THE
GRAND
COSMIC
STORY

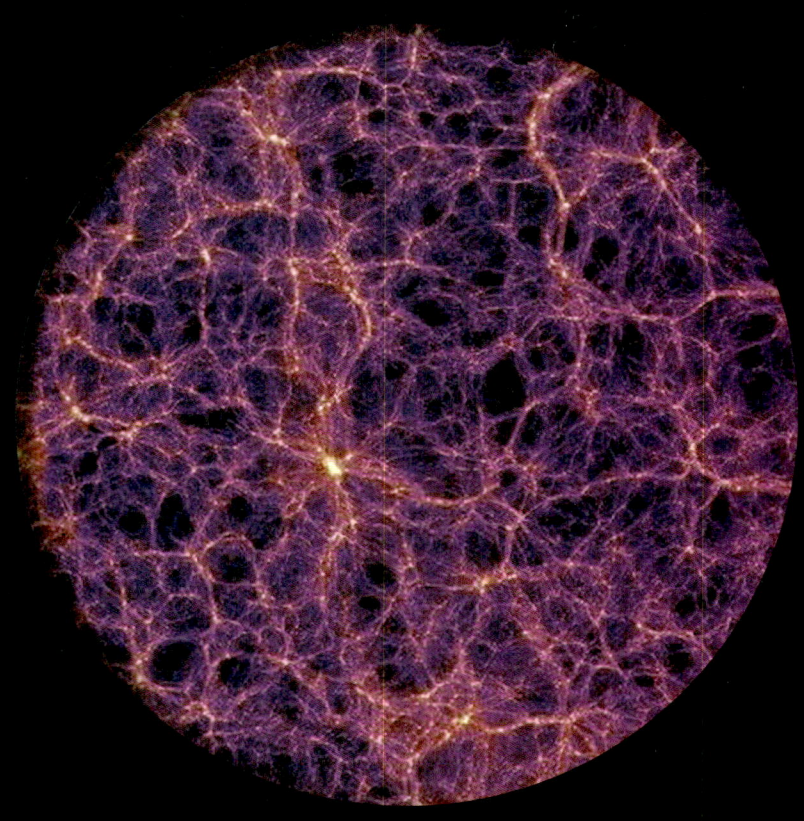

This image depicts a god's-eye view of our Universe from the outside—a likely impossible vantage for humans—revealing its large-scale structure. The porthole reveals just the spherical portion of the Universe that we can observe, 93 billion light-years in diameter. We are near the observable center, offset slightly by 0.017 billion light-years, the result of us being able to see farther in the direction of our cumulative galactic motion. At this scale, the Laniakea Supercluster, home to our galaxy and approximately 100,000 other nearby galaxies, is too small to see—smaller than a single pixel.

THE GRAND COSMIC STORY

AN ILLUSTRATED TIMELINE
o····· 13.8 BILLION YEARS AND BEYOND ·····▸

ETHAN SIEGEL

MARK A. GARLICK

JON LOMBERG

WILLIAM LIDWELL

WASHINGTON, D.C.

CONTENTS

years into
the future

This view of the Universe is not of today's skies as seen from Earth, but rather transports us across space and time. As seen from above the cloud tops of a distant planet long ago, this view transcends the confines of our limited perspective to far-flung stars, nebulae, galaxies, and more. By investigating the Universe observationally, we discover the cosmos and, in turn, come to better understand ourselves.

IMAGE BY MARK A. GARLICK

THE GRAND COSMIC STORY

INTRODUCTION

—— *13.8 billion years and beyond* ——

here is a remarkable story out there that's universal to us all: the story of the cosmos. For most of history, this story was a marvelous mystery that filled us with awe and wonder, with seemingly endless possibilities that stretched the limits of our imaginations. But over the course of the 20th and 21st centuries, novel discoveries led humanity to replace speculation with science, allowing us—for the first time—to recount the real story that the cosmos tells us about itself.

That's the endeavor of *The Grand Cosmic Story*. One page spread and 100 million years at a time, we start from the earliest times that science can say anything definitive about our Universe at all and step through our shared cosmic history. Each spread focuses on a piece of the cosmic story that takes place during that time interval, informed by the best available scientific information that humanity presently possesses wherever possible, filling in any remaining gaps with scenarios that are firmly within the realm of what's scientifically plausible.

Throughout this volume, we've adhered to three overarching principles: first, that the Universe itself unfolds naturally, according to the physical laws of nature and without requiring any sort of supernatural intervention; second, that those laws apply everywhere and at all times, and are not "special" to humans, Earth, our Solar System, or our specific here and now in the Universe; and third, wherever direct evidence is lacking to support one scientific scenario over another, we offer informed speculation, demanding that it be fully consistent with all available knowledge and not in conflict with our best theories of physical reality.

We consider, in calculating statistics for what lies within our Universe, only the components of the cosmos whose signals could, at some point throughout history, be detectable by someone or something within the Milky Way. We do not include the unobservable universe, whose full extent is unknown, or other universes that may be out there beyond this one. In realms where we do not yet possess meaningful data to guide us—such as questions concerning the abundance of life beyond Earth—we provide ranges to reflect the broad possibilities that are still scientifically admissible.

The scope of this book is to provide an overview of all that exists, from subatomic scales to the grandest cosmic ones. It covers the formation of atoms, the lives and deaths of stars, the formation of rocky planets, the origin of life, the growth of galaxies and cosmic structure, and the eventual end of everything that now exists. It is all at once a story that is both wonderful and terrifying, awe-inspiring and humbling, grand and yet somehow mundane. Our story brings us up to the present day with the emergence of humanity on planet Earth: a bit player in the grand cosmic story encompassing our Universe's history, but to us, perhaps the most wondrous occurrence of all. We conclude with three speculative scenarios concerning humanity's future, followed by our ultimate, inevitable fate.

Our ambitious goal is to make this Universe, in all of its splendor, accessible to each and every one of us. Through stories, statistics, and gloriously imaginative illustrations, the story of our Universe takes on a life of its own. While no signal within our cosmos can travel faster than the speed of light, our imaginations know no such limit, traversing billions of light-years or unfathomable eons of time with the flip of a page. We hope that you enjoy this book over and over again, and that you leave with a sense that you don't just exist within this Universe but that you yourself are an important and inextricable part of the cosmic story that unites us all.

ABOUT THIS BOOK

With each turn of the page in this book, you step forward another 100 million years in the cosmic story that unites us all, each phase represented by narrative text and an imaginative illustration. In addition to the text and artwork, each page includes a table with vital information about the changing cosmos during that epoch.

The information in these three columns—Universe, Astronomical Objects, and Worlds with Life—expresses a picture of the cosmos during each phase of evolution, according to current scientific understanding. The range of numbers on each page represents the changes thought to have occurred during that spread's 100-million-year time period. Since significant change occurs within each lengthy time period, each category contains not a single number but a numerical range. The changes you see, page to page, reveal the evolution of those phenomena through longer sweeps of time.

The numbers calculated reflect our best current model of the Universe, including modern measurements of the expansion rate, estimates for the contents of our Universe (including dark matter and dark energy), and the latest data about the cosmic microwave background and the large-scale clustering and distribution of galaxies. Here are additional details about each entry.

DESCRIPTION OF KEY EVENT OR FEATURE

AGE OF THE UNIVERSE

Red dwarf stars, the lowest-mass stars in the Universe, have the longest lifetimes and display the greatest amounts of stellar activity. While this may render most planets orbiting them inhospitable for tens or even hundreds of billions of years, perhaps in the far future these stars will stabilize, creating potentially habitable conditions long after the less massive stars have died away

IMAGE BY MARK A. GARLICK

7.3–7.4 BILLION YEARS

THE LIVES OF RED DWARFS

the most common and longest-lived stars

UNIVERSE	ASTRONOMICAL OBJECTS	WORLDS WITH LIFE
Diameter: $2.17 \times 10^{23} \rightarrow 2.20 \times 10^{23}$ ly	Galaxies: $2.83 \times 10^{11} \rightarrow 2.83 \times 10^{11}$	Simple Life: $6.62 \times 10^{21} \rightarrow 2.61 \times 10^{23}$
Observable: $5.44 \times 10^{23} \rightarrow 5.49 \times 10^{23}$ ly	Stars: $1.65 \times 10^{21} \rightarrow 1.67 \times 10^{21}$	Complex Life: $1.31 \times 10^{21} \rightarrow 1.35 \times 10^{21}$
Expansion: $100.31 \rightarrow 99.27$ km/s/Mpc	Black Holes: $1.34 \times 10^{18} \rightarrow 1.35 \times 10^{18}$	Intelligent Life: $3.47 \times 10^{8} \rightarrow 8.74 \times 10^{14}$
Density: $1.20 \rightarrow 1.16$ p/m³	Planetary Systems: $2.23 \times 10^{20} \rightarrow 2.26 \times 10^{20}$	Technological Life: $9.18 \times 10^{4} \rightarrow 9.03 \times 10^{10}$
Temperature: $4.62 \rightarrow 4.58$ K	Worlds: $2.53 \times 10^{21} \rightarrow 2.56 \times 10^{21}$	Interstellar Life: $4 \rightarrow 6.00 \times 10^{1}$

pon developing the capacity to perceive light, in any form, a biological organism's detection abilities will inherently be biased towards the brightest-shining, most easily visible objects. In any night sky, no matter which set of ultraviolet, visible, or infrared wavelengths an organism's sight is adapted to, the brightest, hottest, most luminous stars will shine most bright. Organisms that are more sensitive to bluer, shorter-wavelength light will preferentially see the youngest, highest-mass stars, as well as evolved blue supergiants. Organisms more sensitive to redder, longer-wavelength light will see large numbers of evolved red giant and supergiant stars: the ones entering the final stages of their lives. But the most common, enduring stars will remain invisible to all but the most sensitive infrared receptors: red dwarfs.

Somewhere between 70% and 82% of all stars that will ever be born in the Universe are red dwarf stars, which still cross that critical ~4 million K temperature in their cores, enabling them to fuse hydrogen into helium inside. However, the low masses of these stars—they can be no more than 40% as massive as a Sun-like star—ensure that only a relatively small volume of the star is undergoing nuclear fusion at any one time, leading to low luminosities and low temperatures but also very long lifetimes. Whereas the brightest stars might live only for millions or a few billion years at most, the longest-lived red dwarf stars can endure for tens of trillions of years.

Recall that the famed saying, "the candle that burns twice as bright burns half as long," is even worse for stars: a star possessing double the mass of another lives approximately just one-eighth as long. For the lowest-mass stars, however, this holds a fascinating implication: The slow rate of fusion, even with smaller amounts of fuel overall, ensures that long after those high-mass, bright stars evolve into stellar remnants, these red dwarfs will still endure, fusing hydrogen into helium in their cores for trillions of years to come.

The lowest-mass red dwarf stars might shine only 0.1% as brightly as a Sun-like star, which themselves are hundreds to millions of times fainter than giant or supergiant stars. But what nature fails to provide in brightness, it makes up for in sheer numbers and in duration. In those galaxies that have exhausted their supplies of gas and stopped forming stars several billion years ago—those "red and dead" elliptical galaxies—the more massive stars will simply run out of fuel and die, but the longer-lived, lower-mass stars will endure.

Inside the red dwarf stars—the lowest-mass, longest-lived, most numerous class of star of all—the nuclear processes that cause the star to shine are so slow that a new process can occur inside of them: whole-star convection. In the higher-mass, shorter-lived stars, spent nuclear fuel simply builds up in the star's core, causing the star to contract, heat up, and evolve when the rate of nuclear reactions drops below a threshold. But in these low-mass red dwarf stars, the nuclear processes powering them are so slow that the spent nuclear fuel, in the form of helium, can actually get displaced towards the outer layers and replenished by new, unburnt hydrogen fuel. This convective process includes the entire star's mass for red dwarfs, ensuring that when they do eventually die, practically the entire star will have converted into helium.

The longest-lived red dwarfs might endure for over 100 trillion years, or more than 10,000 times the present age of the Universe. With such long life spans, who knows what might arise on worlds around them?

DEPICTION OF KEY EVENT OR FEATURE

TIME FROM PRESENT

DATA TABLE
— estimates per 100 million years —

NOTE: While the UNIVERSE and ASTRONOMICAL OBJECTS columns detail the evolution of each specific quantity over a 100-million-year period, the WORLDS WITH LIFE column displays a confidence interval, complete with uncertainties, for the mean value of each type of world listed.

UNIVERSE
Diameter: $\emptyset_1 \rightarrow \emptyset_2$ ly
  Observable: $\emptyset_1 \rightarrow \emptyset_2$ ly
Expansion: $V_1 \rightarrow V_2$ km/s/Mpc
Density: $D_1 \rightarrow D_2$ p/m³
Temperature: $T_1 \rightarrow T_2$ K

ASTRONOMICAL OBJECTS
Galaxies: $G_1 \rightarrow G_2$
Stars: $S_1 \rightarrow S_2$
Black Holes: $BH_1 \rightarrow BH_2$
Planetary Systems: $PS_1 \rightarrow PS_2$
Worlds: $W_1 \rightarrow W_2$

WORLDS WITH LIFE
Simple Life: min – max
Complex Life: min – max
Intelligent Life: min – max
Technological Life: min – max
Interstellar Life: min – max

UNIVERSE

Diameter	The change in the diameter of the entire Universe—unobservable and observable combined—during this time interval, expressed in light-years (ly).
Observable	The change in the end-to-end diameter of the observable Universe during this time interval, expressed in terms of a speed (kilometers per second, or km/s) per unit distance (megaparsec, or Mpc).
Expansion	The changing expansion rate—which slows as the Universe expands—during this time interval, expressed in kilometers per second per megaparsec.
Density	The changing amount of matter present, on average, per unit volume of the Universe, as measured in protons per cubic meter (p/m³).
Temperature	The changing temperature of the cosmic background radiation as the Universe expands and cools, measured in Kelvin (K).

NOTE: While we can only directly measure the properties of our observable Universe, the lack of spatial curvature tells us that the entire cosmos, including the parts that are both observable (where we can detect signals emitted at the speed of light) and unobservable (beyond the limits of where light can reach in the finite time that has elapsed since the Big Bang), must be at least a few hundred times the extent (and tens of millions of times the volume) of the part we can observe.

ASTRONOMICAL OBJECTS

Galaxies	The changing number of large collections of stars, containing anywhere from thousands to quadrillions of stars. The number of galaxies may grow as new ones form, but it also may shrink as existing ones collide and merge together.
Stars	The changing number of astronomical bodies with a mass of at least 7.5 percent of our Sun's that are actively fusing light elements into heavier ones in their cores.
Black Holes	The changing number of dense regions of space that contain so much mass within them that nothing, not even light, can escape from within. They often, but not always, arise as remnants of stars.
Planetary Systems	The changing number of star systems that contain at least one orbiting planet.
Worlds	The total number of planets, moons, dwarf planets, and other spheroidal objects that either orbit a parent star or are orphaned bodies without a parent star of their own.

WORLDS WITH LIFE

Simple Life	Any world that possesses an organism, no matter how primitive, with a metabolism and the ability to reproduce. Includes Earth beginning at least 3.8 billion years ago.
Complex Life	A world containing at least one organism that has evolved multiple internal components that perform specialized functions and contribute to the organism's overall functioning and survival. Includes Earth for at least the past 900 million years.
Intelligent Life	A world containing life-forms that exhibit traits of advanced reasoning and substantial problem-solving skills. Includes Earth for at least the past 100 million years.
Technological Life	A world with life that develops and utilizes tools beyond mere biological structures, including bone, rock, metal, electronics, and nanotechnology. Detectable via broadcast signals, artificial lighting, nuclear activity, and/or spacecraft. Examples include humans on Earth, particularly in recent centuries.
Interstellar Life	A world from which life has journeyed to a destination beyond its home planetary system. Does not include Earth today, but may in time to come.

Before the Beginning, random quantum processes were a fundamental component of space-time, just as they are today. Even in the absence of particles, quantum fluctuations persist throughout empty space, with at least one such region experiencing an inflationary process. When inflation ends, these now-stretched fluctuations remain imprinted on the Universe as the initial seeds of structure, persisting even through the hot Big Bang's inferno-like beginnings.

IMAGE BY WILLIAM LIDWELL

BEFORE THE BEGINNING

—— *all is void and randomness* ——

UNIVERSE	ASTRONOMICAL OBJECTS	WORLDS WITH LIFE
Diameter: 0	Galaxies: 0	Simple Life: 0
Observable: 0	Stars: 0	Complex Life: 0
Expansion: 0	Black Holes: 0	Intelligent Life: 0
Density: 0	Planetary Systems: 0	Technological Life: 0
Temperature: 0	Worlds: 0	Interstellar Life: 0

efore the Beginning, only one thing was certain to exist: space-time. It may have been devoid of particles; it may have been filled with various forms of energy; conditions may have been chaotic beyond comprehension. But within this space-time, random processes continuously occurred. In at least one region within this volume, these random processes brought about a particular quantum state that led to a rapid, relentless expansion of space known as cosmic inflation. This inflating region of space-time increases its volume more rapidly than all others, swiftly dominating all other quantum states. Practically the entire cosmic volume becomes devoid of particles, but still obeying the fundamental laws of physics, including the quantum rules governing all of physical reality.

Part of the quantum rules that govern reality include the existence, and omnipresence, of quantum fields. These quantum fields not only permeate all of space but also have inherent fluctuations to them: uncertainties to whatever values they may take on at any moment. This has a profound and fascinating consequence for space itself: It implies that the zero-point energy of space, or the amount of energy inherent to a specific volume of even completely empty space, may not be zero. Instead, the total amount of energy that empty space intrinsically possesses can be finite and positive, and—unlike the small finite and positive value we observe today—it could have been quite large early on.

If space itself contains a positive amount of energy, then it cannot be static but must evolve with time. As it evolves by expanding, new space is continuously created along with that same value of energy inherent to it, compelling the expansion to continue relentlessly. The only

variations that occur, from region to region, arise from the inevitable fluctuations in those quantum fields themselves. This expansion continues relentlessly, with each tiny region of space containing its own quantum fields. Although those fields fluctuate, the expansion stretches those fluctuations to progressively larger scales as time goes on.

In most of those regions, this expansion continues indefinitely and without letting up: a continuation of cosmological inflation. But every once in a while, the values in one of these inflating regions of space give rise to an unlikely fluctuation that causes inflation to become unstable. Whenever and wherever this occurs, inflation comes to an end, and most of the energy that had hitherto been inherent to the fabric of space itself—the driving force behind inflation—gets converted into particles and antiparticles by the equation governing matter-energy equivalence: $E = mc^2$. Abruptly, in these regions of space, inflation stops, replaced by a hot big bang.

In one of these regions, some 13.8 billion years ago, the end of inflation gave rise to the start of our Universe; and the hot Big Bang that ensued was the start of our cosmic history as we know it. But outside of that region, no matter how large it may be, our Universe is surrounded by more space that's still inflating, peppered with individual regions where inflation comes to an end and gives rise to completely independent hot big bangs.

The full suite of all of these individual universes make up a multiverse, forever separated by relentlessly inflating space. But in one particular universe—our Universe—we came into existence some 13.8 billion years after the hot Big Bang that signaled the end of our Universe's inflation. What follows is the cosmic story that unites us all.

The Universe as we know it begins with the hot Big Bang, with energetic radiation, particles, and antiparticles permeating all of the space within it. As it both expands and gravitates, the ultimate cosmic race begins.

IMAGE BY JON LOMBERG

THE BEGINNING

—— from randomness, order ——

UNIVERSE
Diameter: ~10^{-10} → 1.19×10^{12} ly
  Observable: ~10^{-12} → 2.97×10^{9} ly
Expansion: ~10^{46} → 6,564.59 km/s/Mpc
Density: ~10^{68} → 7,292.88 p/m³
Temperature: ~10^{23} → 84.47 K

ASTRONOMICAL OBJECTS
Galaxies: 0
Stars: 0 → 1.7×10^{10}
Black Holes: 0 → 6.81×10^{9}
Planetary Systems: 0
Worlds: 0

WORLDS WITH LIFE
Simple Life: 0
Complex Life: 0
Intelligent Life: 0
Technological Life: 0
Interstellar Life: 0

In the beginning, there is energy, with every permissible quantum state imaginable realized in some form. The rapidly expanding Universe suddenly becomes very hot, as particles and antiparticles of every flavor, color, charge, and variety pop into existence, all with tremendous energy. In the dense environment of the early Universe, they travel at nearly the speed of light, crashing into one another quadrillions of times in the first nanosecond alone. All the different species of matter, antimatter, and radiation gravitate, attempting to pull the Universe back together, while the initial expansion rate attempts to drive the very fabric of space itself apart.

If the Universe had begun a tenth of a percent more dense at the outset, gravitation would have won out over the expansion, recollapsing the Universe before a single second had elapsed. If the Universe had begun a tenth of a percent less dense, the expansion would have won so thoroughly that after a single second had elapsed, no two particles would have ever met again.

And yet, these two phenomena—the initial expansion rate and the gravitational effects the matter, antimatter, and radiation present—balance one another. As the Universe expands, it not only gets less dense, but also cooler, as every quantum of energy gets stretched to longer, cooler wavelengths with the expansion of space. And as the energy density drops, so does the expansion rate. As time goes on, the Universe continues expanding, getting less dense, and cooling, with the expansion rate and energy density remaining in perfect balance to neither recollapse nor run away into oblivion.

As the Universe cools, less energy becomes available to create new particle-antiparticle pairs, but things remain dense enough for particle-antiparticle pairs to annihilate into radiation: particles of pure energy. The more the Universe cools, the more easily the heavy, unstable particles and antiparticles decay into less energetic, more stable species. As annihilations and decays continue, a small asymmetry emerges, leaving the Universe with one excess matter particle for about every one billion particle-antiparticle pairs that annihilate or decay away.

As time continues to pass, the Universe becomes less symmetric. The Higgs symmetry breaks, and particles gain large, nonzero rest masses. The electroweak force splits irrevocably into the electromagnetic and the weak nuclear force. The quark-gluon plasma becomes unstable and fragments into individual protons and neutrons. Neutrinos cease interacting with the leftover particles, and then the last of the charged antimatter—positrons, the counterpart of electrons—annihilate into photons. All of this takes less than a second since the start of the hot Big Bang: the most consequential second in all of cosmic history.

At this point, only a hot, dense, primordial soup of photons, neutrinos, electrons, protons, and neutrons remain, alongside the mysterious dark matter and dark energy. Atomic nuclei and neutral atoms are both still impossible, as the energetic early Universe blasts such bound states apart instantaneously. After about three minutes have elapsed, the remaining protons and neutrons fuse together to make the first stable nuclei, including deuterium, helium, and lithium, amid a sea of bare protons: hydrogen nuclei. After another 380,000 years, the nuclei can bind with electrons without getting blasted apart, creating the first neutral atoms. The leftover radiation from the Big Bang, at last, is free to travel without smashing into free, charged particles.

And finally, as it expands, the Universe gravitates. The tiny imperfections in density that were seeded by inflation, at just the 1-part-in-30,000 level, begin to gravitationally grow. The densest clumps of matter grow the fastest; after the first 100 million years, the very first stars only just begin to form.

The first stars begin to shine, as large clouds of mass gather together under the influence of gravity, and then cool and collapse. These so-called Population III stars are hot, massive, blue, and very short-lived.

IMAGE BY MARK A. GARLICK

CANDLES IN THE DARKNESS

—— *the first stars* ——

UNIVERSE
Diameter: $1.19 \times 10^{12} \rightarrow 1.89 \times 10^{12}$ ly
  Observable: $2.97 \times 10^9 \rightarrow 4.73 \times 10^9$ ly
Expansion: $6,564.59 \rightarrow 3,267.48$ km/s/Mpc
Density: $7,292.88 \rightarrow 1,811.59$ p/m³
Temperature: $84.47 \rightarrow 53.00$ K

ASTRONOMICAL OBJECTS
Galaxies: $0 \rightarrow 6.67 \times 10^{10}$
Stars: $1.7 \times 10^{10} \rightarrow 7.34 \times 10^{16}$
Black Holes: $6.81 \times 10^9 \rightarrow 1.27 \times 10^{16}$
Planetary Systems: $0 \rightarrow 2.94 \times 10^{16}$
Worlds: $0 \rightarrow 1.48 \times 10^{17}$

WORLDS WITH LIFE
Simple Life: 0
Complex Life: 0
Intelligent Life: 0
Technological Life: 0
Interstellar Life: 0

here's an ongoing struggle between gravitational growth and cosmic expansion, and it's now that the first victors in that conflict begin to emerge. Our Universe is born at the start of the hot Big Bang with almost perfectly equal amounts of energy everywhere. From one region of space to the next, the density varies only slightly, with the densest regions being about 100.003% of the average density and the least dense regions being only about 99.997% as dense as average. But as long as you start with even the tiniest imperfection, gravitation will take these minuscule differences as seeds and grow them into cosmic structures that span the entire Universe.

It's the imperfections on small cosmic scales that begin to gravitationally grow at the earliest times, with the initially overdense regions drawing more and more matter into them, while the underdense regions preferentially give up their matter to their denser surroundings. The more matter accumulates in a specific region, the greater its relative gravitational pull will be with respect to all the other regions around it. Gravitation is an unforgiving, runaway process, and the more successful one region is at drawing greater amounts of matter into it, the faster and more massive it grows with time.

For the densest initial regions that the Universe possessed at the start of the hot Big Bang, it takes many tens to (much more commonly) hundreds of millions of years for enough matter to accumulate—dark matter and normal matter alike—in order for those very first molecular clouds to begin gravitationally collapsing. With 99.9999999% of the normal matter in the Universe made up exclusively of hydrogen and helium atoms, these molecular clouds have tremendous difficulty in cooling,

and so they must reach enormous masses for gravitational collapse to occur.

When that critical threshold is finally reached, however, each massive gas cloud gains its own imperfections and collapses at different rates in different regions. Just as before, the components that collect the greatest amounts of matter the most rapidly tend to pull matter from the surrounding regions into it, leading to an even more rapid onset of gravitational growth and collapse. As the density of atoms increases, hydrogen atoms collide and begin forming molecular hydrogen gas (H_2), which is the primary way that the densest clumps of gas radiate away their heat.

At last, the next stage in the evolution of the cosmos can occur: the formation of the very first stars. Composed only of hydrogen and helium, these stars are very different from the ones we see today. They're overwhelmingly massive, with the smallest stars weighing in at about 10 times the mass of our Sun, while the largest can reach hundreds or even thousands of times the Sun's mass. Without heavier elements around, there are no planets around any of these stars, only clouds of gas, which rapidly evaporate from the intense radiation emanating from these first stars.

Because they're so massive, these stars live for only very short periods of time—a few million years at most— before they burn through the nuclear fuel in their cores and die in catastrophic stellar cataclysms like supernovae. Wherever our Universe has pristine gas that's never formed stars before, only these ultramassive, hot, blue, short-lived stars can arise. But the deaths of these first stars are, perhaps, the most special event of all, as they produce the first copious amounts of the heavy elements—including oxygen, carbon, nitrogen, phosphorous, silicon, sulfur, iron, neon, calcium, magnesium, and more—that will make possible all that's to come as the cosmic story unfolds.

Amid a sea of neutral atoms in the
early Universe, dense clumps of matter
gravitationally collapse, forming stars in
clusters that grow and merge into the first
protogalaxies. As starlight streams out into
the Universe, it ionizes neutral hydrogen
atoms, creating a pinkish glow as electrons
cascade down the energy levels of hydrogen
atoms when they recombine, while dense
clouds enshroud these luminous regions.

IMAGE BY MARK A. GARLICK

STARS DANCE WITH STARS

—— *the first galaxies* ——

UNIVERSE
Diameter: $1.89 \times 10^{12} \rightarrow 2.48 \times 10^{12}$ ly
  Observable: $4.73 \times 10^{9} \rightarrow 6.21 \times 10^{9}$ ly
Expansion: $3,267.48 \rightarrow 2,174.76$ km/s/Mpc
Density: $1,811.59 \rightarrow 803.16$ p/m³
Temperature: $53.00 \rightarrow 40.49$ K

ASTRONOMICAL OBJECTS
Galaxies: $6.67 \times 10^{10} \rightarrow 3.26 \times 10^{11}$
Stars: $7.34 \times 10^{16} \rightarrow 5.28 \times 10^{17}$
Black Holes: $1.27 \times 10^{16} \rightarrow 1.79 \times 10^{16}$
Planetary Systems: $2.94 \times 10^{16} \rightarrow 5.44 \times 10^{17}$
Worlds: $1.48 \times 10^{17} \rightarrow 2.74 \times 10^{18}$

WORLDS WITH LIFE
Simple Life: 0
Complex Life: 0
Intelligent Life: 0
Technological Life: 0
Interstellar Life: 0

On larger cosmic scales than ever before, the cosmic race between gravitation and cosmic expansion now leads to the formation of the earliest galaxies. Regions of space that are denser than average not only attract the surrounding matter into them, leading to the formation of dense molecular gas clouds that will form stars and star clusters, but independent star clusters begin to attract one another as well. As these overdense regions attract one another and begin accelerating towards one another, they can interact, collide, and merge together. Driven by both dark matter and normal matter, enough mass gathers together in one place that the Universe forms the earliest protogalaxies.

Inside these growing clumps of matter, stars form at a significantly increased rate during this time. In many locations, the matter that's coalescing has neither participated in any star formation previously, nor has it been enriched by the ejecta from dying stars in the surrounding space. These pristine clouds of gas must grow to be many millions of times the mass of the Sun before they collapse, as they have no heavy elements inside them to help them cool. In these locations, the stars that form are tens, hundreds, or possibly even thousands of times as massive as the Sun, making them bright, blue, luminous, and very short-lived.

Although a large fraction of the stars that form during this era are still pristine—made of the elements left over from the Big Bang alone—the heavy elements produced from the very first stars have already begun to substantially enrich the Universe. Wherever these originally pristine stars have already lived and died, the surrounding space that's in between the various clumps of stars, known as the interstellar medium, will be filled with the ejecta from the death throes of these stellar corpses. As the first generation of stars dies, they make way for the formation of new stars that are, in many ways, more evolved than their predecessors.

When a molecular cloud of gas contains even a small amount of heavy elements—as little as 0.001%—it behaves very differently from a cloud made of hydrogen and helium alone. These heavier elements are superior at radiating heat away, and they enable gas clouds to collapse both more rapidly and at much lower masses as compared to pristine ones. With heavy elements inside, second-generation stars not only form more easily than first-generation ones but obtain a much wider range of masses. Whereas the first pristine stars are typically 10 solar masses or more, the average star that contains some amount of heavy elements has less than half of the Sun's mass. For the first time, the Universe not only forms short-lived, hot blue stars but also stars that span the full range of stellar colors, brightnesses, temperatures, and masses: from hot, bright, and blue to cool, dim, and red, along with all of the whites, yellows, and oranges in between.

Additionally, with many element types present, these second-generation stars can form planets for the first time. Most of the planets that form this early on will either be large and gaseous or small and icy; the Universe will have to wait to become much more enriched before rocky planets are possible. Nonetheless, for the very first time, our Universe possesses planetary systems around stars of all different colors, sizes, and masses.

Supermassive black holes take shape and grow, formed from dying stars and the direct collapse of large amounts of matter. Today, practically every massive galaxy has a supermassive black hole at its center.

IMAGE BY MARK A. GARLICK

HERE BE DRAGONS

— *the first supermassive black holes* —

UNIVERSE
Diameter: $2.48 \times 10^{12} \rightarrow 3.01 \times 10^{12}$ ly
  Observable: $6.21 \times 10^{9} \rightarrow 7.52 \times 10^{9}$ ly
Expansion: $2{,}174.76 \rightarrow 1{,}630.20$ km/s/Mpc
Density: $803.16 \rightarrow 451.36$ p/m³
Temperature: $40.49 \rightarrow 33.41$ K

ASTRONOMICAL OBJECTS
Galaxies: $3.26 \times 10^{11} \rightarrow 5.77 \times 10^{11}$
Stars: $5.28 \times 10^{17} \rightarrow 1.40 \times 10^{18}$
Black Holes: $1.79 \times 10^{16} \rightarrow 2.67 \times 10^{16}$
Planetary Systems: $5.44 \times 10^{17} \rightarrow 1.68 \times 10^{18}$
Worlds: $2.74 \times 10^{18} \rightarrow 8.52 \times 10^{18}$

WORLDS WITH LIFE
Simple Life: 0
Complex Life: 0
Intelligent Life: 0
Technological Life: 0
Interstellar Life: 0

By the time 300 million years have elapsed since the Big Bang, the number of pristine, first-generation stars is already in decline. A strong majority of all the new stars that form now possess some substantial amount of heavy elements, leading to a wide diversity of stars in star clusters and in young, growing galaxies.

Whenever the Universe forms stars, they not only vary in color, brightness, mass, and lifetime, but also in what they'll wind up leaving behind. Stars that are born with less than about eight times the Sun's mass are long-lived, lasting anywhere from tens of millions to hundreds of trillions of years; when they run out of fuel, their cores contract to form white dwarfs. But more massive stars are shorter-lived and usually end their lives in a cataclysmic supernova explosion, and when they do, they leave either a neutron star or—if they were massive enough at birth—a black hole behind. While fewer than 1% of second-generation (and later) stars will leave behind black holes, a quarter or more of the first, pristine stars will form black holes.

Black holes are not only the densest objects in the Universe, packing more mass into a tiny volume of space than anything else, but they also help determine how young galaxies grow up and evolve. As black holes orbit around the centers of their home galaxies, they experience the gravitational tug of other stars, gas, dust, and even dark matter. Because they're both dense and massive, these cumulative interactions cause the most massive black holes to cluster together in the centers of these young galaxies. These black holes not only feed on the matter they encounter but can merge together, creating the seeds of what will eventually grow into supermassive black holes.

Perhaps surprisingly, this growth happens very quickly. Even as the Universe approaches the 400-million-year mark, the largest black holes have already grown to be upwards of one million solar masses, with event horizons that exceed the physical size of our Sun. As each individual galaxy grows larger and more massive, through mergers with neighboring star clusters and protogalaxies, the central black holes accumulate more matter and grow more massive as well. The more intensely the galaxy gains gas and forms new stars, the more rapidly the central black hole's mass increases.

As these early galaxies continue to form and grow, the total number of stars continues to rapidly increase. The largest, most massive galaxies now shine with the brightness of more than a billion Suns, with most of those stars possessing numerous giant planets. A very small number of stellar systems might contain rocky worlds, but the ones that exist will either be airless, like Mercury, or frozen, like Pluto. The Universe, despite all that's developed, cannot yet form Earth-like planets.

The big downside, however, is that from the vantage point of any star or protogalaxy, most of the Universe cannot yet be seen. There's simply too much leftover neutral matter in interstellar and intergalactic space, and too few ultraviolet photons, even from all the stars that have formed, to ionize them all. The Universe, by the time it reaches the age of 400 million years, is more evolved than ever but remains opaque to the types of light that human eyes can perceive.

Although many stars and galaxies have already formed, they are mostly obscured by a still-persistent fog of neutral atoms permeating intergalactic space. Here, two bright sources of light illuminate the darkness for a brief extent but fade like headlights in the fog with greater distances.

IMAGE BY MARK A. GARLICK

A BLEAK, UNKINDLY FOG

—— *most of the universe is still opaque* ——

UNIVERSE
Diameter: $3.01 \times 10^{12} \rightarrow 3.50 \times 10^{12}$ ly
  Observable: $7.52 \times 10^{9} \rightarrow 8.73 \times 10^{9}$ ly
Expansion: $1,630.20 \rightarrow 1,303.7$ km/s/Mpc
Density: $451.36 \rightarrow 288.60$ p/m³
Temperature: $33.41 \rightarrow 28.78$ K

ASTRONOMICAL OBJECTS
Galaxies: $5.77 \times 10^{11} \rightarrow 8.65 \times 10^{11}$
Stars: $1.40 \times 10^{18} \rightarrow 2.80 \times 10^{18}$
Black Holes: $2.67 \times 10^{16} \rightarrow 4.08 \times 10^{16}$
Planetary Systems: $1.68 \times 10^{18} \rightarrow 3.56 \times 10^{18}$
Worlds: $8.52 \times 10^{18} \rightarrow 1.82 \times 10^{19}$

WORLDS WITH LIFE
Simple Life: 0
Complex Life: 0
Intelligent Life: 0
Technological Life: 0
Interstellar Life: 0

On the scales of stars, galaxies, and black holes, the Universe is beginning to grow up. The most massive, early galaxies continue to draw additional matter into them, growing via both accretion of intergalactic matter and also through mergers with star clusters and other protogalaxies. At the centers of these objects, the seeds of supermassive black holes continue to rapidly grow. From the corpses of the most massive, shortest-lived stars, new generations of stars and stellar systems arise, more enriched than ever before. In this brief 100-million-year interval, the number of stars in the Universe doubles over its previous value. Although many of these new stars will have planets, including a few with rocky, Earth-sized planets, the Universe is not yet ready for life to emerge. Additionally, no matter which galaxy you choose to look out from, the overwhelming majority of the Universe is still obscured.

The reasons the Universe is still too immature to either be fully transparent to starlight or to begin developing the first instances of life are one and the same: There simply haven't been enough stars that have lived and died to allow the next stage of cosmic evolution to take place. Perhaps surprisingly, it all comes down to what's occurring on the smallest scales of all: at the level of atoms and molecules.

Whenever stars form, they do so from a cloud of neutral atoms: molecular clouds of gas. There are still plenty of pristine stars forming—stars made up exclusively from the hydrogen and helium remaining from the early moments of the hot Big Bang—and they're predominantly massive, hot, blue, and short-lived. But once those stars complete their life cycles, enriching the Universe with heavier elements, the subsequent generations of stars that form will be, on average, much lower in mass. For molecular

clouds to gravitationally collapse, they must radiate away enough heat. But atoms and molecules can only experience very particular energy transitions, defined by the quantum mechanical orbitals that electrons can occupy. Unless the energy of a particular transition precisely matches the energy imparted to the atom in question, a transition will not occur.

But there's an exception to that rule: If you impart enough energy to an atom or molecule to ionize it—that is, to fully unbind an electron—it can absorb any amount of energy that meets or exceeds its ionization threshold. And here's the kicker: The heavier your atom or molecule is, in general, the easier it is to ionize an electron. This is why clouds of gas that are rich in heavier elements cool so much more efficiently, and can form lower-mass stars and planets, than their more pristine counterparts. However, even in the densest regions of space that have experienced the most star formation, possessing the greatest fraction of heavy elements, there's still too little of them to lead to the formation of Sun-like stars and Earth-like planets. Further generations of stars living and dying, via a variety of processes, are necessary precursors for life.

In the depths of intergalactic space, however, the opposite side of the story is unfolding. In locations where no stars or galaxies have yet formed, the matter is still pristine: made of hydrogen and helium. Until enough high-energy, ultraviolet photons are created, those intergalactic atoms will remain neutral, rendering the Universe still opaque to visible light. Greater numbers of stars, including in these hard-to-reach locations, must still form for the Universe to become transparent to starlight. The atoms within the Universe, after finally becoming neutral, must become reionized once again for us to see through it.

With enough stars having cumulatively formed, the total number of ultraviolet photons is at last sufficient to ionize practically every atom in the intergalactic medium. At last, the Universe has become transparent to starlight.

LET THERE BE LIGHT

—— *reionization clears the fog* ——

UNIVERSE
Diameter: $3.50 \times 10^{12} \rightarrow 3.95 \times 10^{12}$ ly
  Observable: $8.73 \times 10^{9} \rightarrow 9.86 \times 10^{9}$ ly
Expansion: $1,303.7 \rightarrow 1,085.85$ km/s/Mpc
Density: $288.60 \rightarrow 200.11$ p/m³
Temperature: $28.78 \rightarrow 25.48$ K

ASTRONOMICAL OBJECTS
Galaxies: $8.65 \times 10^{11} \rightarrow 1.21 \times 10^{12}$
Stars: $2.80 \times 10^{18} \rightarrow 4.89 \times 10^{18}$
Black Holes: $4.08 \times 10^{16} \rightarrow 6.16 \times 10^{16}$
Planetary Systems: $3.56 \times 10^{18} \rightarrow 6.36 \times 10^{18}$
Worlds: $1.82 \times 10^{19} \rightarrow 3.30 \times 10^{19}$

WORLDS WITH LIFE
Simple Life: 0
Complex Life: 0
Intelligent Life: 0
Technological Life: 0
Interstellar Life: 0

The number of stars within the Universe continues to grow at a tremendous rate, nearly doubling again during this period. At last, it finally happens: More than 99% of the neutral atoms in the Universe become reionized. In most directions and locations in space, light of optical wavelengths can at last travel through the Universe without being absorbed and scattered by the intervening matter. As the last remaining neutral matter becomes ionized again in most places, the veil of obscurity is finally lifted, revealing the rich structures of stars and young galaxies that permeate the visible Universe.

Simultaneously, cosmic "winners" and "losers" in the great gravitational race are beginning to clearly emerge. The earliest galaxies—and the earliest regions where star formation took place in general—are now growing up into substantial, quite large structures all on their own. Many of these young galaxies have already, in such a short time, grown to have more than a billion stars inside of them. The ones with the largest, fastest-growing black holes at their centers have achieved supermassive black hole masses of over 100 million solar masses. And the first major mergers, where two large, roughly equal-sized galaxies collide, are occurring, triggering new waves of star formation and a flurry of stellar cataclysms, including supernovae. For the first time, these early galaxies are beginning to clump together on larger cosmic scales.

At the start of this time interval, the average photon released from a star will travel through interstellar space until it leaves its host galaxy, continuing on its way. Before it gets very far, however—after just a few million light-years—it's more likely than not that the photon will strike a neutral atom, preventing it from traveling farther in the same direction. As more and more stars are created, however, including hot, blue, short-lived massive stars, more and more ultraviolet photons travel through the intergalactic medium, striking and ionizing these neutral atoms.

By this moment in time, the Universe has expanded enough, and the density of particles is now low enough that when these atoms do get struck by an ultraviolet photon, they don't just ionize; they stay ionized. In other words, when an electron gets kicked off of a neutral atom, the electron is exceedingly unlikely to encounter another ionized nucleus to re-form a neutral atom, and the newly ionized nucleus is similarly unlikely to encounter a free electron. With just one charged particle for every 1,000 cubic centimeters of space at this moment, the Universe slowly becomes transparent to optical light.

At 550 million years, a milestone is reached: From the vantage point of any random galaxy, you'll be able to look out across the Universe and see stars and galaxies stretching for billions of light-years in all directions. Although there will remain a few select lines of sight where an insufficient number of stars have formed to fully reionize the matter occupying that space—including regions that are themselves only beginning to collapse, where pristine, first-generation stars will subsequently form—starlight can now travel freely throughout the cosmos. What was once a shrouded, shadowy abyss suddenly experiences the removal of its dusty cosmic veil, exposing the glittering riches of the Universe lying in the far reaches of deep space forevermore.

As the dust enshrouding the earliest galaxies is at last blown away, the active supermassive black holes at the centers of those galaxies propel energetic jets of particles and radiation throughout the Universe. These quasars are among the most energetic phenomena in all the Universe.

IMAGE BY JON LOMBERG

BETWEEN DRAGONS AND THEIR WRATH

—— the first quasars ——

UNIVERSE
Diameter: $3.95 \times 10^{12} \rightarrow 4.37 \times 10^{12}$ ly
 Observable: $9.86 \times 10^{9} \rightarrow 1.09 \times 10^{10}$ ly
Expansion: $1,085.85 \rightarrow 932.05$ km/s/Mpc
Density: $200.11 \rightarrow 147.33$ p/m³
Temperature: $25.48 \rightarrow 23.01$ K

ASTRONOMICAL OBJECTS
Galaxies: $1.21 \times 10^{12} \rightarrow 1.59 \times 10^{12}$
Stars: $4.89 \times 10^{18} \rightarrow 7.72 \times 10^{18}$
Black Holes: $6.16 \times 10^{16} \rightarrow 9.00 \times 10^{16}$
Planetary Systems: $6.36 \times 10^{18} \rightarrow 1.02 \times 10^{19}$
Worlds: $3.30 \times 10^{19} \rightarrow 5.34 \times 10^{19}$

WORLDS WITH LIFE
Simple Life: 0
Complex Life: 0
Intelligent Life: 0
Technological Life: 0
Interstellar Life: 0

Even as the Universe ages, gravitation steadfastly remains the most important force. On small cosmic scales, regions that were born just slightly denser than average have at last collapsed, forming waves of pristine stars in places that have never experienced it before. On slightly larger scales, galaxies are beginning to come into their own, as most of the early, isolated star clusters have now grown and merged together. On even grander scales, the earliest primitive skeleton of the cosmic web—a tenuous outline of dark matter—begins to form, with intersecting filaments hinting at the locations where the first galaxy clusters will arise.

But it's the early galaxies that deserve the limelight at this stage of cosmic evolution. Now that the intergalactic matter has been more than 99% reionized, the light from galaxies can travel all throughout the Universe, with only sparse clumps of neutral, pristine gas interspersed throughout the space between those galaxies. The largest, most massive galaxies are the ones that appear the brightest, as they've been forming stars continuously for around half a billion years at this point. As each prior generation of stars begins to die out, creating both supernova explosions and, later on, planetary nebulae as the most massive Sun-like stars reach the end of their life cycles, it triggers the formation of the next generation. As the newest stars form, they're richer in heavy elements than each of the prior generations.

In the core regions of these galaxies, you can find the greatest number and densities of stars. The rate of star formation continues to increase throughout the Universe, and the supermassive black holes at the centers of galaxies continue to grow extremely rapidly. For the very first time, the heaviest supermassive black holes cross the one-billion-solar-mass threshold: a milestone most Milky Way–sized galaxies will never reach. These arise in the most massive galaxies that have been starbursting—where the entire galaxy itself is forming new stars all throughout it—for several hundreds of million of years.

With each new episode of star formation, gas and dust are consumed, as that's the very material required to make new stars. Even though galaxies are growing by accruing matter from the surrounding space, and even though smaller clumps and clusters of stellar material fall onto them, the galaxies that have formed stars the fastest and grown by the greatest amounts have finally started to run out of star-forming material. The growing population of stars helps evaporate away additional dust, while the supermassive, central black hole also works to swallow up much of the most central material, funneling it into an accretion disk. Overall, these starburst galaxies are finally starting to deplete their dust—required material for creating new stars—which means the initial starburst phase of their lives is coming to a close.

As the dust clears, the full power of the supermassive black hole at the centers of these galaxies is finally revealed. Even as thousands of new stars form every year, an accretion disk forms surrounding the supermassive black hole. The black hole's gravitational forces accelerate and heat up the matter inside the accretion disk, drawing some of it across the event horizon but also shooting out material, close to the speed of light, in two jets that form perpendicular to the disk itself.

At long last, the first quasars have been born. From X-rays to visible light and into the radio wave spectrum, these ultra-energetic features common to quasars and active galactic nuclei have finally emerged.

A variety of stellar cataclysms, including core-collapse supernovae, exploding white dwarfs, and merging neutron stars, all expel vast amounts of matter back into the interstellar medium. This ejected material will be richer in heavy elements than before the cataclysm, seeding the Universe with heavy elements.

IMAGE BY JON LOMBERG

THE SEEDS OF COMPLEXITY

heavy elements form worlds

UNIVERSE
Diameter: $4.37 \times 10^{12} \rightarrow 4.78 \times 10^{12}$ ly
  Observable: $1.09 \times 10^{10} \rightarrow 1.20 \times 10^{10}$ ly
Expansion: $932.05 \rightarrow 814.66$ km/s/Mpc
Density: $147.33 \rightarrow 112.46$ p/m³
Temperature: $23.01 \rightarrow 21.02$ K

ASTRONOMICAL OBJECTS
Galaxies: $1.59 \times 10^{12} \rightarrow 2.03 \times 10^{12}$
Stars: $7.72 \times 10^{18} \rightarrow 1.15 \times 10^{19}$
Black Holes: $9.00 \times 10^{16} \rightarrow 1.28 \times 10^{17}$
Planetary Systems: $1.02 \times 10^{19} \rightarrow 1.53 \times 10^{19}$
Worlds: $5.34 \times 10^{19} \rightarrow 8.14 \times 10^{19}$

WORLDS WITH LIFE
Simple Life: 0
Complex Life: 0
Intelligent Life: 0
Technological Life: 0
Interstellar Life: 0

Within the expanding Universe, it's not just light emitted from stars and galaxies that reaches farther and farther into deep space but their gravitational influence as well. Light and gravity, despite their tremendous differences, have something remarkable in common: They can only travel at the speed of light. Early on after the hot Big Bang, only the smallest scales, leading to individual star clusters, will gravitationally collapse. Significantly later, larger structures, like protogalaxies and then full-fledged galaxies, emerge. But it's only now, when the Universe is some three-quarters of a billion years old, that the first cosmic structures on scales larger than an individual galaxy are beginning to form.

Up until this point, the gravitational effects of individual galaxies haven't been able to pull neighboring, surrounding galaxies together into a larger, bound structure. Driven by an invisible, constantly forming web of dark matter, however, these galactic clumps of matter find themselves caught up as part of a larger entity. For the first time, individual galaxies are being drawn together into groups and clusters, while the influence of dark matter begins to shape what will grow into cosmic structure on larger scales.

Quasars are beginning to become abundant, and the recycled material from previous generations of stars leads to regions that are now relatively rich in heavy elements, particularly towards the central cores of the largest galaxies. Inside those galaxies, we also find the remnants of the earliest stars: black holes, neutron stars, and white dwarfs. Whenever new stars form, about half of them form similarly to our Sun: in isolation, as the only stellar object in their star system. But the other half of them have companion stars as well and are part of binary, trinary, or even richer multi-star systems.

Over time, the most massive stars burn through their fuel and leave stellar remnants behind. The ones in binary (or greater) systems can interact, inspiral, and even merge together, producing their own unique sets of cosmic fireworks when they do. Black hole mergers emit enormous amounts of energy in the form of gravitational waves: ripples in the fabric of space-time. Neutron star mergers create kilonova events, enabling the formation of the heaviest elements of all found in nature, all while creating a new black hole at the center of the merger. And white dwarf mergers create a type of supernova that's entirely different from the core-collapse variety that arises from the most massive star; instead of producing a neutron star or a black hole, white dwarf mergers trigger a runaway fusion reaction, destroying both objects entirely and enriching the Universe with copious amounts of heavy elements.

On the largest scales, galaxy groups and galaxy clusters are forming for the first time, while on much smaller cosmic scales, the most massive, most evolved galaxies are building up substantial abundances of the heavy elements found all over the periodic table. Most stellar systems that are forming at this moment will now have a variety of planets—gas giants, ice giants, and rocky, terrestrial-like planets—and with just a few more generations, the first worlds with chances for life will soon emerge. It's even possible, perhaps, that in some very central region of the most enriched galaxies of all, the first planets with a chance at habitability are emerging right now.

Among the many young galaxies and protogalaxies forming in the Universe, an early merger creates the seed for what will eventually grow into our familiar Milky Way galaxy. Although it is not born with a spiral structure inside, this will develop over time.

IMAGE BY JON LOMBERG

THE BACKBONE OF NIGHT

—— *the Milky Way galaxy* ——

UNIVERSE
Diameter: $4.78 \times 10^{12} \rightarrow 5.17 \times 10^{12}$ ly
  Observable: $1.20 \times 10^{10} \rightarrow 1.29 \times 10^{10}$ ly
Expansion: $814.66 \rightarrow 725.06$ km/s/Mpc
Density: $112.46 \rightarrow 88.98$ p/m³
Temperature: $21.02 \rightarrow 19.45$ K

ASTRONOMICAL OBJECTS
Galaxies: $2.03 \times 10^{12} \rightarrow 2.52 \times 10^{12}$
Stars: $1.15 \times 10^{19} \rightarrow 1.63 \times 10^{19}$
Black Holes: $1.28 \times 10^{17} \rightarrow 1.76 \times 10^{17}$
Planetary Systems: $1.53 \times 10^{19} \rightarrow 2.18 \times 10^{19}$
Worlds: $8.14 \times 10^{19} \rightarrow 1.18 \times 10^{20}$

WORLDS WITH LIFE
Simple Life: 0
Complex Life: 0
Intelligent Life: 0
Technological Life: 0
Interstellar Life: 0

The Universe continues to expand, cool, and gravitate, forming stars and growing galaxies just as before. Only now, more than 800 million years after the start of the hot Big Bang, some truly remarkable features begin to emerge. The fastest-growing, most prolific star-forming galaxies continue to draw matter into them, but the rate at which stars form continues to grow even faster. As the process of star formation consumes more and more gas and dust, and the radiation from these numerous hot blue stars continues to evaporate away the remaining gas and dust, these early galaxies start to run into trouble. While the rate of star formation in the overall Universe continues its ascent, many of the most massive galaxies lose the ability to continue forming stars at such a breakneck pace. As they lose the ability to form stars from gas and dust in their core regions, their central black holes draw more and more of that material in: turning on and becoming active galaxies.

Although the expansion rate is still remarkably high—the Universe, at this time, is expanding at a full 10 times the rate it is today—it's smaller compared to what it was not so long ago. As a result, the radiation emitted by stars and galaxies continues to stream through the Universe, and the gravitational influence of massive objects now extends farther than ever before. Isolated star clusters are now rare, as most of them have merged and become components of early galaxies. The most massive individual star clusters tend to evolve into globular clusters: spheroidal collections of anywhere from tens of thousands to tens of millions of stars, found orbiting throughout the halos of spiral and elliptical galaxies. And the young galaxies themselves, despite the cosmic distances between them, have influenced one another gravitationally to begin the formation of galaxy groups and clusters in earnest. It's no longer just a few outliers but, rather, the start of an ongoing process that will shape the Universe for billions of years to come.

In one unremarkable, nondescript, and relatively secluded spot amid the great expanse of space, a sufficient amount of matter—including star clusters, gas, and dust—accumulates around a protogalaxy to evolve into a full-fledged galaxy for the first time. Many of the earliest stars to form have already died, but the lowest-mass ones, below about 75% of the Sun's mass, will continue to persist even to the present day. These earliest of the low-mass stars are red in color, low in brightness, and fuse hydrogen into helium in their cores at a slow burn, ensuring a long lifetime. Found in a mix of globular clusters and all throughout the young galactic disk as star clusters are torn apart by gravitational interactions within different portions of this galaxy, these stars persist even as their more massive, shorter-lived siblings die out. The young Milky Way, complete with many of the oldest stars we can still find today, has finally come into its own.

Even though starlight now streams freely through the expanse of space, it still encounters the occasional clump of neutral matter in the sparsest regions. As the Universe approaches 900 million years of age, more and more of the remaining neutral matter in intergalactic space becomes ionized. As this chapter in the Universe's history draws to a close, the veil of dust is finally lifted between distant galaxies and quasars. The Universe, at last, is fully visible.

Now that there are sufficient amounts of heavy elements present in the Universe, the first solid-surfaced, planet-sized objects, composed of icy, rocky, and metallic materials, are beginning to form. Only due to prior episodes of stellar enrichment are these worlds now possible.

IMAGE BY MARK A. GARLICK

RISE OF THE GENESIS PLANETS

—— the elements for life are in place ——

UNIVERSE
Diameter: $5.17 \times 10^{12} \rightarrow 5.50 \times 10^{12}$ ly
  Observable: $1.29 \times 10^{10} \rightarrow 1.39 \times 10^{10}$ ly
Expansion: $725.06 \rightarrow 652.18$ km/s/Mpc
Density: $88.98 \rightarrow 71.90$ p/m³
Temperature: $19.45 \rightarrow 18.11$ K

ASTRONOMICAL OBJECTS
Galaxies: $2.52 \times 10^{12} \rightarrow 3.06 \times 10^{12}$
Stars: $1.63 \times 10^{19} \rightarrow 2.23 \times 10^{19}$
Black Holes: $1.76 \times 10^{17} \rightarrow 2.35 \times 10^{17}$
Planetary Systems: $2.18 \times 10^{19} \rightarrow 2.99 \times 10^{19}$
Worlds: $1.18 \times 10^{20} \rightarrow 1.64 \times 10^{20}$

WORLDS WITH LIFE
Simple Life: 0
Complex Life: 0
Intelligent Life: 0
Technological Life: 0
Interstellar Life: 0

One aspect of the evolving Universe, from the Big Bang onward, appears to remain constant: More stars are forming now than ever before, with the older, more massive stars continuing to die in spectacular fashion. The most massive ones will have their cores collapse after burning through their nuclear fuel, triggering a supernova explosion that creates copious abundances of elements up to zirconium: the 40th element (with 40 protons in its nucleus) in the periodic table. When they die, they leave behind either neutron stars or black holes. Stars that are less massive but still bright and blue when they're born will also run out of fuel relatively quickly, dying a much gentler death as they blow off their outer layers while their cores—now made largely of carbon and oxygen—contract to form white dwarfs.

Although these processes have been going on for hundreds of millions of years at this point, the Universe still lacks some very important elements. There's almost no boron or beryllium present, as those are only created when heavier elements are struck by high-energy cosmic particles that blast them apart. There's practically no iodine, cesium, silver, gold, platinum, or iridium in existence, as stars that go supernova don't form them at all and the "gently dying" stars only produce tiny amounts of most of the heavier elements. And many of the intermediate elements—including sulfur, calcium, titanium, iron, cobalt, nickel, copper, and zinc—are only created in relatively small abundances by supernovae.

What is needed to form sufficient amounts of these elements—many of which are essential to the biological processes we find on Earth—is for stellar remnants, the corpses of prior generations of stars, to merge together. When two neutron stars, the corpses of stars that go supernova, merge together, they produce a spectacular explosion known as a kilonova. While approximately 95% of the total mass of merging neutron stars goes into producing either a heavier neutron star or a black hole, the remaining 5% primarily produces the heaviest elements of all. The majority of elements in the Universe heavier than zirconium—that is, with more than 40 protons in their nucleus—are produced by merging neutron stars.

White dwarfs, the corpses of Sun-like stars that aren't massive enough to go supernova, can either merge together or draw in sufficient quantities of matter from a companion star until they become too massive, at which point they explode. Unlike merging neutron stars, this will trigger a runaway fusion reaction within the white dwarf(s), causing their complete destruction. The majority of many elements in the Universe, including most of the elements from sulfur through zinc, are produced by these exploding white dwarfs, a special type of supernova that differs entirely from core-collapse events.

In the richest regions of the Universe, where all of these events have occurred with great enough frequency, a new class of stars can form: stars with not only rocky planets but also the full suite of elements that make up the periodic table. Although it will still take additional time for the first life-forms to arise, the most essential of the primitive ingredients—all of the stable atoms—now exist in the most enriched galaxies and can be found on the surfaces and in the interiors of rocky planets.

Now that a full one billion years have passed since the hot Big Bang, the Universe has formed more than 20 quintillion stars over its history. For the first time, more than 1% of all the stars that have ever formed are now present in the Universe.

Chemical reactions occur inside nebulae, within planet-forming systems, and from the outflows surrounding giant stars, resulting in the synthesis of both simple and complex organic molecules. These molecules include sugars, amino acids, and nucleobases, seeding the Universe with the precursor molecules needed for life.

IMAGE BY JON LOMBERG

THE BUILDING BLOCKS ASSEMBLE

—— the first chemical precursors of life ——

UNIVERSE
Diameter: $5.50 \times 10^{12} \rightarrow 5.91 \times 10^{12}$ ly
  Observable: $1.39 \times 10^{10} \rightarrow 1.48 \times 10^{10}$ ly
Expansion: $652.18 \rightarrow 593.76$ km/s/Mpc
Density: $71.90 \rightarrow 59.52$ p/m³
Temperature: $18.11 \rightarrow 17.01$ K

ASTRONOMICAL OBJECTS
Galaxies: $3.06 \times 10^{12} \rightarrow 3.62 \times 10^{12}$
Stars: $2.23 \times 10^{19} \rightarrow 2.93 \times 10^{19}$
Black Holes: $2.35 \times 10^{17} \rightarrow 3.06 \times 10^{17}$
Planetary Systems: $2.99 \times 10^{19} \rightarrow 3.94 \times 10^{19}$
Worlds: $1.64 \times 10^{20} \rightarrow 2.21 \times 10^{20}$

WORLDS WITH LIFE
Simple Life: 0
Complex Life: 0
Intelligent Life: 0
Technological Life: 0
Interstellar Life: 0

From the ashes of stars that live and die in the Universe, the material available to form future generations of stars and stellar systems becomes progressively richer. This includes not only the building blocks of matter—the atoms and their atomic nuclei—but also the structures that form when various building blocks bind together. The most significantly enriched regions of space, which is where the greatest fractions of heavy elements are found, are in the central regions of the most active, star-forming galaxies over the first one billion years of cosmic history.

The gas clouds in these enriched regions now possess the full suite of stable elements across the periodic table, as well as significant populations of unstable ones that will undergo radioactive decay. These atoms, as soon as the gas cools sufficiently, will begin to spontaneously link up, forming a vast array of molecules. While most of them will be simple molecules, such as methane, water, ammonia, hydrogen cyanide, and carbon monoxide, many of the atoms will link up and react—particularly in the presence of the radiation from nearby, massive stars and evolved stellar remnants—to produce significant abundances of complex, carbon-containing molecules: what we know today as organic molecules.

"Organic molecules" aren't only generated and used in biological processes, as the name might suggest. Instead, any molecule with carbon-hydrogen bonds in it is defined as an organic molecule. Given that hydrogen is the most abundant element in the Universe, while carbon is fourth, there are too many possibilities to count when it comes to the types of molecular combinations that form in our Universe. Even in this early stage of our cosmic history, once the raw ingredients are in place, we fully expect that there will be millions of such molecular configurations realized across these rich, young, rapidly evolving galaxies.

When you include the other abundant elements—including oxygen, nitrogen, phosphorus, sulfur, iron, copper, and more—a large number of chemical compounds will spontaneously arise in these stellar and interstellar environments. Simple sugars can be created very easily, as can a wide variety of amino acid precursors: methylamine, formamide, glycolonitrile, and aminoacetonitrile. Hydrocarbons and alcohols will spontaneously arise, and so will acids, aldehydes, and ketones. Additionally, ringed carbon structures, known as aromatic molecules, are created in great abundances in the interstellar environment. This includes not just simple ones, like benzene, but also complex, interlocking rings known as polycyclic aromatic hydrocarbons.

In the same regions where stars are forming, living, and dying most prolifically, these molecules appear in the greatest abundances. They can be found in Herbig-Haro objects, which occur when young, massive stars, shrouded by gas and dust, emit jets of light and particles that smash into the interstellar material. Those collisions wind up illuminating the regions in visible and infrared light and trigger chemical reactions in the stellar nurseries where new stars and stellar systems are forming.

As those new stars and stellar systems form, their protoplanetary disks will contain not only the same molecules found in interstellar space but more complex, evolved ones as well. It's highly likely that the same chemical compounds found in modern meteorites—including more than 80 species of amino acids, all five known nucleobases used in life processes, and a wide variety of organic compounds—will form in great abundances in new stellar and planetary systems. At last, the precursor ingredients for life are in place.

With organic ingredients at last present on rocky worlds in orbit around parent stars, life finally emerges from nonlife. This occurrence doesn't just happen once but likely in trillions of locations where the right conditions exist for molecules to extract energy from their environment and also to replicate.

IMAGE BY JON LOMBERG

LIFE, LIFE EVERYWHERE

—— the first proto-organisms ——

UNIVERSE
Diameter: $5.91 \times 10^{12} \rightarrow 6.26 \times 10^{12}$ ly
Observable: $1.48 \times 10^{10} \rightarrow 1.57 \times 10^{10}$ ly
Expansion: $593.76 \rightarrow 544.69$ km/s/Mpc
Density: $59.52 \rightarrow 50.01$ p/m³
Temperature: $17.01 \rightarrow 16.05$ K

ASTRONOMICAL OBJECTS
Galaxies: $3.62 \times 10^{12} \rightarrow 4.25 \times 10^{12}$
Stars: $2.93 \times 10^{19} \rightarrow 3.78 \times 10^{19}$
Black Holes: $3.06 \times 10^{17} \rightarrow 3.91 \times 10^{17}$
Planetary Systems: $3.94 \times 10^{19} \rightarrow 5.08 \times 10^{19}$
Worlds: $2.21 \times 10^{20} \rightarrow 2.91 \times 10^{20}$

WORLDS WITH LIFE
Simple Life: $1.04 \times 10^{13} - 1.06 \times 10^{14}$
Complex Life: 0
Intelligent Life: 0
Technological Life: 0
Interstellar Life: 0

s the Universe becomes progressively more enriched—as massive stars die, as stellar remnants merge, and as heavy atoms and atomic nuclei get ejected back into interstellar space—the new stars and star systems that form reap the benefits. For the first time, stars that are forming where prior generations of stars have enriched the interstellar material most significantly have it all: enough heavy elements and complex chemical compounds to make life a realistic possibility.

In these regions, the stars that are forming are not the pristine, hydrogen-and-helium-only stars that formed from material left over from the Big Bang: what astronomers call Population III stars. Nor are they the slightly enriched stars—the ones that struggle to make large, rocky planets containing a wide diversity of complex chemicals—that form subsequently: what are known as Population II stars. Instead, for perhaps the very first time in cosmic history, there are regions where sufficient numbers of stars have lived and died to produce stellar systems that have significant fractions of heavy elements, and those systems contain large, rocky planets and will be rife with organic molecules. At last, we're forming stars of the same type as our modern-day Sun: Population I stars.

Many of these first Population I stars will form with rocky planets around them, or with giant planets that have large, rocky moons orbiting them. Many of those worlds will be large and massive, like Earth, but will only have thin atmospheres. Many of those will happen to lie at the right distance from their parent star to have liquids such as water on their surfaces. And, as their outer layers—crust, ocean, and atmosphere—form, a non-negligible fraction of them will start producing chemical reactions never before seen in the Universe in an explosion of life.

In the liquid reservoirs on the surfaces of these worlds, including near deep-sea hydrothermal vents, in tide pools, in the shallows of continental shelves, and in lakes heated by underground volcanic activity (i.e., hydrothermal fields), something remarkable is occurring. The organic molecules present in these aqueous environments continuously collide, combine, and cleave one another apart. Amino acids synthesize into peptides and are broken apart again, producing innumerable configurations of transient molecules. Ions combine with those peptides, transforming them into enzymes. Some of these peptides and enzymes, just by random chance, turn out to be useful in some fashion and may even perform some type of metabolic function. Amid this primordial chemical soup, nucleic acids begin aligning with the various amino acids. These nucleic acids and peptides evolve together: the peptides driving metabolism and the nucleic acids causing reproduction of the original.

Although it's far more primitive than anything extant on Earth, these replicating molecules likely represent the earliest life-forms in the cosmos. Remarkably, in the span of under 100 million years, the Universe has gone from lifeless to having trillions of worlds where biological activity takes place.

Most of them, for better or worse, don't last very long. Any slight nudge in the wrong direction—a stray collision with another peptide, a temperature change of just a few degrees, or an encounter with a random electrical discharge or lava flow, for example—extinguishes primitive life-forms before they ever truly begin thriving. The opportunity to create life is at last present, but the opportunities for total extinction are vast. In the harsh, turbulent regions of active star formation where the Universe's earliest forms of life arise, extinction is the norm; survival is the exception.

Although life arises frequently, it also frequently goes extinct. Here, a massive star reaches the end of its life, exploding in a supernova and bringing an end to any organic life-forms that existed on the planets and moons around it.

IMAGE BY MARK A. GARLICK

DEATH, DEATH EVERYWHERE

—— *the first mass extinctions* ——

UNIVERSE
Diameter: $6.26 \times 10^{12} \rightarrow 6.61 \times 10^{12}$ ly
  Observable: $1.57 \times 10^{10} \rightarrow 1.65 \times 10^{10}$ ly
Expansion: $544.69 \rightarrow 502.72$ km/s/Mpc
Density: $50.01 \rightarrow 42.52$ p/m³
Temperature: $16.05 \rightarrow 15.20$ K

ASTRONOMICAL OBJECTS
Galaxies: $4.25 \times 10^{12} \rightarrow 4.92 \times 10^{12}$
Stars: $3.78 \times 10^{19} \rightarrow 4.78 \times 10^{19}$
Black Holes: $3.91 \times 10^{17} \rightarrow 4.90 \times 10^{17}$
Planetary Systems: $5.08 \times 10^{19} \rightarrow 6.44 \times 10^{19}$
Worlds: $2.91 \times 10^{20} \rightarrow 3.76 \times 10^{20}$

WORLDS WITH LIFE
Simple Life: $5.47 \times 10^{14} - 5.85 \times 10^{15}$
Complex Life: 0
Intelligent Life: 0
Technological Life: 0
Interstellar Life: 0

Even in a Universe where life is common, persistent survival is a challenge. Although it takes just over one billion years for the first ultraprimitive forms of life to arise—proto-organisms capable of gathering and metabolizing resources and then reproducing copies of themselves—the environments where that life appears are among the most violent in the young Universe. The only way to get all the raw ingredients in place, including copious amounts of the full suite of heavy elements that can rapidly become incorporated into new stars and new worlds, is to look to the richest regions of stellar birth and death over the early history of the Universe.

That means it's only the locations that underwent the greatest amount of star formation, historically, that will admit the possibility of life arising so early on. These are also the most chaotic regions of all: the centers of the richest, most massive, quickest-to-evolve galaxies in the young Universe. Only in these locations can the ingredients for life—and then, with the next generation of stars, life itself—arise in such record time. In these regions, however, the combination of ongoing star formation, a large number of stars in a small volume of space, high densities of stellar remnants, and high rates of stellar cataclysms all contribute to an incredibly high extinction rate.

Life, particularly the earliest forms of life, is an extremely fragile thing. The difference between survival and extinction can be a tiny difference in the temperature, acidity, or saltiness of an environment. Although planetary orbits might be stable around their stars, on average, over cosmically long timescales, the central environments of galaxies provide a large number of extinction factors to contend with.

When matter accretes onto a galaxy's central black hole, copious amounts of radiation and cosmic particles can flood the surrounding environments, making it difficult for life to survive. They can irradiate life-forms, destroy protective planetary ozone layers, and even strip away atmospheres. When white dwarfs or neutron stars merge together, they can create energetic explosions that sterilize the environments around them for multiple light-years. When massive stars go supernova or even hypernova, they can completely wipe out life in a multiple-light-year radius around them and can sterilize worlds for thousands of light-years if they happen to unluckily align with an energetic, cataclysmic jet of particles.

Furthermore, having large numbers of stellar and planetary systems all in the same region of space is bound to lead to disastrous gravitational encounters: where large, massive objects collide with one another or where the planetary and lunar orbits of living worlds are significantly altered. If the collisions or orbital changes are great enough—and many of them will be—the world will freeze, fry, or lose its atmospheres and/or oceans.

Most of the life that gets created in the earliest stages will rapidly go extinct, with only a few lucky worlds escaping catastrophe within the next ~100 million years of life arising in the first place. But some of the living worlds out there—a minority for sure, but an important one—will escape such a gruesome fate. With so many chances—so many stars, with planets and massive moons, with the raw ingredients for life, and where chemical metabolism and replication begins—even a small chance of survival leads to a great many living worlds. Just 10% of the way into our cosmic history, there are likely quadrillions of worlds where life, at least once, has appeared.

In the early stages of cosmic history, the star-formation rate rises and rises for the first few billion years without ever taking a break. New galaxies form, quasars appear and evolve, and galaxy groups and even larger-scale structures begin to form. Over time, even the internal structure of galaxies becomes more intricate, complex, and ordered.

IMAGE BY JON LOMBERG

LANTERNS OF THE NIGHT ALIGHT

—— stellar birth intensifies ——

UNIVERSE
Diameter: $6.61 \times 10^{12} \rightarrow 6.94 \times 10^{12}$ ly
  Observable: $1.65 \times 10^{10} \rightarrow 1.74 \times 10^{10}$ ly
Expansion: $502.72 \rightarrow 467.51$ km/s/Mpc
Density: $42.52 \rightarrow 36.71$ p/m³
Temperature: $15.20 \rightarrow 14.48$ K

ASTRONOMICAL OBJECTS
Galaxies: $4.92 \times 10^{12} \rightarrow 5.62 \times 10^{12}$
Stars: $4.78 \times 10^{19} \rightarrow 5.91 \times 10^{19}$
Black Holes: $4.90 \times 10^{17} \rightarrow 6.01 \times 10^{17}$
Planetary Systems: $6.44 \times 10^{19} \rightarrow 7.96 \times 10^{19}$
Worlds: $3.76 \times 10^{20} \rightarrow 4.76 \times 10^{20}$

WORLDS WITH LIFE
Simple Life: $2.56 \times 10^{15} - 2.87 \times 10^{16}$
Complex Life: 0
Intelligent Life: 0
Technological Life: 0
Interstellar Life: 0

While the earliest forms of life continue to arise and struggle for long-term survival, the stellar story of the Universe is still rising to a cosmic crescendo. Ever since the first stars began to form, the Universe has been increasing its overall star-formation rate. Despite the effects of cosmic expansion, where the space between galaxies and other cosmically bound structures increases with time, the atom-based matter that exists in the Universe is increasingly drawn into the regions where gravity is the strongest.

The earliest, most massive galaxies not only draw in matter from their surroundings but also pull the nearest galaxies to them, leading to galactic mergers on long enough timescales. As galaxies gravitationally attract one another, however, something simple occurs that leads to something remarkable.

The simple thing is that, from a gravitational perspective, the closest part of one galaxy to another experiences a greater gravitational attraction than the farthest part: an example of tidal forces. But when you pull, preferentially, on one part of a galaxy versus another, it causes the gas within it to compress together at specific locations. Wherever this gas compresses, it can lead to enough mass collecting in one location to lead to gravitational collapse: exactly what we need to trigger the formation of new stars.

In other words, even as the Universe continues to both expand and get less dense, it also continues to gravitate, leading to progressively more complex, denser, and richer structures in the Universe. As these structures grow, they exert larger and larger tidal forces on their environments, causing star formation to intensify wherever a large enough amount of hydrogen-rich gas exists in one place. Overall, this leads to a progressive increase in the rate of star formation that has yet to abate; the star-formation rate at this moment is 10 times as great as it was just one billion years ago.

But the stars that are forming are also more evolved than they've ever been before. Only a very small fraction of new stars—less than 1 in 100,000—are made of pristine material that's never formed stars previously. Even though most of the newly forming stars are still too poor in heavy elements to ever give rise to living worlds, a substantial fraction are precisely what we need. In fact, more than 20% of the stars that form during this 100-million-year interval are Population I stars: the same class of star as our modern-day Sun. Not only are we forming more stars than ever before, but more and more of them are of the right type to give rise to life in the Universe.

At the same time that all of this is occurring, galaxies that are separated by as much as tens of millions of light-years are gravitationally tugging on one another, resisting the expansion of the Universe. In the richest large-scale regions, hundreds or even thousands of young galaxies are beginning the slow process of clumping together into enormous galaxy clusters: a process that will take billions of years to complete. Meanwhile, at the centers of massive galaxies in possession of supermassive black holes, particularly when two large galaxies merge together, quasars form. Just as the number of stars increases, so does the number of quasars, as well as the sizes and masses of gravitationally bound structures. The Universe, despite all that exists within it at the present, is just getting started.

Throughout the cosmos, galaxies fight for space, colliding with their neighbors and triggering new waves of star formation. Galaxies that interact will typically merge together to form larger cosmic behemoths, with large numbers of major mergers eventually leading to the formation of giant elliptical galaxies.

IMAGE BY JON LOMBERG

GALAXIES DEVOUR GALAXIES

—— *galaxies merge with other galaxies* ——

UNIVERSE
Diameter: $6.94 \times 10^{12} \rightarrow 7.28 \times 10^{12}$ ly
  Observable: $1.74 \times 10^{10} \rightarrow 1.82 \times 10^{10}$ ly
Expansion: $467.51 \rightarrow 436.12$ km/s/Mpc
Density: $36.71 \rightarrow 31.88$ p/m³
Temperature: $14.48 \rightarrow 13.81$ K

ASTRONOMICAL OBJECTS
Galaxies: $5.62 \times 10^{12} \rightarrow 6.40 \times 10^{12}$
Stars: $5.91 \times 10^{19} \rightarrow 7.24 \times 10^{19}$
Black Holes: $6.01 \times 10^{17} \rightarrow 7.28 \times 10^{17}$
Planetary Systems: $7.96 \times 10^{19} \rightarrow 9.76 \times 10^{19}$
Worlds: $4.76 \times 10^{20} \rightarrow 5.98 \times 10^{20}$

WORLDS WITH LIFE
Simple Life: $6.90 \times 10^{15} - 8.10 \times 10^{16}$
Complex Life: 0
Intelligent Life: 0
Technological Life: 0
Interstellar Life: 0

On cosmic scales that now extend for over a billion light-years, the great gravitational dance continues. On supergalactic scales, galaxies and other clumps of matter attract one another. If there's enough mass—including both dark matter and normal matter combined—even well-separated galaxies can overcome the expansion of space between them. Given enough time and enough mutual mass to gravitationally attract, they'll cease their recession from one another, will accelerate towards each other, and will eventually merge.

In most cases, the galaxies are mismatched in size and mass, and the larger one will simply cannibalize the smaller one. Although this triggers a new wave of star formation, potentially even causing a starburst—where the entire galaxy forms new stars in great abundances—it's a short-lived phenomenon that leaves the original, large galaxy roughly intact. Once the galaxy settles down from the disruption caused by the merger, it will maintain its spiral shape and gas-rich properties. Its potential for forming new stars remains high.

But in rarer cases, particularly in the densest, richest regions where multiple large galaxies of comparable sizes exist, a phenomenon known as major mergers will occur. When two similarly sized galaxies merge together—complete with gas, dust, stars, stellar systems, and supermassive black holes—the entire shape, behavior, and composition of the galaxy can be transformed.

Individual stars and stellar systems will largely be undisturbed; the number of star-star and planet-planet collisions that will newly arise, even for a major galactic merger, can typically be counted on one hand. The gas, just as for minor mergers, will trigger starbursts and waves of novel star formation. And matter will accumulate in the environments around the central, supermassive black holes,

causing them to turn on and become active. It may even cause the galaxy to transform into a quasar.

But in the case of major mergers, three additional, spectacular phenomena can occur. First, the supermassive black holes, especially if they're of comparable masses to one another, can receive a substantial gravitational "kick" when they merge. That kick can be strong enough to lead to their ejection, entirely, from the post-merger galaxy that arises. It all depends on the relative masses, orbits, spin speeds, and spin orientations of the two supermassive black holes that will merge together. Although most galactic mergers, even major mergers, will grow a retained central black hole, a small fraction will be ejected. Consequently, many rogue supermassive black holes rove wantonly throughout the Universe.

A second thing that happens, owing to complex gravitational interactions, is that the motions of the constituent stars can become randomized. Instead of primarily orbiting in a plane, many, most, or even practically all of the stars inside can be thrown into randomly oriented orbits, creating a swarm-like structure instead of one that's primarily disk-like. Major mergers, particularly if there are many of them in rapid succession, can transform spiral galaxies into elliptical ones.

And finally, with a large enough merger and a massive enough starburst, the central regions of the galaxy can emit enough radiation and exert enough pressure to expel the galaxy's gas—the potential star-forming material—entirely. Many of the elliptical galaxies that we see in the modern Universe have no such material left and haven't formed any new stars in many billions of years. These giant cosmic smashups lead to the most advanced, evolved structures present in the Universe, and these gas-depleted ellipticals are already arising: just 1.5 billion years after the hot Big Bang.

At the heart of nearly every galaxy lies a supermassive black hole, and if there's a "food source" for these black holes in the form of infalling matter, these black holes will not only feed on them, but accelerate the matter within them and eject them in jetlike structures. These active galactic nuclei, or quasars, are among the most energetic phenomena in the Universe.

THE DENS OF DRAGONS

—— black holes are at the centers of galaxies ——

UNIVERSE
Diameter: $7.28 \times 10^{12} \rightarrow 7.60 \times 10^{12}$ ly
   Observable: $1.82 \times 10^{10} \rightarrow 1.90 \times 10^{10}$ ly
Expansion: $436.12 \rightarrow 409.30$ km/s/Mpc
Density: $31.88 \rightarrow 28.02$ p/m³
Temperature: $13.81 \rightarrow 13.23$ K

ASTRONOMICAL OBJECTS
Galaxies: $6.40 \times 10^{12} \rightarrow 7.17 \times 10^{12}$
Stars: $7.24 \times 10^{19} \rightarrow 8.70 \times 10^{19}$
Black Holes: $7.28 \times 10^{17} \rightarrow 8.65 \times 10^{17}$
Planetary Systems: $9.76 \times 10^{19} \rightarrow 1.17 \times 10^{20}$
Worlds: $5.98 \times 10^{20} \rightarrow 7.37 \times 10^{20}$

WORLDS WITH LIFE
Simple Life: $1.38 \times 10^{16} - 1.70 \times 10^{17}$
Complex Life: 0
Intelligent Life: 0
Technological Life: 0
Interstellar Life: 0

 ince the deaths of the very first stars within practically every galaxy, a supermassive black hole has emerged and persisted. Whenever you have a large collection of collapsed objects— planetary bodies, brown dwarfs, stars, and stellar corpses—they're all going to interact gravitationally, but objects of different mass will respond in different ways to the same types of force. The largest masses are the hardest to accelerate, while the smallest masses can easily achieve high speeds. As a result, over long periods, the most massive objects tend to sink to the centers of large, gravitationally bound collections, while the smallest tend to get ejected into intergalactic space: a process known as "violent relaxation."

As a result, the most massive, earliest black holes merge together in the galactic center, leading to the supermassive black holes that we see all throughout the Universe's cosmic history. So long as a galaxy has copious amounts of gas, the material that's essential for forming new stars, new star clusters are continuously going to be formed, with the central regions of galaxies serving as the greatest hotbeds of star formation. Much of this material will wind up being consumed by the central, supermassive black hole, where it not only grows the black hole's mass but also gets accreted and accelerated by it, leading to energetic radiation and even collimated jets of particles.

Every galaxy, once formed sufficiently, is thought to start out with conditions like this. The dust-rich and gas-rich environment forms stars rapidly, additional matter falls into the galaxy, the star-formation rate increases, and the central black hole grows and accretes matter, emitting radiation and particles. But there's a limit to how long the galaxy can remain like this, because there's only so much energy that the dust can absorb. Not only will it re-radiate that energy away, but eventually, it will boil away and become completely ionized. This transformation not only ends new star formation, but allows the radiation and particles from the central black hole to escape into intergalactic space.

Whenever we see a galaxy whose central black hole is energetically emitting particles and radiation, we see either an active galactic nucleus or—for the most energetic sources—a quasar. As the structures within the Universe continue to grow and evolve, not only do we see the star-formation rate continuously rising, but the abundance of quasars steadily increases as well. So long as there continues to be material falling onto these central black holes, the black holes will accrete and accelerate that matter, and the galactic core will continue to remain active: a state they can maintain for hundreds of millions of years. In all the time we've been observing the Universe and over all the quasars we've ever observed, we've never seen a single one turn "on" or "off" on human timescales.

As galaxies grow and evolve, a greater fraction of new stars will possess heavy elements and can form worlds around them where life has the potential to arise. About a third of new stars formed during this time period will be enriched enough where life's emergence becomes a realistic possibility. For the first time, there are potentially inhabited planets forming not only in the centers of the richest galaxies but also in the planes of spiral galaxies thousands of light-years away from the galactic centers. Even as more energy is released into the Universe from stars and black holes, life's emergence only becomes more plentiful.

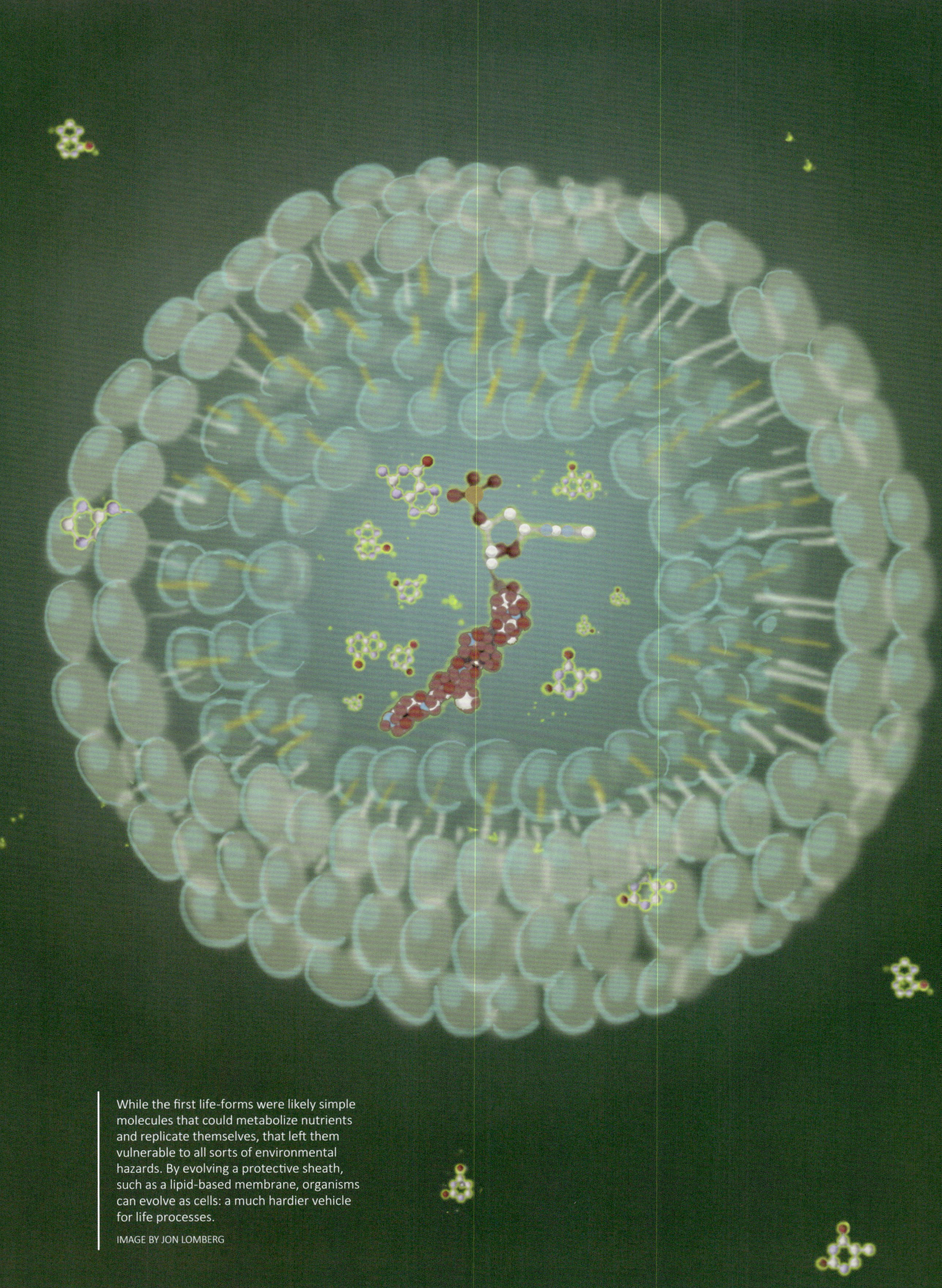

While the first life-forms were likely simple molecules that could metabolize nutrients and replicate themselves, that left them vulnerable to all sorts of environmental hazards. By evolving a protective sheath, such as a lipid-based membrane, organisms can evolve as cells: a much hardier vehicle for life processes.

IMAGE BY JON LOMBERG

RISE OF THE REPLICATORS

the first cells

UNIVERSE
Diameter: $7.60 \times 10^{12} \rightarrow 7.92 \times 10^{12}$ ly
  Observable: $1.90 \times 10^{10} \rightarrow 1.98 \times 10^{10}$ ly
Expansion: $409.30 \rightarrow 385.33$ km/s/Mpc
Density: $28.02 \rightarrow 24.77$ p/m³
Temperature: $13.23 \rightarrow 12.70$ K

ASTRONOMICAL OBJECTS
Galaxies: $7.17 \times 10^{12} \rightarrow 8.04 \times 10^{12}$
Stars: $8.70 \times 10^{19} \rightarrow 1.04 \times 10^{20}$
Black Holes: $8.65 \times 10^{17} \rightarrow 1.02 \times 10^{18}$
Planetary Systems: $1.17 \times 10^{20} \rightarrow 1.40 \times 10^{20}$
Worlds: $7.37 \times 10^{20} \rightarrow 9.04 \times 10^{20}$

WORLDS WITH LIFE
Simple Life: $2.36 \times 10^{16} - 3.02 \times 10^{17}$
Complex Life: 0
Intelligent Life: 0
Technological Life: 0
Interstellar Life: 0

The Universe is now teeming with worlds with all the raw ingredients needed for life to arise. Whenever a new stellar system forms, if there's a great enough abundance of heavy elements, there's a chance that rocky planets or moons will arise at the right distance from their parent star—with the right atmospheric and surface conditions—so that liquid water can come to exist in copious amounts on that world's surface. From the material that forms these worlds, many molecules that we recognize as the precursors of life, including sugars, amino acids, nucleobases, and other complex molecules, find themselves floating in naturally occurring aqueous environments on these worlds. At this point in time, just 1.6 to 1.7 billion years after the hot Big Bang, there are over 10 quintillion (10^{19}) stars whose worlds have the capacity to give rise to life.

The first major steps towards what we recognize today as a living organism likely occur many times over: the development of a peptide that could metabolize one of the abundant resources in its environment, alongside a set of nucleotides that could align with the various amino acids and trigger reproduction. Although most of these primitive metabolizing, reproducing molecules will die out for a variety of reasons, a small but significant subset of them will not only survive for long periods of time but will fill their niche with copies of themselves.

Over time and many generations, variety will arise within these molecules, as they collide with and assimilate information from other peptides and also as mutations occur. But the outside environment is a dangerous place, filled with large energy gradients and subject to wildly changing temperatures. Unless these primitive "metabolizing replicators" acquire some sort of ability to protect themselves from these potentially deadly conditions, their ability to survive and thrive over the long term becomes an extraordinary long shot.

However, the molecules present in these primordial, planetary environments include highly nonpolar molecules, including many of the lipids. Lipids, in both polar and nonpolar varieties, arise naturally through the combination of various acids—particularly large acid molecules—with a simple, naturally occurring alcohol: glycerol. When polar and nonpolar lipids combine, they can spontaneously arrange to form a membrane, capable of partitioning the world into an interior and exterior environment.

This membrane, in an aqueous environment rich in biomolecules, can serve two incredibly useful functions. First, it can keep complex, sensitive molecules shielded from the outside environment, protecting them from varying and sometimes aggressive external pressures. And second, it can confine biochemical activity inside the membrane, enabling the development of biocomplexity and conferring an evolutionary advantage onto any such replicating molecule that produces a better enzyme, protein, or other beneficial molecule.

Whenever these two phenomena combine—when metabolizing, self-replicating molecules make use of a membrane and develop the ability to isolate themselves from external hazards, protecting the critical material inside—the Universe will have newly created a proto-cell. All extant life on Earth can be traced back to such a cell, and such entities should have arisen naturally in the prebiotic environments of potentially habitable worlds. With so many chances arising so rapidly, the emergence of the first cells is all but inevitable.

A massive enough molecular cloud of gas, when it collapses under its own gravity, won't simply form small star clusters that will dissociate within their host galaxy but, rather, an enormous globular cluster, with hundreds of thousands to tens of millions of stars inside. These structures, unlike the smaller star clusters that are more common today, can survive for tens of billions of years.

IMAGE BY MARK A. GARLICK

CLOUDS COLLAPSE AND STARS SWARM

—— *the first globular clusters* ——

UNIVERSE
Diameter: $7.92 \times 10^{12} \rightarrow 8.22 \times 10^{12}$ ly
 Observable: $1.98 \times 10^{10} \rightarrow 2.06 \times 10^{10}$ ly
Expansion: $385.33 \rightarrow 364.41$ km/s/Mpc
Density: $24.77 \rightarrow 22.10$ p/m³
Temperature: $12.70 \rightarrow 12.22$ K

ASTRONOMICAL OBJECTS
Galaxies: $8.04 \times 10^{12} \rightarrow 8.80 \times 10^{12}$
Stars: $1.04 \times 10^{20} \rightarrow 1.21 \times 10^{20}$
Black Holes: $1.02 \times 10^{18} \rightarrow 1.18 \times 10^{18}$
Planetary Systems: $1.40 \times 10^{20} \rightarrow 1.63 \times 10^{20}$
Worlds: $9.04 \times 10^{20} \rightarrow 1.08 \times 10^{21}$

WORLDS WITH LIFE
Simple Life: $3.56 \times 10^{16} - 4.76 \times 10^{17}$
Complex Life: 0
Intelligent Life: 0
Technological Life: 0
Interstellar Life: 0

osmic expansion is a relentless phenomenon, demanding that the matter present within our observable Universe occupy increasingly larger and larger volumes. By the time 1.8 billion years have elapsed since the onset of the hot Big Bang, every location in space is receiving light from material that's up to as much as 10 billion light-years away.

And yet, despite how spread out all of the matter in the Universe is, individual galaxies are only growing: accreting intergalactic matter and merging together at even greater rates than they've ever experienced before. The overall star-formation rate, even this late in the cosmic game, continues to rise. The number and abundance of quasars and active galaxies increases as well, as does the total energy output from starlight. The heaviest supermassive black holes now approach a whopping 10 billion solar masses, with event horizons that are some 20 times as large as a typical stellar system's Kuiper belt.

There are literally billions of galaxies at this time undergoing a stellar "baby boom," where hundreds or even thousands of new stars are being formed with each passing year in that galaxy alone. Something that we consider a large star-forming region in the modern Milky Way, like the Orion Nebula, only contains about 2,000 times the mass of the Sun total. By contrast, the star-forming regions present in these glittering galaxies can weigh in at more than 10 million solar masses each, far outpacing anything in the nearby, late-time Universe.

Most star-forming regions are relatively small and low in mass: of a few thousand solar masses or less. When these gas-rich regions gravitationally collapse, they form dense collections of young stars typically known as open star clusters. With hundreds or even thousands of new stars inside, they normally persist for a few hundred million years before the combined effect of stellar deaths and the gravitational interactions of the stars inside result in the cluster breaking apart. When that occurs, the constituent stars are strewn throughout the galaxy in a mix of singlet (51%), binary (35%), trinary (10%), and quaternary or higher (4%) star systems.

But when extremely large, massive clouds of gas gravitationally collapse, they can form anywhere from tens of thousands to many millions of stars, all within a volume of space just a few tens of light-years across. These collections of stars are bound together much more strongly than their open star cluster counterparts and are created not just in the planes or central regions of galaxies but all throughout the galactic halo. The resulting swarm of millions of stars—all tightly bound together in a fashion that will remain stable for billions of years—forms what's known as a globular star cluster.

While the modern Milky Way contains about 150 of these globular clusters, the most massive elliptical galaxies we know of possess upwards of 10,000 of them apiece. Almost all of the globular clusters we know of formed within the first two to three billion years of our cosmic history, and almost all of the stars inside are extremely metal-poor, as though only a few generations of stars had previously enriched the material that gave rise to the globular cluster's stars. When we see a typical globular cluster today, we have every right to expect that its constituent stars will be around 12 billion years old. If you want to feast your eyes on a "living fossil" from our ancient cosmic history, you need look no further than your typical, run-of-the-mill globular cluster, including the ones right in our own cosmic backyard.

In a stellar system with multiple stars, it will be the most massive, bluest, hottest stars that burn through their fuel the fastest and reach the ends of their life cycles first. Even if life were to develop on the rocky world shown here, the short lifetime of the blue star will ensure that all life will be extinguished in just a few hundred million years at most.

IMAGE BY MARK A. GARLICK

CANDLES THAT BURN TWICE AS BRIGHT

—— large, hot, blue stars live short lives ——

UNIVERSE
Diameter: $8.22 \times 10^{12} \rightarrow 8.52 \times 10^{12}$ ly
  Observable: $2.06 \times 10^{10} \rightarrow 2.13 \times 10^{10}$ ly
Expansion: $364.41 \rightarrow 345.69$ km/s/Mpc
Density: $22.10 \rightarrow 19.84$ p/m³
Temperature: $12.22 \rightarrow 11.79$ K

ASTRONOMICAL OBJECTS
Galaxies: $8.80 \times 10^{12} \rightarrow 9.68 \times 10^{12}$
Stars: $1.21 \times 10^{20} \rightarrow 1.41 \times 10^{20}$
Black Holes: $1.18 \times 10^{18} \rightarrow 1.35 \times 10^{18}$
Planetary Systems: $1.63 \times 10^{20} \rightarrow 1.90 \times 10^{20}$
Worlds: $1.08 \times 10^{21} \rightarrow 1.29 \times 10^{21}$

WORLDS WITH LIFE
Simple Life: $4.96 \times 10^{16} - 6.92 \times 10^{17}$
Complex Life: 0
Intelligent Life: 0
Technological Life: 0
Interstellar Life: 0

Whenever the Universe gathers together enough normal, atom-based matter in one place to trigger gravitational collapse, the inevitable result is the formation of new stars. But the new stars that wind up being produced always come in a great variety of sizes, masses, and colors. The stars that accumulate the greatest amounts of matter become high-mass, large, hot, blue stars, and these are the ones that shine the most luminously, as they burn through their core fuel the fastest. The ones with less matter become lower-mass, smaller, cooler, redder stars; their intrinsic brightnesses are significantly lower, and their lifetimes are much longer than their hot, blue, short-lived counterparts.

While most of the stars in the Universe are on the cool, red, low-mass end, with some 95% of all stars coming in at or under the mass of the Sun, the greatest amount of starlight is produced by those rare stars that are more massive than our Sun is. These stars are larger in radius than our Sun, shine at hotter temperatures, and can produce up to millions of times as much light, apiece, as our Sun does. With starlight spanning a greater temperature and energy range shining on their planets, there's every reason to believe that any Earth-sized planets orbiting one of these hot, bright, blue stars at the right distance for liquid water will be more likely to develop life on it than around a cooler, fainter, redder star like our own.

But these stars have a major disadvantage when it comes to the survival of life on inhabited planets surrounding them: Their lifetimes are incredibly short. There's a famous saying that "the candle that burns twice as bright burns half as long," but for stars, it's actually worse than that. A star that's born with twice the mass of the Sun will live just one-eighth as long, while a star that's born with 10 times the Sun's mass will live only one-thousandth as long. Instead of lasting for more than 10 billion years, as our Sun will, these brighter, more massive stars might live only for one to two billion years, for a few hundred million years, or even more briefly.

Yes, the conditions will frequently be right for life to arise on planets around these stars, but there simply won't be enough time for those life-forms to evolve into something beyond a single cell in the short times these planets retain conditions that are friendly to life. Even before the cores of these massive stars run out of fuel, the inner temperatures will rise as they begin to burn through the hydrogen in their inner cores, causing the rates of fusion inside to increase substantially and increasing the total energy output of these stars. Planets that once had the right conditions to have liquid water on their surfaces will, in a relatively short amount of cosmic time, see their oceans boil, all but eliminating the possibility of life continuing to survive and thrive in these environments.

Wherever new stars form with potentially habitable planets, life's emergence is possible. But if your star is too massive, its lifetime will be too short for any planetary life to advance beyond the simplest forms. Around the most massive, shortest-lived stars, life is likely to arise; but its rapid extinction is all but inevitable. For the first planets where life survives, thrives, and becomes complex and differentiated, we must not only wait longer; we must look elsewhere.

Although galaxies come in a great variety of shapes and sizes at this time, and with a wide variety of stellar populations, this epoch marks the end of Population III stars: stars made only of the pristine material left over from the Big Bang. All across the Universe, there are no lingering locations free of the pollutive ashes from prior generations of stars.

IMAGE BY JON LOMBERG

THE DYING OF A GENERATION

— *the era of first-generation stars ends* —

UNIVERSE
Diameter: $8.52 \times 10^{12} \rightarrow 8.83 \times 10^{12}$ ly
  Observable: $2.13 \times 10^{10} \rightarrow 2.21 \times 10^{10}$ ly
Expansion: $345.69 \rightarrow 328.46$ km/s/Mpc
Density: $19.84 \rightarrow 17.86$ p/m³
Temperature: $11.79 \rightarrow 11.39$ K

ASTRONOMICAL OBJECTS
Galaxies: $9.68 \times 10^{12} \rightarrow 1.06 \times 10^{13}$
Stars: $1.41 \times 10^{20} \rightarrow 1.63 \times 10^{20}$
Black Holes: $1.35 \times 10^{18} \rightarrow 1.55 \times 10^{18}$
Planetary Systems: $1.90 \times 10^{20} \rightarrow 2.20 \times 10^{20}$
Worlds: $1.29 \times 10^{21} \rightarrow 1.54 \times 10^{21}$

WORLDS WITH LIFE
Simple Life: $6.59 \times 10^{16} - 9.58 \times 10^{17}$
Complex Life: 0
Intelligent Life: 0
Technological Life: 0
Interstellar Life: 0

Imagine what a well-equipped observer would see as they look out at the Universe, even just two billion years after the Big Bang: stars and galaxies scattered throughout space in all directions, extending for billions of light-years. But in between these islands of luminous stellar activity are vast expanses of space where no stars or galaxies have yet formed. Even in these sparse locations that generate no light of their own, however, a variety of conditions persist.

In the regions close to large, luminous collections of mass, you're likely to find that there's much less matter present than you'd expect, cosmically, on average. The reason that cosmically complex structures like stars, galaxies, and clusters of galaxies can form is that gravitation is a runaway force. The more matter you draw in towards one location, the more successful that location will be at drawing in additional matter from its surroundings. The overdense regions will, over time, get progressively more and more overdense, while the surrounding regions will tend to give up their matter to the greatest local source of gravitational attraction. The more gas, dust, and even dark matter gets drawn into the nearest large "clump" of material, the less remains in the surrounding intergalactic medium.

But far away from any stars, galaxies, or clusters of galaxies, the matter that's present there proves difficult to move. Without a large, initial overdensity to serve as a seed for structure to form around it, the material found in these great cosmic voids struggles to gravitationally collapse. Gravitational growth, in these locations, can only occur over very long timescales. By this point in time, some two billion years after the hot Big Bang, the final collections of pristine material—material that's never formed stars and never been polluted by material from stars that formed elsewhere—are disappearing.

Even though the Universe is expanding, the expansion rate has been slowing as the Universe has evolved, all while gravitation continues its relentless pull. As a result, the Universe has been getting "clumpier" over time: Galaxies draw more matter into them, new stars and star clusters and galaxies form as gravitational collapse occurs for the first time, and cosmically underdense regions give up their matter to the relatively denser regions in their vicinity. While many intergalactic clumps of matter remain—including clumps of neutral, gaseous matter— almost all of them now possess some substantial fraction of heavy elements. Pristine populations of material are increasingly rare.

This marks the era where the last truly pristine clumps of matter have been discovered to exist: in the form of isolated clouds of gas that haven't yet finished collapsing to form stars and yet have remained far away enough from star-forming regions that they haven't yet been polluted by the recycled material from prior generations of stars. Composed of hydrogen and helium alone, these pristine clouds of gas are the last ones capable of producing those ultra-high-mass stars made of hydrogen and helium alone.

Although the Universe began some 13.8 billion years ago, and the normal matter within it quickly came to exist in the forms of hydrogen (75%) and helium (25%) and almost nothing else, truly pristine populations of material don't survive for long. By the time we pass the two-billion-year mark in our cosmic history, the last of it has either collapsed to form stars or has been polluted by the material from stars that formed elsewhere. The last of the "first stars" are being born, and dying, just two billion years after the Big Bang.

As the Universe becomes further enriched with heavy elements, particularly within massive galaxies that have undergone large quantities of star formation already, great numbers of stellar systems rich in rocky planets arise. This illustration shows a protoplanetary disk around a young star, with protoplanets already having formed within the disk of a spiral galaxy: a common disk-within-a-disk configuration.

IMAGE BY JON LOMBERG

RISE OF THE ROCKY WORLDS

rocky planets and moons become common

UNIVERSE
Diameter: $8.83 \times 10^{12} \rightarrow 9.12 \times 10^{12}$ ly
  Observable: $2.21 \times 10^{10} \rightarrow 2.28 \times 10^{10}$ ly
Expansion: $328.46 \rightarrow 313.05$ km/s/Mpc
Density: $17.86 \rightarrow 16.18$ p/m³
Temperature: $11.39 \rightarrow 11.01$ K

ASTRONOMICAL OBJECTS
Galaxies: $1.06 \times 10^{13} \rightarrow 1.15 \times 10^{13}$
Stars: $1.63 \times 10^{20} \rightarrow 1.86 \times 10^{20}$
Black Holes: $1.55 \times 10^{18} \rightarrow 1.75 \times 10^{18}$
Planetary Systems: $2.20 \times 10^{20} \rightarrow 2.51 \times 10^{20}$
Worlds: $1.54 \times 10^{21} \rightarrow 1.81 \times 10^{21}$

WORLDS WITH LIFE
Simple Life: $8.38 \times 10^{16} - 1.27 \times 10^{18}$
Complex Life: 0
Intelligent Life: 0
Technological Life: 0
Interstellar Life: 0

Newborn stars and stellar systems all come along with a remarkable possibility: two scenarios by which planets and moons can form. Universally, each new protostar will develop a circumstellar disk around it, and that disk provides the seed material necessary for planet formation. Wherever an imperfection in the disk arises, material will progressively accumulate around these protoplanetary "seeds," as any overdensity will draw more and more matter into it compared to the surrounding regions. These initial overdensities can give rise to protoplanetesimals, which can merge, grow, and accrete ever-increasing amounts of matter over time, eventually growing up to form full-fledged planets in mature stellar systems.

The main mechanism by which planets form around stars is known as the core-accretion scenario. In this picture, there are enough heavy elements—things like not only carbon and oxygen but also silicon, sulfur, magnesium, and iron—so that a dense planetary core can form even close to a parent star. With such heavy elements to "anchor" a potential new planet, even the intense winds and radiation coming off of a central protostar can't blow this material apart; they will persist even in the face of such extreme conditions. Once a sufficiently large, massive planetary core forms, it will continue to accrete additional material, giving rise to both rocky planets and gas giant worlds. Via core accretion, planets can form interior to, exterior to, and most important, within each star's zone of potential habitability. As a result of formation in this manner, rocky planets and rocky moons around gas giant worlds both offer opportunities for life to arise.

However, even around stars where core accretion is impossible, gravitational instabilities should still arise within those protoplanetary disks. If those instabilities are far enough away from the parent protostar, the winds and radiation won't be able to blow them away, and so giant planets without rocky cores can still form. While the core-accretion scenario seems necessary in order to form potentially habitable planets around most stars, every new star is capable of forming distant, gas-rich worlds around it via the gravitational instability mechanism.

Ever since the very first stars formed, the rate of star formation throughout the Universe has steadily increased: a trend that continues even now. However, for the first time, the star-formation rate across the cosmos now begins increasing at a slower rate than previously. Even though there are more stars than ever before—we've finally formed 10% of all the stars that exist in the Universe at present—the radiation from such excessively high star-formation rates begins to put the brakes on unchecked growth. Star formation, after more than two billion years, is finally approaching its cosmic peak.

It turns out that in order to form the potentially habitable planets that only arise via the core-accretion scenario, a stellar system needs to be born from material that's at least 25–30% as rich in heavy elements as our modern-day Solar System is. During this sliver of cosmic history, for the very first time, as many stars with these potentially habitable planets are being born as stars without them. Overwhelmingly, these potentially habitable new systems are found in the planes of galaxies with prominent disks and/or spiral arms, and in the very centers of actively star-forming galaxies.

At long last, the Universe has grown up enough that there are some locations out there where not only can we form Earth-like planets around Sun-like stars, but the conditions on those planets may be stable enough for life to survive and thrive for billions of years into the future, with the potential for complex, differentiated, and eventually, intelligent life to arise.

Although our own "planet Earth" didn't arise until more than 9 billion years after the Big Bang, the conditions were in place as early as ~2 billion years after the Big Bang for such worlds to form. Somewhere in the Universe, just 2.1 billion years after the Big Bang, an Earth-like world forms around a Sun-like star, with all the ingredients necessary for life to arise, survive, and thrive for billions of years.

IMAGE BY JON LOMBERG

THE BIRTH OF TERRA PRIMA

—— *the first Earth-like planet* ——

UNIVERSE
Diameter: $9.12 \times 10^{12} \rightarrow 9.41 \times 10^{12}$ ly
  Observable: $2.28 \times 10^{10} \rightarrow 2.35 \times 10^{10}$ ly
Expansion: $313.05 \rightarrow 299.26$ km/s/Mpc
Density: $16.18 \rightarrow 14.74$ p/m^3
Temperature: $11.01 \rightarrow 10.68$ K

ASTRONOMICAL OBJECTS
Galaxies: $1.15 \times 10^{13} \rightarrow 1.24 \times 10^{13}$
Stars: $1.86 \times 10^{20} \rightarrow 2.11 \times 10^{20}$
Black Holes: $1.75 \times 10^{18} \rightarrow 1.96 \times 10^{18}$
Planetary Systems: $2.51 \times 10^{20} \rightarrow 2.85 \times 10^{20}$
Worlds: $1.81 \times 10^{21} \rightarrow 2.11 \times 10^{21}$

WORLDS WITH LIFE
Simple Life: $1.03 \times 10^{17} - 1.62 \times 10^{18}$
Complex Life: 0
Intelligent Life: 0
Technological Life: 0
Interstellar Life: 0

More than two billion years after the Big Bang takes place, the Universe has evolved sufficiently so that it finally happens: The first truly Earth-like planet in the Universe forms. A molecular cloud of gas in the disk of a spiral galaxy—close to the central galactic bulge but not within it and with approximately the same fraction of heavy elements as our Solar System possesses today—collapses. Within that collapsing cloud, gravitational instabilities form, leading to the formation of protostars with protoplanetary systems around them.

Of the stars that form, 92% will have less mass than our Sun does, with four out of every five stars becoming red dwarfs. About 4% of the stars will be more massive than our Sun—hotter, bluer, and shorter-lived than our own—while the remaining 4% will be comparable in mass, temperature, and lifetime to our own Sun. Overall, about half of the new stars will be part of a multi-star system— binaries, trinaries, and even richer systems—while the other half will be like our own Solar System, possessing just a single star.

Only around 2% of these new stars will be about the same mass as our Sun and not have another stellar companion, with each such star forming its own circumstellar disk. Owing to the heavy elements inside this disk, a series of protoplanetary cores will form, with gravitation causing them to attract, grow, merge, and accumulate material from their surroundings. While some protoplanets will get ejected as others get redirected into their parent star, the survivors will gravitationally grow. The ones that grow the largest cores will become giant planets with their own moon-forming circumplanetary disks; the ones that fail to cross a critical threshold fast enough will become rocky, terrestrial planets, similar to Mercury, Venus, Earth, and Mars.

These conditions will persist until the protostar becomes a full-fledged star and the remaining protoplanetary disk evaporates entirely. What's then left behind is a fully formed stellar system, complete with rocky planets (some of which have moons from protoplanetary collisions), giant planets (all of which should have moons and ring systems), asteroids, Kuiper belt objects, and large, diffuse Oort cloud at the outskirts. Any rocky planets or large, rocky moons around giant planets that happen to be located at the right distance from their parent stars to have liquid water on their surfaces are prime candidates for life to arise.

Around one such star that's newly formed in a young, rich, evolved disk galaxy, a potentially habitable planet containing molten surface—rife with volcanic activity— begins to cool. Simultaneously, comets, asteroids, and some of the remaining protoplanetesimals impact the young planet, creating a primitive atmosphere rich in simple gases. While the hydrogen and helium boil away quickly, the surviving, heavier gases include methane, water vapor, nitrogen, ammonia, as well as carbon dioxide and sulfur dioxide. With a sufficiently dense atmosphere, liquid water begins to pool on the surface, creating continents, salty oceans, as well as more variable lakes, rivers, streams, and other sources of fresh water driven by weather events.

In many of these environments, particularly in concert with early volcanic activity, all of the necessary conditions and precursor molecules are in place to support the emergence of life on that world. As soon as that life arises, survives, reproduces, and begins thriving, we know we'll have formed the Universe's first truly Earth-like planet: Terra Prima.

Around nearly all massive galaxies, anywhere from 30 to 100 dwarf galaxies, as well as hundreds to tens of thousands of globular clusters, can cluster in the nearby environment. In what will become our own Local Group, the now-ancient globular cluster Omega Centauri has already formed, with many of its stars having planets, as shown here.

IMAGE BY MARK A. GARLICK

RELICS OF THE EARLY UNIVERSE

— globular clusters encircle galaxies —

UNIVERSE
Diameter: $9.41 \times 10^{12} \rightarrow 9.70 \times 10^{12}$ ly
  Observable: $2.35 \times 10^{10} \rightarrow 2.42 \times 10^{10}$ ly
Expansion: $299.26 \rightarrow 286.54$ km/s/Mpc
Density: $14.74 \rightarrow 13.47$ p/m³
Temperature: $10.68 \rightarrow 10.36$ K

ASTRONOMICAL OBJECTS
Galaxies: $1.24 \times 10^{13} \rightarrow 1.34 \times 10^{13}$
Stars: $2.11 \times 10^{20} \rightarrow 2.38 \times 10^{20}$
Black Holes: $1.96 \times 10^{18} \rightarrow 2.19 \times 10^{18}$
Planetary Systems: $2.85 \times 10^{20} \rightarrow 3.21 \times 10^{20}$
Worlds: $2.11 \times 10^{21} \rightarrow 2.46 \times 10^{21}$

WORLDS WITH LIFE
Simple Life: $1.24 \times 10^{17} - 2.01 \times 10^{18}$
Complex Life: 0
Intelligent Life: 0
Technological Life: 0
Interstellar Life: 0

For every large, luminous, modern galaxy that exists in the Universe today, there are anywhere between 30 and 100 tiny dwarf galaxies that are found gravitationally bound to its outskirts, as well as hundreds to tens of thousands of globular clusters scattered throughout its galactic halo. When we examine most of these minuscule galaxies and compact collections containing anywhere from a few thousand to a few million stars, we find that they overwhelmingly consist of very old stars: stars that formed within the first ~2.5 billion years after the Big Bang.

In fact, most of the globular clusters that we know of had already finished forming their stars by this point in time. Others, including some of the closest, densest collections of stars to us—like globular clusters Messier 62 and NGC 6752—are rapidly forming all of the stars inside of them right now: all at once. But unlike the stars forming in the central regions of galaxies or within the disks of more massive galaxies, there are practically no planets at all around them.

There's an explanation for this, however, for both low-mass galaxies and also for globular clusters. In both cases, there are gravitationally bound collections of matter initially, and the normal matter cools and gravitationally contracts. In both cases, the normal, atom-based matter within these collections has been enriched by previous generations of stars, but typically only at the 1–10% level compared to the fraction of heavy elements that our Sun presently possesses. With such a minuscule amount of heavy elements present, there isn't enough material present to form the metal-and-rock protoplanets that are required to build up either a rocky planet or a gas giant world close to a parent star. Within both globular clusters and low-mass galaxies, planets are exceedingly rare.

In the case of low-mass galaxies, the normal matter exists alongside dark matter: the silent, invisible, gravitational "glue" that forms the backbone of the cosmic web. The dark matter in any individual bound structure, such as a small, low-mass galaxy, will remain distributed in a large-volume, diffuse halo. The normal matter, however, being made of atoms, can gravitationally collapse, forming a burst of stars all at once. When a large wave of star formation occurs, however, the hot, young stars that form can easily—through radiation and the emission of particles—blow the remaining gas clear out of the galaxy. Stars form all at once in these objects, and then never again, as the material needed to form future generations of stars gets expelled in the immediate aftermath.

In the case of globular clusters, the normal matter that collapses is simply a cloud of gas in the halo of a larger galaxy. When gravitational collapse occurs, a tremendous collection of stars gets created: from tens of thousands of stars up to several million. Again, the remaining gas is expelled, but unlike the scenario for low-mass galaxies, there is no dark matter inherent to these globular clusters; there's only the "background" dark matter that's distributed in a halo around the parent galaxies.

These ancient, low-mass relics can persist until the present day, but they'll evolve differently. Over time, the hotter, bluer, more massive stars will die first, leaving only the low-mass, cooler, red stars behind. In globular clusters, the more massive stars sink to the center, forming a dense stellar core. Within that core, many stars will merge, forming heavier, bluer stars known as blue stragglers. But in low-mass galaxies, where dark matter dominates, the stars remain diffusely distributed and only rarely merge. More than 11 billion years later, these stellar collections will still remain, glittering around every modern galaxy.

With liquid water precipitating on a rocky planet orbiting a still-living star, the conditions for life to arise are likely all in place. If a nearby star ends its life in a supernova, it may drive many organisms to extinction, but as long as even a small percentage of organisms survive, life will continue on this world for a long time to come.

IMAGE BY MARK A. GARLICK

LIFE IS BUT A WALKING SHADOW

—— the ephemerality of life ——

UNIVERSE
Diameter: $9.70 \times 10^{12} \rightarrow 9.98 \times 10^{12}$ ly
  Observable: $2.42 \times 10^{10} \rightarrow 2.50 \times 10^{10}$ ly
Expansion: $286.54 \rightarrow 274.80$ km/s/Mpc
Density: $13.47 \rightarrow 12.35$ p/m³
Temperature: $10.36 \rightarrow 10.07$ K

ASTRONOMICAL OBJECTS
Galaxies: $1.34 \times 10^{13} \rightarrow 1.43 \times 10^{13}$
Stars: $2.38 \times 10^{20} \rightarrow 2.66 \times 10^{20}$
Black Holes: $2.19 \times 10^{18} \rightarrow 2.42 \times 10^{18}$
Planetary Systems: $3.21 \times 10^{20} \rightarrow 3.59 \times 10^{20}$
Worlds: $2.46 \times 10^{21} \rightarrow 2.83 \times 10^{21}$

WORLDS WITH LIFE
Simple Life: $1.46 \times 10^{17} - 2.46 \times 10^{18}$
Complex Life: 0
Intelligent Life: 0
Technological Life: 0
Interstellar Life: 0

espite the tremendous number of galaxies that are strewn all throughout the Universe—even at this early stage, there are already trillions—the overwhelming majority of stars are locked up inside the most massive 1% of all galaxies. Not only are most of the stars found in the largest galaxies, but the overwhelming majority of the most enriched stars are found there as well. These are the stars with the greatest fractions of heavy elements around them, as well as the stars most likely to possess planets and the ingredients necessary for life on their surfaces.

Among the planets that have numerous similarities to Earth—including in terms of mass, size, atmospheric thicknesses, surface temperatures, and abundances of water—many of them will see life arise on them. In most cases, however, life is expected to be extremely fragile; a tiny shift in environmental conditions can render a once-fertile location completely inhospitable to the life that exists there. A tiny change in the pH of an aqueous environment, a change in salinity, or a small amount of added (or removed) heat can all mean the difference between survival and extinction for primitive life-forms.

On the overwhelming majority of worlds where life develops, due to the fragility of early life, we have every reason to think that complete extinction is the norm and that long-term survival is the exception. Even if an organism happens to occupy an environment that experiences relative stability, all sorts of events can drive them to extinction. They can poison their own environments with their metabolic waste products, rendering it inhospitable for future generations. (Yeast cells do this in sugar-rich environments here on Earth.) A flare or outburst from their parent star can annihilate life on that world, as can even a minor gravitational interaction or orbital change. Severe weather events, asteroid/comet strikes, volcanic activity, nearby stellar cataclysms, jets from black holes, and many other phenomena can all rapidly sterilize an otherwise thriving biosphere.

But in many cases, an imperfect extinction event—one that destroys most, but not all, of the species living on a world at a particular time—can be precisely what's needed to trigger an explosion in the diversity of life-forms found in any environment. In all cases where life is abundant, we expect there to be a finite set of resources, a set of physical and chemical properties inherent to the environment, and a set of living organisms that are well adapted to succeed, evolutionarily, in that particular ecosystem.

Without an extinction event or a change in conditions, the change in the genomes of the organisms present will be slow, driven largely by random mutations. With too large of an extinction event, everything dies, and there's nothing left to occupy the now-unfilled ecological niches. But with a large, but imperfect, extinction event, life-forms that existed on the fringes can now rise to dominance. Organisms that survive these mass extinctions can reproduce and have access to resources that were monopolized by a previously dominant species; with their extinction, a rapid change in the biological balance of an inhabited world will surely ensue.

A 100% successful mass extinction event is always game over for life on any inhabited world. But when at least some life-forms survive, it can herald a giant evolutionary leap. After all, the same biological rules should apply to life all across the Universe, and the species that survive are always the ones most adaptable to change.

Interstellar space might not be the first place you'd think of looking for living organisms, but as life arises on a variety of worlds and those worlds then experience impact events, life-forms can be ejected into interstellar space. If that life then lands on another world and successfully reproduces, it can seed biological activity there: an example of the process known as panspermia.

IMAGE BY JON LOMBERG

AND YET, LIFE FINDS A WAY

—— life can spread across worlds ——

UNIVERSE
Diameter: $9.98 \times 10^{12} \rightarrow 1.03 \times 10^{13}$ ly
 Observable: $2.50 \times 10^{10} \rightarrow 2.56 \times 10^{10}$ ly
Expansion: $274.80 \rightarrow 264.33$ km/s/Mpc
Density: $12.35 \rightarrow 11.39$ p/m^3
Temperature: $10.07 \rightarrow 9.80$ K

ASTRONOMICAL OBJECTS
Galaxies: $1.43 \times 10^{13} \rightarrow 1.52 \times 10^{13}$
Stars: $2.66 \times 10^{20} \rightarrow 2.95 \times 10^{20}$
Black Holes: $2.42 \times 10^{18} \rightarrow 2.66 \times 10^{18}$
Planetary Systems: $3.59 \times 10^{20} \rightarrow 3.98 \times 10^{20}$
Worlds: $2.83 \times 10^{21} \rightarrow 3.24 \times 10^{21}$

WORLDS WITH LIFE
Simple Life: $1.69 \times 10^{17} - 2.93 \times 10^{18}$
Complex Life: 0
Intelligent Life: 0
Technological Life: 0
Interstellar Life: 0

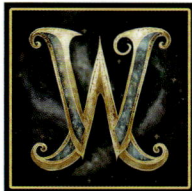

What happens on a living world, as far as we can tell, doesn't always remain on a living world. In a Universe rife with cosmic collisions, gravitational interactions, asteroid and comet strikes, and all sorts of violent, energetic cataclysms, life's persistence is constantly challenged. And yet, that very same phenomenon—of wild, unstable energy gradients across atmospheres, oceans, and continental landmasses—can be exactly the thing needed to give rise to life-sustaining environments, even in the most counterintuitive of places.

It's easy to look at our own planet, where life clearly arose, as the blueprint for how life successfully arises and then thrives over long timescales in the Universe. But there's a problem inherent to making this assumption: Until we take a biological census of the Universe, all we can state with confidence is that our home planet emerged as a winner in some sort of great cosmic lottery. If each potentially habitable world is a lottery ticket, however, there are two unknowns that remain: We don't know what the odds of getting a winning ticket are, and we don't know what the other prizes are. It's possible that our own world, despite how successful life has been here, doesn't even represent the grand prize.

There are exomoons to consider: rocky moons orbiting gas-rich, giant planets. Perhaps, driven by the energy inputs from both their parent star and from tidal heating from their parent planet, life would have an even easier time of arising on such a world. There could be worlds containing much more water than Earth that are completely continent-free; with the right ingredients, the entire planet-wide ocean could be a laboratory for primitive life. Super-Earth planets, larger in size and with greater masses than Earth, could potentially be habitable if their atmospheres can remain thin enough, while cooler planets may harbor life in the subsurface waters beneath an icy crust. Where there's liquid water, the right biomolecules, and an energy gradient, life's possibility cannot be so easily dismissed.

Moreover, once a world becomes inhabited—where life arises and begins thriving—the possibilities for how that life evolves and persists might become greater than even our imaginations permit. On a world where a runaway greenhouse effect ensues, perhaps life won't go extinct but can take refuge in the cloud-rich layers of the upper atmosphere. On a world where atmospheric stripping leads to the loss of liquid water on its surface, perhaps life can thrive in a subterranean environment where liquid water persists underground. And perhaps, as a parent star warms and increases its energy output over its lifetime, once-sterile, lifeless worlds can suddenly gain the conditions needed for habitability.

And perhaps, if life arising at all from nonlife is a rare event, it may be spread from world to world, even across interstellar distances, through the natural activity of planetary bombardment. Here on Earth, a fraction of the meteorites that fall from the sky actually originate from other planets, moons, and asteroids in our Solar System; massive strikes can kick up debris and send planetary fragments hurtling throughout, and even out of, their system of origin. Perhaps the intergalactic medium in nearly all galaxies is populated with dormant, ejected life-forms that have stowed away aboard these fragments; and perhaps, when they land in just the right environment, life will survive and thrive there, too. This idea—known as panspermia—could be responsible for a significant fraction of inhabited worlds within our Universe.

While the earliest galaxies emerged after only a few hundred million years, the first full-fledged galaxy clusters take billions of years to form. As gravitational effects on large cosmic scales, tens of millions of light-years across, rapidly begin to add up, clusters of galaxies, containing hundreds to even thousands of galactic members, finally take shape.

IMAGE BY JON LOMBERG

THE TIE THAT BINDS

— the first mature galactic clusters —

UNIVERSE
Diameter: $1.03 \times 10^{13} \rightarrow 1.05 \times 10^{13}$ ly
  Observable: $2.56 \times 10^{10} \rightarrow 2.63 \times 10^{10}$ ly
Expansion: $264.33 \rightarrow 254.30$ km/s/Mpc
Density: $11.39 \rightarrow 10.50$ p/m³
Temperature: $9.80 \rightarrow 9.54$ K

ASTRONOMICAL OBJECTS
Galaxies: $1.52 \times 10^{13} \rightarrow 1.61 \times 10^{13}$
Stars: $2.95 \times 10^{20} \rightarrow 3.26 \times 10^{20}$
Black Holes: $2.66 \times 10^{18} \rightarrow 2.92 \times 10^{18}$
Planetary Systems: $3.98 \times 10^{20} \rightarrow 4.40 \times 10^{20}$
Worlds: $3.24 \times 10^{21} \rightarrow 3.69 \times 10^{21}$

WORLDS WITH LIFE
Simple Life: $1.95 \times 10^{17} - 3.47 \times 10^{18}$
Complex Life: 0
Intelligent Life: 0
Technological Life: 0
Interstellar Life: 0

Ever since the Big Bang first occurred, every massive particle has experienced the gravitationally attractive force from every other massive particle within its vicinity. However, this force isn't instantaneous across the vast cosmic distances: It can only propagate at the speed of light. Because the Universe is also expanding, that means that, at any given moment in cosmic history, there's a finite distance over which any massive concentration of matter can influence the other matter in the Universe.

It took tens to hundreds of millions of years for the very first clumps of matter to collapse on stellar scales, leading to the very first star clusters. It took hundreds of millions of years more for those star clusters to merge together into protogalaxies, and then nearly a billion years (total) for those protogalaxies to grow and evolve into full-fledged galaxies. It was only in the second billion years of our shared cosmic history that galactic groups began to form, with dark matter composing the rudimentary bones of what will eventually grow into the larger-scale structures that make up our cosmic web.

It's only ~2.5 billion years after the Big Bang that gravitation can extend its influence over the large distances that separate galaxies and individual galactic groups from one another, leading to the first galaxy clusters. Although proto-clusters have existed previously, this epoch corresponds to the very first time that full-fledged galaxy clusters have come into existence.

Collections of massive galaxies—on par with or even larger than the modern Milky Way—get drawn into the cluster's center, where they then proceed to pull on one another, exerting large tidal forces. These forces lead to a galaxy-wide burst of star formation, triggering the conversion of gas into new stars. Meanwhile, large amounts of gas exist between the individual galaxies, found in what's known as the intracluster medium. This gas itself, within the cluster, is hot and emits X-rays all on its own. X-ray emissions are how we find and identify galaxy clusters today, including the most distant one ever found: CL J1001+0220.

The gas found throughout the cluster acts like a frictional medium as galaxies pass through it. The intracluster gas causes each speeding galaxy to heat up and emit X-rays, all while stripping out a portion of their gas, leaving a trail of newly forming stars behind them. The concentration of large, massive galaxies in a cluster's core is the telltale signifier of the very first mature galaxy clusters. Although they're just coming into existence now, more will join them as time goes on. As the great gravitational dance continues, more and more galaxies and galaxy groups will be drawn into the earliest clusters; they may consist of only a few dozen large galaxies right now, but that number will soon grow into the hundreds—and later into the thousands—for the most massive, fastest-growing ones of all.

Although the galaxies closest to the cluster's center will form stars the most quickly, these are also the environments where large, massive galaxies go to die. As the galaxies collect and merge in the cluster's core, they trigger additional galaxy-wide bursts of star formation. The new stars can blow the remaining gas out of the galaxy entirely, while simultaneously changing their shapes from spirals into ellipticals. Without gas, no new stars form; as the hottest, shortest-lived stars evolve and die, only the cooler, redder ones remain. At the centers of galaxy clusters, "red and dead" elliptical galaxies will soon come to dominate.

As galaxies gravitationally tug on each other, they frequently collide and merge. If one galaxy punches through the central disk of another, not only will streams of gas and stars arise, but ripples will propagate outward through the disk, creating a stable ring of stars and the phenomenon of a rare ring galaxy.

IMAGE BY JON LOMBERG

BOUND UPON WHEELS OF FIRE

—— the emergence of cosmic rings ——

UNIVERSE
Diameter: $1.05 \times 10^{13} \rightarrow 1.08 \times 10^{13}$ ly
  Observable: $2.63 \times 10^{10} \rightarrow 2.70 \times 10^{10}$ ly
Expansion: $254.30 \rightarrow 245.39$ km/s/Mpc
Density: $10.50 \rightarrow 9.74$ p/m³
Temperature: $9.54 \rightarrow 9.30$ K

ASTRONOMICAL OBJECTS
Galaxies: $1.61 \times 10^{13} \rightarrow 1.70 \times 10^{13}$
Stars: $3.26 \times 10^{20} \rightarrow 3.57 \times 10^{20}$
Black Holes: $2.92 \times 10^{18} \rightarrow 3.18 \times 10^{18}$
Planetary Systems: $4.40 \times 10^{20} \rightarrow 4.82 \times 10^{20}$
Worlds: $3.69 \times 10^{21} \rightarrow 4.17 \times 10^{21}$

WORLDS WITH LIFE
Simple Life: $2.20 \times 10^{17} - 4.02 \times 10^{18}$
Complex Life: 0
Intelligent Life: 0
Technological Life: 0
Interstellar Life: 0

With the ongoing passage of time, the cosmic structures that have formed continue to evolve. More and more of the newly forming star systems have enough heavy elements in them to make planets, with rock-and-metal cores forming the foundation for gas giant planets. Galaxies continue to form, grow, and accrete matter from the intergalactic medium, while gravity pulls them together into pairs, groups, and clusters. And even among the relatively longer-lived stars, the ones that are about two to eight times the Sun's mass rapidly run out of hydrogen and helium fuel in their cores, leading to a gentle death for the star. In all three of these cases, there's the potential for one of the most striking features ever to appear in our Universe to arise: a cosmic ring.

When giant planets form, they most frequently arise from a scenario known as core accretion. A dense collection of heavy elements, formed out of the imperfections in a young star's protoplanetary disk, forms a rock-and-metal core. Once that core grows massive enough, it begins scooping up all of the volatile material around it: light gases like hydrogen and helium that would otherwise be blown away by this hot, energetic environment. Each planet that forms this way develops its own circumplanetary disk, which possesses its own gravitational instabilities, leading to a system of moons. When any moon experiences an energetic enough collision, it will experience utter destruction, with the collisional debris getting stretched out into a ring. All of the gas giant planets in our own Solar System formed this way, with each acquiring a unique ring system. The giant planets existing even at this early stage should possess both moons and rings as well, with the rings alternately experiencing creation and destruction over time.

Galaxies, meanwhile, not only attract one another in the depths of intergalactic space but also fall into larger structures like groups and clusters. Over time, the more massive galaxies will sink to the centers of these clusters, where they have the potential to interact and merge together. Most frequently, galactic collisions will produce a burst of new stars in both members, with the post-merger remnant settling down into a more evolved spiral or elliptical galaxy. But every once in a while, a collision with just the right geometry occurs. If one fast-moving galaxy passes through the center of a gas-rich, massive galaxy, it will create a shock wave that ripples outward. The rippling gas collapses at various locations, forming stars within the main galaxy, in streams trailing behind each galaxy, and also—quite spectacularly—in a ring encircling the galaxy that was "punched through." Only about 1 in 10,000 galaxies becomes a ring galaxy, but in such a rich cosmos, that means there are about a billion of them strewn throughout the observable Universe.

And finally, when Sun-like stars reach the end of their life cycles, they blow off their outer layers into the interstellar medium. As the central core contracts into a white dwarf, it heats up, ionizing the outer material and creating a planetary nebula. About 80% of planetary nebulae wind up with a bipolar structure: forming a shape like an hourglass. The other 20% form a ring-like structure: where dense knots of gas around the dying star's equator are illuminated more brightly than all of the surrounding regions. There are three ways—for planets, for galaxies, and for dying Sun-like stars—that the Universe makes rings, and this epoch in cosmic history sees them all created in great abundances.

Throughout the history of the Milky Way, an enormous number of smaller, low-mass galaxies have been absorbed. However, a few major mergers have occurred, the largest of which was with a galaxy known as "the Kraken," which occurs during this time period.

CLASH OF THE TITANS

the Milky Way merges with the Kraken

UNIVERSE
Diameter: $1.08 \times 10^{13} \rightarrow 1.11 \times 10^{13}$ ly
  Observable: $2.70 \times 10^{10} \rightarrow 2.77 \times 10^{10}$ ly
Expansion: $245.39 \rightarrow 236.81$ km/s/Mpc
Density: $9.74 \rightarrow 9.03$ p/m³
Temperature: $9.30 \rightarrow 9.07$ <

ASTRONOMICAL OBJECTS
Galaxies: $1.70 \times 10^{13} \rightarrow 1.79 \times 10^{13}$
Stars: $3.57 \times 10^{20} \rightarrow 3.91 \times 10^{20}$
Black Holes: $3.18 \times 10^{18} \rightarrow 3.45 \times 10^{18}$
Planetary Systems: $4.82 \times 10^{20} \rightarrow 5.28 \times 10^{20}$
Worlds: $4.17 \times 10^{21} \rightarrow 4.71 \times 10^{21}$

WORLDS WITH LIFE
Simple Life: $2.46 \times 10^{17} - 4.63 \times 10^{18}$
Complex Life: 0
Intelligent Life: 0
Technological Life: 0
Interstellar Life: 0

he Milky Way galaxy, our cosmic home in this vast Universe, was not always as it is today. Like most galaxies in the Universe, it was much smaller, lower in mass, and yet was forming stars much more rapidly 11 billion years ago than it is today. For the first part of its existence, the proto–Milky Way grew slowly, accreting matter from the intergalactic medium, growing up from a small, initial set of star clusters that merged together to form a protogalaxy. Over time, gas fell onto the Milky Way, small collections of matter were drawn in from the intergalactic medium, and gravitational infall dominated where the Milky Way got most of its matter from.

But the young Universe is filled with collections of matter, and even 2.7 billion years after the Big Bang, the observable Universe possesses just 2.5% of its present volume. Galaxies, although much smaller back then than they are at present, on average, are also much closer together in space than they are today. With so many clumps of matter competing to draw in additional material from the intergalactic medium, it's only a matter of time before two major clumps—that is, two young galaxies of comparable mass—draw one another in due to their mutual gravity.

When an event such as this occurs, the two galaxies gravitationally attract, accelerate towards one another, exert tidal forces on one another (which triggers new star formation), and most often, eventually collide and merge together. From the slow gravitational growth of accreting intergalactic matter, the young Milky Way had just about 10% of the total number of stars it possesses today. But right about now, between 2.7 and 2.8 billion years after the Big Bang, a spectacular event occurs: The Milky Way merges with the largest galaxy, relative to its mass at the time, that it will ever experience.

This long-ago-devoured galaxy, known as "the Kraken," might have only contained 3–4% of the stars that the Milky Way did before they merged but possessed anywhere from 9–20% of the Milky Way's mass. This is the closest event to a "major merger" our galaxy will experience until the far future: when the Milky Way and Andromeda galaxies eventually merge together. Of all the galaxies that the Milky Way has merged with, from its initial birth up until the present day, no galaxy played a larger role in shaping the evolutionary history of our cosmic home than the Kraken did.

This must have triggered a tremendous burst of star formation 11 billion years ago; if we could take a census of the faint, long-lived stars in our galaxy, we'd likely find evidence for a spike in the population of stars that dates back to this epoch. When we examine the globular clusters found in the halo of the Milky Way—of which there are about 150—we find that at least 13 of them, or approximately 10%, originated from an outside galaxy that was devoured by the Milky Way from precisely this epoch in cosmic history.

Just as fossils in Earth's sedimentary rock reveal the organisms that roamed our planet back in its ancient history, these astronomical fossils can help us conduct our own sort of "galactic archaeology," teaching us how the Milky Way grew up. Although the gradual accumulation of matter from the intergalactic medium is the most important factor in growing our home galaxy, mergers played major roles throughout our cosmic past. The Kraken—the largest of them all—quickly became our most significant victim of galactic cannibalism.

When two black holes of comparable mass but with large spins merge together, the post-merger black hole can often receive a high-velocity kick that boosts the black hole up to speeds of a few percent the speed of light. This can be sufficient to eject black holes, even supermassive black holes, from galaxies.

IMAGE BY MARK A. GARLICK

THERE CAN BE ONLY ONE

—— *extra black holes merge or get ejected* ——

UNIVERSE
Diameter: $1.11 \times 10^{13} \rightarrow 1.13 \times 10^{13}$ ly
  Observable: $2.77 \times 10^{10} \rightarrow 2.84 \times 10^{10}$ ly
Expansion: $236.81 \rightarrow 228.90$ km/s/Mpc
Density: $9.03 \rightarrow 8.41$ p/m³
Temperature: $9.07 \rightarrow 8.86$ K

ASTRONOMICAL OBJECTS
Galaxies: $1.79 \times 10^{13} \rightarrow 1.88 \times 10^{13}$
Stars: $3.91 \times 10^{20} \rightarrow 4.25 \times 10^{20}$
Black Holes: $3.45 \times 10^{18} \rightarrow 3.73 \times 10^{18}$
Planetary Systems: $5.28 \times 10^{20} \rightarrow 5.74 \times 10^{20}$
Worlds: $4.71 \times 10^{21} \rightarrow 5.28 \times 10^{21}$

WORLDS WITH LIFE
Simple Life: $2.73 \times 10^{17} - 5.28 \times 10^{18}$
Complex Life: 0
Intelligent Life: 0
Technological Life: 0
Interstellar Life: 0

 ractically every galaxy, as far as we can tell, has a supermassive black hole residing at its core. Once an initial seed black hole forms, gravitational interactions between objects of different masses cause the lightest objects to get ejected, while the heaviest ones become more tightly gravitationally bound, sinking to the galactic center. These supermassive black holes grow rapidly, often punctuated by bursts of activity followed by dormant periods, quickly achieving masses that rise into the millions or even billions of solar masses.

When two galaxies of comparable size and/or mass merge together, however, it can lead to a cosmically interesting scenario: one where a single galaxy winds up with multiple supermassive black holes. When this occurs, the black holes both sink to the center, driven by the gravitational interactions between each black hole and the individual stars, gas, dust, and stellar remnants that they encounter. Over time, as more material gets drawn into the galactic center, the two supermassive black holes begin orbiting one another: first from far away, and then more and more tightly as time goes on. Eventually, they'll get locked into a gravitational dance, where they don't simply orbit one another but where their mutual orbits begin to appreciably decay.

This is a feature of gravity that's absent in Newtonian gravity but that is absolutely inextricable from Einstein's General Relativity. Whenever a massive object moves and accelerates through the gravitational field generated by another nearby mass, its orbit begins to decay. It loses energy, because that energy gets carried away by the gravitational field itself: in the form of gravitational radiation, also known as gravitational waves.

Although we have yet to detect the gravitational waves emitted by supermassive black holes—our current technology is only sensitive to merging black holes with masses below that of a few hundred Suns—the physics underlying them is very well understood. They will orbit, emit gravitational waves, inspiral, and merge. Whenever you have a galactic merger, where each of the merging galaxies possesses a supermassive black hole, those black holes themselves will meet somewhere close to the center of the galaxy. When they do, they'll merge together; when this occurs is only a matter of time.

Whenever black holes merge together, they form an even more massive black hole. Although some of the total mass will be radiated away in the form of gravitational waves, the maximum amount of mass that can be lost to this form of energy—via $E = mc^2$—is about 11% of the smaller-mass black hole. The remainder goes into the post-merger black hole, which is always more massive than either of the progenitors was on its own.

But there's an important caveat here that's particularly relevant for our Milky Way: When two black holes merge, depending on how fast each one is spinning and what the relative orientations of their mutual orbits are, they can radiate gravitational waves preferentially in one direction, rather than equally in all directions. When this occurs, the post-merger black hole recoils in the opposite direction, at speeds up to thousands of kilometers per second. Even supermassive black holes can receive large enough "kicks" to get ejected from their host galaxy.

Today, the Milky Way's supermassive black hole is only four million solar masses: the smallest known among comparably sized galaxies. Most likely, it originally possessed a larger, more massive one, but a colossal merger ejected it. In the aftermath of our merger with the Kraken, we likely lost our original supermassive black hole. What remains is all that the cosmos could regrow in the time that's passed since.

Galaxies come in a great variety of sizes, masses, and shapes, with spirals and ellipticals being the most common, followed by irregulars. Towards the centers of galaxy clusters, giant ellipticals are most common; on their outskirts and in relative isolation, large spirals and dwarf irregular and dwarf elliptical galaxies are more common.

IMAGE BY MARK A. GARLICK

A DIVERSITY OF GALAXIES

—— galaxies come in many shapes and sizes ——

UNIVERSE
Diameter: $1.13 \times 10^{13} \rightarrow 1.16 \times 10^{13}$ ly
  Observable: $2.84 \times 10^{10} \rightarrow 2.90 \times 10^{10}$ ly
Expansion: $228.90 \rightarrow 221.59$ km/s/Mpc
Density: $8.41 \rightarrow 7.85$ p/m³
Temperature: $8.86 \rightarrow 8.65$ K

ASTRONOMICAL OBJECTS
Galaxies: $1.88 \times 10^{13} \rightarrow 1.96 \times 10^{13}$
Stars: $4.25 \times 10^{20} \rightarrow 4.59 \times 10^{20}$
Black Holes: $3.73 \times 10^{18} \rightarrow 4.01 \times 10^{18}$
Planetary Systems: $5.74 \times 10^{20} \rightarrow 6.20 \times 10^{20}$
Worlds: $5.28 \times 10^{21} \rightarrow 5.89 \times 10^{21}$

WORLDS WITH LIFE
Simple Life: $3.01 \times 10^{17} - 5.95 \times 10^{18}$
Complex Life: 0
Intelligent Life: 0
Technological Life: 0
Interstellar Life: 0

It's finally time: Three billion years after the Big Bang, the adolescent Universe is maturing. Yes, there are still plenty of relatively primitive structures out there, including stars too poor in heavy elements to have planets, isolated and immature galaxies, and large populations of molecules incapable of encoding complex information. But there are also rich regions of space that have been forming stars and growing galaxies for more than 90% of the Universe's history so far, and the densest of those locations have drawn so much matter into them that we're already finding galaxies that are entering the final stages of their lives.

When galaxies form, they typically result from numerous clumps of matter gravitating and merging together from random directions in our three-dimensional Universe. As gravity pulls this matter together, regardless of what shape or orientation it started off possessing, inevitably one direction will collapse more quickly than the other two, leading to the creation of a galactic disk. After only a few hundred million years, the familiar shape of a gas-rich spiral galaxy can begin to emerge.

It takes a significantly longer amount of time—a few billion years—for gravitation to begin drawing individual galaxies together into much larger structures, like rich galaxy clusters. It's right around now that not only are galaxy clusters becoming common, but large, individual galaxies are also interacting and merging together in the cores of these clusters. If you were to take a snapshot of one of these rich clusters, you'd find a collection of features that had never existed in the Universe prior to right now.

As galaxies approach the cluster's center, they speed through the intracluster medium at breakneck speeds: up to 2% the speed of light. The gas within them can get stripped out by collisions with the sparse amounts of matter found between the galaxies, leading to trailing streams of newly forming stars. Collisions and mergers of galaxies can create all sorts of irregularly shaped structures, including stretched and distended spiral arms, jets of matter, and hybrid galaxies containing features of both spirals and ellipticals all at once. Supermassive black holes will get fed, activating the central engines in these galaxies. And in the most extreme cases of galactic mergers, star formation inside of them can become so intense that it winds up expelling practically all remaining gas inside.

Galaxies that run out of gas can no longer form new stars; they can only shine with the light emitted from the already-formed, still-surviving stars they possess inside. These gas-poor, giant elliptical galaxies can now be found at the centers of the earliest, richest galaxy clusters, surrounded by a variety of spirals, ellipticals, as well as hybrid and irregular galaxies. As they continue to age, the hottest, bluest stars will be the shortest-lived, resulting in their rapid deaths. Over time, it will only be the cooler, redder, lower-mass stars that survive, as they're the longest-lived ones. No new stars will form inside of them, as without gas, there can be no new stars. For the first time, the richest, densest, most massive structures in the Universe have "red and dead" galaxies at their cores, surrounded by a gaggle of galaxies, all gloriously diverse in their structure and contents. The Universe is growing up, and it's the richest clusters that provide the most fertile ground for the rapid evolution of galaxies.

Averaged over all galaxies across the Universe, star formation now reaches its peak: with more stars forming in this 100-million-year interval than over any other in cosmic history. Within a galaxy like the young Milky Way, up to 10 billion stars can form at this time, with perhaps 100 million of those stellar systems serving as locations with all the right ingredients for life to form on a rocky planet.

IMAGE BY JON LOMBERG

100 QUINTILLION POINTS OF LIGHT

—— *the peak of star formation* ——

UNIVERSE
Diameter: $1.16 \times 10^{13} \rightarrow 1.19 \times 10^{13}$ ly
  Observable: $2.90 \times 10^{10} \rightarrow 2.97 \times 10^{10}$ ly
Expansion: $221.59 \rightarrow 214.85$ km/s/Mpc
Density: $7.85 \rightarrow 7.35$ p/m³
Temperature: $8.66 \rightarrow 8.47$ K

ASTRONOMICAL OBJECTS
Galaxies: $1.96 \times 10^{13} \rightarrow 2.04 \times 10^{13}$
Stars: $4.59 \times 10^{20} \rightarrow 4.94 \times 10^{20}$
Black Holes: $4.01 \times 10^{18} \rightarrow 4.29 \times 10^{18}$
Planetary Systems: $6.20 \times 10^{20} \rightarrow 6.67 \times 10^{20}$
Worlds: $5.89 \times 10^{21} \rightarrow 6.54 \times 10^{21}$

WORLDS WITH LIFE
Simple Life: $3.28 \times 10^{17} - 6.65 \times 10^{18}$
Complex Life: 0
Intelligent Life: 0
Technological Life: 0
Interstellar Life: 0

Ever since the very first stars formed just a few tens of millions of years after the Big Bang, there's been one constant in the Universe's evolution: Overall, the star-formation rate has only continued to rise. Within each individual galaxy, star-formation rates rise and fall over time, dependent on what types of interactions, mergers, and accretion events a galaxy experiences, but on the large-scale cosmic average, more stars have been forming with each successive time interval than they had been in the previous one. At long last, a little over three billion years after the Big Bang, that finally begins to change.

The Universe, finally, has reached the peak of star formation: where it forms new stars more rapidly than at any other time in cosmic history.

Star formation, like many phenomena in the Universe, can only proceed under the right conditions, and the very conditions that give rise to it will, if left unchecked, bring it to a screeching halt. In order to form new stars, you need a collection of gas massive enough that it can collapse under its own gravity. That gas also has to efficiently radiate away heat so that it can form clumps of matter dense enough to create stars; if you can't radiate away your heat quickly enough, you won't be able to achieve the needed densities to initiate nuclear fusion: the litmus test for whether you've formed a star.

Once nuclear fusion commences, however, these newborn stars emit large amounts of radiation and also create winds, both of which are forms of feedback that work to impede the formation of additional stars.

In other words, the greater the rate of star formation, the more severe the feedback from star formation is, and that works to slow and even prevent the formation of additional stars in these environments thereafter. Within individual galaxies, many spurts of star formation have been interspersed with periods of relative quiet already, and there's tremendous variation between galaxies as far as when and where they do and don't form stars.

While one galaxy might be just ramping up the number of stars it's forming, another one might be falling from its star-forming peak, entering a relatively quiet period. The Universe, driven by the first three billion years of cosmic evolution, is no longer anywhere close to a uniform place.

But if we consider the Universe as a whole, summed up over all the stars and galaxies inside, this one brief, hundred-million-year time interval is the one where more stars form than at any other time. Looking back from the present, our Universe's current star formation measures just 3% of what it was during this epoch when star formation was maximized. On average, a young Milky Way–like galaxy forms somewhere upwards of 10 billion new stars during this time, including perhaps a hundred million of which might someday give rise to a potentially inhabited, Earth-sized planet in orbit around it.

Given how enriched the Universe has already become, around half of these newly forming stars represent a chance for rocky planets with liquid water and organic molecules on their surface. And wherever those conditions exist, so too does the opportunity for new life to arise.

While there are many geological, chemical, and physical processes that can transform a world's surface, oceanic, and atmospheric features, the presence of biological activity can also be transformative. Here, life in the waters beneath an icy crust produces gases that cause fissures in the ice, altering the contents of the planet's atmosphere.

IMAGE BY MARK A. GARLICK

BREATHING LIFE INTO WORLDS

—— *life terraforms worlds* ——

UNIVERSE
Diameter: $1.19 \times 10^{13} \rightarrow 1.21 \times 10^{13}$ ly
  Observable: $2.97 \times 10^{10} \rightarrow 3.03 \times 10^{10}$ ly
Expansion: $214.85 \rightarrow 205.33$ km/s/Mpc
Density: $7.35 \rightarrow 6.88$ p/m³
Temperature: $8.47 \rightarrow 8.23$ K

ASTRONOMICAL OBJECTS
Galaxies: $2.04 \times 10^{13} \rightarrow 2.12 \times 10^{13}$
Stars: $4.94 \times 10^{20} \rightarrow 5.31 \times 10^{20}$
Black Holes: $4.29 \times 10^{18} \rightarrow 4.58 \times 10^{18}$
Planetary Systems: $6.67 \times 10^{20} \rightarrow 7.17 \times 10^{20}$
Worlds: $6.54 \times 10^{21} \rightarrow 7.17 \times 10^{21}$

WORLDS WITH LIFE
Simple Life: $3.56 \times 10^{17} - 7.39 \times 10^{18}$
Complex Life: 0
Intelligent Life: 0
Technological Life: 0
Interstellar Life: 0

The very first worlds where life arises likely came into existence more than a billion years prior to this moment in cosmic history, but in most of those instances, life will almost certainly quickly peter out. Life, for all of its ubiquity here on Earth, is tremendously fragile, as most organisms can only withstand minuscule changes to their environments and ecological niches before they are forced to either adapt or go extinct. In its primitive, early stages, it's likely that life was even more sensitive to variations in its early conditions, and tiny changes in temperature, acidity, or nutrient availability could mean the difference between life continuing or ending.

But on at least a small fraction of the worlds where life arises, it not only persists through its difficult infancy but also begins to spread, diversify, and thrive. While many events can lead to its extinction—both internally, such as a self-poisoning of its environment or destruction of life's needed resources, as well as externally, including from collisions, solar variations, or astrophysical cataclysms—there will inevitably be many worlds where life continues to persist. And wherever life persists, its metabolic by-products will begin to accumulate.

Gradually, over hundreds of millions or even billions of years, the results of life processes will begin to transform the atmospheres, waters, and even landmasses on those worlds. What was once a primitive atmosphere consisting largely of gases like hydrogen, helium, methane, ammonia, water vapor, with carbon dioxide and carbon monoxide also present, will eventually be transformed by the tiny microorganisms that use one or more of those molecules as a source of food, while producing other molecules as waste. No matter how large a planet or moon may be, its resources and atmosphere will be finite, and with enough time and enough living organisms, that initially inorganic atmosphere will eventually reach what's known as "carrying capacity." In short, it will be saturated by the effects of life.

On Earth, the outermost layers of our world that have been changed in this fashion are known as the biosphere and include the solid Earth itself, from the thin surface topsoil all the way down to bedrock, the fresh and salted bodies of water that float atop Earth's crust, and the atmosphere, all of which have been modified over the billions of years where life has persisted on our planet. Elsewhere in the Universe, creatures that have evolved to derive energy from a variety of metabolic pathways—whether thermal, chemical, or electrical—will similarly transform their own worlds: The only relevant questions are precisely how and on what timescales.

Although the very first inhabited worlds likely arose billions of years before this moment in cosmic history, our best methods for detecting a living world all rely on planet-wide changes caused by the sustained presence of life. For the first time, just a little over three billion years after the start of the hot Big Bang, worlds are being completely transformed by tiny, primitive microorganisms that evolved, survived, thrived, and continued to do what living creatures do: metabolize resources and emit waste products. It's the simplest biological story in all the Universe, and with sufficiently advanced technology, we could be able to identify inhabited worlds whose light comes to us from even more than 10 billion years ago.

With galaxy clusters now commonplace,
it's only a matter of time before two
nearby clusters gravitationally attract one
another with sufficient force to lead to
them merging together. The violent events
of galaxy cluster collisions not only lead
to rapid bursts of star formation and the
creation of hot, X-ray-emitting gas but can
also cause dark matter and normal matter
to separate from one another.

IMAGE BY MARK A. GARLICK

MANY TORRENTS BECOME ONE

—— *galactic clusters collide and merge* ——

UNIVERSE
Diameter: $1.21 \times 10^{13} \rightarrow 1.24 \times 10^{13}$ ly
  Observable: $3.03 \times 10^{10} \rightarrow 3.10 \times 10^{10}$ ly
Expansion: $208.33 \rightarrow 202.33$ km/s/Mpc
Density: $6.88 \rightarrow 6.46$ p/m³
Temperature: $8.28 \rightarrow 8.11$ K

ASTRONOMICAL OBJECTS
Galaxies: $2.12 \times 10^{13} \rightarrow 2.19 \times 10^{13}$
Stars: $5.31 \times 10^{20} \rightarrow 5.67 \times 10^{20}$
Black Holes: $4.58 \times 10^{18} \rightarrow 4.87 \times 10^{18}$
Planetary Systems: $7.17 \times 10^{20} \rightarrow 7.66 \times 10^{20}$
Worlds: $7.17 \times 10^{21} \rightarrow 7.76 \times 10^{21}$

WORLDS WITH LIFE
Simple Life: $3.83 \times 10^{17} - 8.15 \times 10^{18}$
Complex Life: 0
Intelligent Life: 0
Technological Life: 0
Interstellar Life: 0

For a long time, the majority of structure was driven by the flow of matter from the intergalactic medium into overdense regions. As the Universe ages, however, those cosmic flows begin to occur on progressively larger and larger scales, with individual star clusters merging into galaxies, with smaller galaxies assembling to make larger ones, and with large collections of galaxies falling into still larger structures: galactic groups and clusters. Although these groups and clusters continue to grow via accretion, they're also capable of being drawn, via the dark matter web that's continuing to form on the largest cosmic scales, into the same vicinity as one another. Owing to the force of their mutual gravitation, they can accelerate towards and eventually collide with one another, creating a magnificent cosmic "bullet" when they strike one another.

Reaching speeds of several thousands of kilometers per second relative to one another—up to ~1–2% the speed of light at their fastest—these galactic groups and clusters are filled not only with individual galaxies but also with large reservoirs of gas in and between most of their component galaxies, as well as approximately five times as much dark matter as the total amount of normal, atom-based matter inside. When these two large-scale structures smash into one another, three important but unique things occur.

The first is that the component galaxies, inside, gravitationally tug on one another and rapidly speed through the intracluster gas. As they do, they'll simultaneously undergo periods of rapid star formation, while also losing a fraction of their gas to the friction that speeding through a gaseous medium creates. The result will be a series of stellar trails that form, getting stripped out of the individual galaxies themselves, creating a population of stars within the cluster that exist in the space between the actual galaxies. Although you might expect many galaxies to collide, given that dozens, hundreds, or even thousands of galaxies exist in each cluster, it turns out that galaxy collisions arising from cluster mergers are rare. The overwhelming majority remain undisturbed.

The second thing that happens is that the normal matter within the colliding galaxy clusters interacts, exchanging energy and heating up to temperatures so high that X-ray emissions are triggered. These colliding clusters cause the gas inside of them to create shock fronts, which slows down the gas, triggers star formation, and creates a large concentration of normal matter at the central collision point. And yet, despite this enormous cosmic collection of gas and matter, this isn't where the majority of the gravitational mass is located.

That's because of the third thing that occurs in a collision such as this: The dark matter within each galaxy cluster simply coasts right through. Dark matter, as far as we can tell, doesn't collide with photons, nor with atom-based normal matter, nor with other dark matter; it's completely noninteracting except for its gravitational influence. As a result, it simply speeds through the collision completely unimpeded, following the individual galaxies that also don't collide. It will take many hundreds of millions of years, if not billions, for the dark matter to come to equilibrium at the central collision point. Although that occurs where the intracluster gas—made of normal matter—went "splat" during the initial smashup, the dark matter and the individual galaxies it carries along for the ride will oscillate about that point for a long cosmic period before eventually settling down. These cosmic "bullets," first appearing now, will be a hallmark of the largest-scale structures that form for billions of years to come.

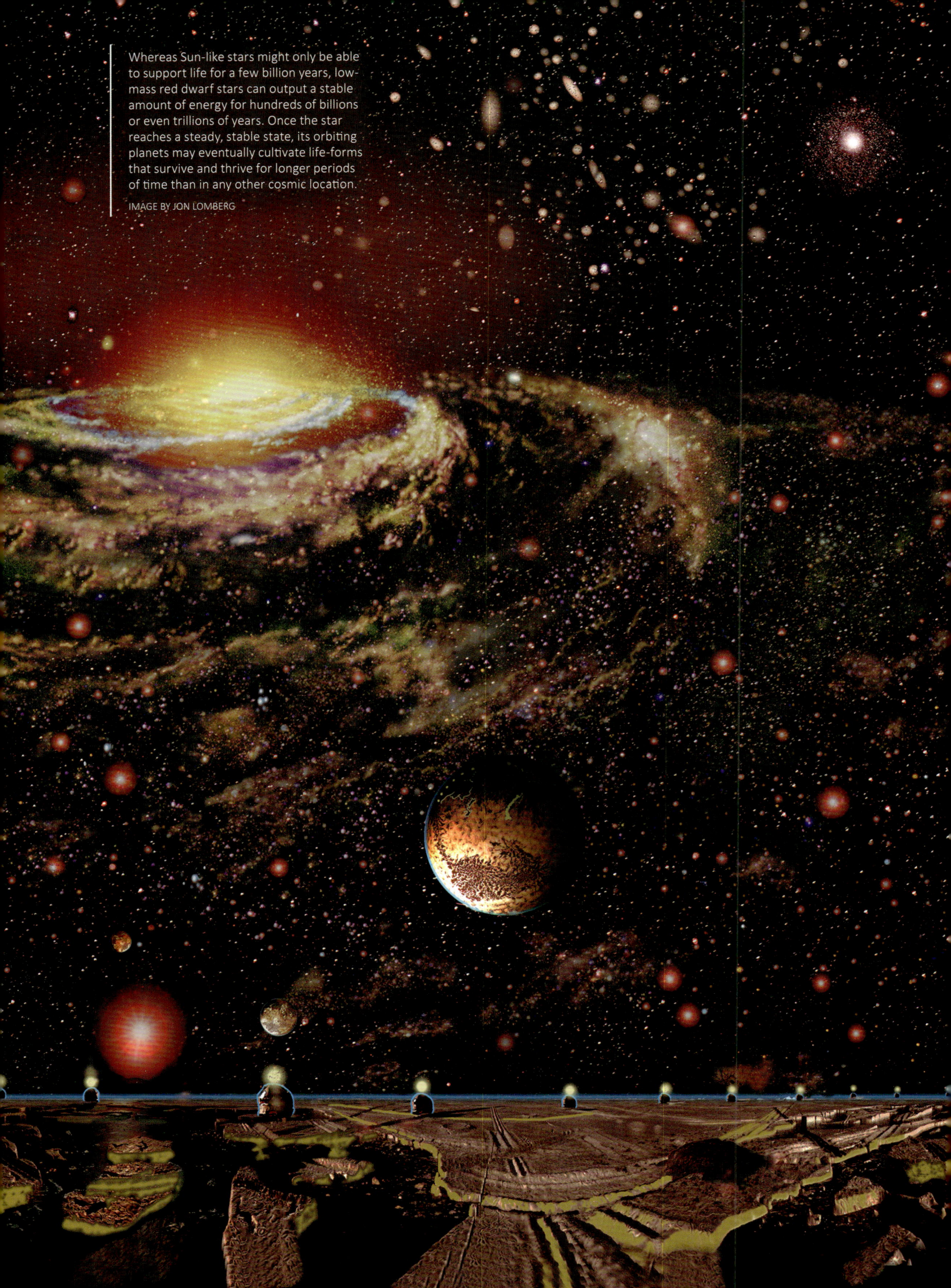

Whereas Sun-like stars might only be able to support life for a few billion years, low-mass red dwarf stars can output a stable amount of energy for hundreds of billions or even trillions of years. Once the star reaches a steady, stable state, its orbiting planets may eventually cultivate life-forms that survive and thrive for longer periods of time than in any other cosmic location.

IMAGE BY JON LOMBERG

CANDLES THAT BURN HALF AS BRIGHT

small, cool, red dwarfs live long lives

UNIVERSE
Diameter: $1.24 \times 10^{13} \rightarrow 1.26 \times 10^{13}$ ly
  Observable: $3.10 \times 10^{10} \rightarrow 3.16 \times 10^{10}$ ly
Expansion: $202.33 \rightarrow 196.80$ km/s/Mpc
Density: $6.46 \rightarrow 6.08$ p/m³
Temperature: $8.11 \rightarrow 7.95$ K

ASTRONOMICAL OBJECTS
Galaxies: $2.19 \times 10^{13} \rightarrow 2.26 \times 10^{13}$
Stars: $5.67 \times 10^{20} \rightarrow 6.02 \times 10^{20}$
Black Holes: $4.87 \times 10^{18} \rightarrow 5.16 \times 10^{18}$
Planetary Systems: $7.66 \times 10^{20} \rightarrow 8.13 \times 10^{20}$
Worlds: $7.76 \times 10^{21} \rightarrow 8.32 \times 10^{21}$

WORLDS WITH LIFE
Simple Life: $4.10 \times 10^{17} - 8.91 \times 10^{18}$
Complex Life: 0
Intelligent Life: 0
Technological Life: 0
Interstellar Life: 0

When we think of life in the Universe, most of us think about planets similar to Earth orbiting stars similar to the Sun. Although that's certainly one way life can arise, it would be incredibly egotistical of us to think those are the only conditions conducive to a living world. Yes, certain ingredients are required: enough heavy elements to form rocky planets and moons, enough liquid water on the surface of those worlds to create an aqueous environment, enough inputted energy to trigger life-sustaining reactions, and enough concentrated chemicals to enable sustained biological reactions. But beyond that, if every world is a lottery ticket in the drawing of life's possibilities, we must remember that we don't know what the prizes up for grabs are on non-Earth-like worlds, and we don't know the odds of winning those other prizes.

It turns out that while Earth-sized planets are incredibly abundant, Sun-like stars aren't actually the norm. Only about 10–15% of all stars have masses, brightnesses, and lifetimes that are comparable to our own parent star. The overwhelming majority of stars—about 80–85% of them, in fact—are actually much lower in mass and brightness, while simultaneously being much longer-lived, than our own. These "red dwarf" stars are known to be abundant with orbiting planets, and in particular with planets that are comparable in size to Earth, and represent the most common type of star found throughout the Universe.

It's true: These stars have some properties that make us question whether life on their orbiting worlds could be possible. These stars have such low luminosities that in order for a planet to be at the right temperature for surface liquid water to be possible, the planet would have to be tidally locked, meaning that the same hemisphere of the planet always faces its parent star, while the opposite hemisphere always faces away. Red dwarf stars are incredibly active, emitting flares that run the risk of frying any life that exists on these worlds, while being astoundingly variable in the amount of energy they output at any given time. And because of the enhanced stellar winds that emerge from stars such as these, it's likely that many of the worlds found orbiting around them will have their atmospheres stripped away entirely.

But we mustn't dismiss life around red dwarf stars as a possibility either. With so many red dwarfs in existence, many of which possess Earth-sized planets in orbit around them, this is where the greatest number of proverbial "lottery tickets" exist for the possibility of life arising. Just as Earth's magnetic field protects us from our Sun's solar wind, so too could the magnetic dynamo at the core of a planet orbiting a red dwarf. The fact that the planet is tidally locked might not be such a disadvantage, especially considering that it would have an orbital period—or a year—that only lasts a few Earth-days. If it can maintain an atmosphere, winds would tend to flow from the "hot" star-facing side to the "cold" space-facing side, with a ring that represents the day/night border perhaps being an ideal location for life to arise.

However, the greatest feature of all in favor of habitability for these red dwarf systems is longevity: Whereas Sun-like stars become uninhabitably hot after only a few billion years, red dwarfs can keep a constant energy output for trillions of years. If life can arise and sustain itself on a planet in a red dwarf system, it just might outlive anything we've ever imagined on a world like Earth.

Whenever two white dwarfs collide and merge, they have the potential to trigger a type Ia supernova explosion: a runaway fusion reaction that destroys the white dwarf and emits a terrifying amount of energetic light. Any planets around those white dwarfs run the risk of being destroyed by these energetic outbursts, with typical galaxies experiencing one such supernova per century.

IMAGE BY MARK A. GARLICK

SECOND-CHANCE SUPERNOVAE

—— some white dwarfs are fated to supernova ——

UNIVERSE
Diameter: $1.26 \times 10^{13} \rightarrow 1.29 \times 10^{13}$ ly
  Observable: $3.16 \times 10^{10} \rightarrow 3.22 \times 10^{10}$ ly
Expansion: $196.80 \rightarrow 191.44$ km/s/Mpc
Density: $6.08 \rightarrow 5.73$ p/m³
Temperature: $7.95 \rightarrow 7.79$ K

ASTRONOMICAL OBJECTS
Galaxies: $2.26 \times 10^{13} \rightarrow 2.32 \times 10^{13}$
Stars: $6.02 \times 10^{20} \rightarrow 6.39 \times 10^{20}$
Black Holes: $5.16 \times 10^{18} \rightarrow 5.45 \times 10^{18}$
Planetary Systems: $8.13 \times 10^{20} \rightarrow 8.63 \times 10^{20}$
Worlds: $8.32 \times 10^{21} \rightarrow 8.92 \times 10^{21}$

WORLDS WITH LIFE
Simple Life: $4.37 \times 10^{17} - 9.70 \times 10^{18}$
Complex Life: 0
Intelligent Life: 0
Technological Life: 0
Interstellar Life: 0

For as long as sufficiently massive stars have formed, supernovae have been exploding throughout the cosmos. Whenever a star is born with more than about 8 to 10 times the mass of the Sun, it's an excellent candidate to become a core-collapse supernova. All stars begin by fusing hydrogen into helium in their cores, with the most massive exhausting their fuel relatively rapidly: running out of it after only a few million years. When that occurs, the stellar core contracts, heats up, and begins fusing helium into carbon as the star swells to transition into a red giant.

When a star's core runs out of helium, there are two general ways for its evolution to proceed. While all such stars will see their cores contract and heat up, only the more massive ones will begin fusing carbon in their core: a critical step on the path to a core-collapse supernova. Stars that don't reach that critical threshold, however, will simply die in a more peaceful fashion: gently blowing off their outer layers to create a planetary nebula, while the core itself contracts to form a white dwarf.

While the majority of the very first stars created in the Universe were overwhelmingly massive, making them excellent candidates for core-collapse supernovae, only about 0.1%–0.2% of stars born in later generations are massive enough to undergo such a core-collapse event. All stars that cannot reach that critical mass threshold will, at the end of the fusion stage of their life cycles, wind up forming a white dwarf instead.

But these white dwarfs—which already, by this stage in the Universe, outnumber the total number of core-collapse supernovae by a factor of 50 or so—aren't done quite yet. Their cores gently contract to form an object that's approximately the size of planet Earth but containing around the mass of the Sun. The end result is a neutral ball of atoms that's roughly hundreds of thousands of times as dense as our world. When you consider, however, that ~50% of all the stars ever created will be members of multi-star systems, you can't help but recognize that many of these white dwarfs will be locked in a gravitational dance with either another star or another white dwarf, both of which give rise to a fascinating possibility: a second-chance supernova.

White dwarfs have an unintuitive property to them: The more mass you add to them, the smaller and denser they become. As a hot, ultradense collection of atoms, it's only a specific rule of quantum physics—the Pauli exclusion principle, which prevents certain classes of identical particles, like electrons, from occupying the same quantum state—that holds these objects up against gravitational collapse. As you add more and more mass to the white dwarf, the atoms at the center of the object themselves get squeezed and compressed. Once they cross a critical threshold, a runaway fusion reaction will occur, tearing the white dwarf apart and creating a "second-chance" stellar cataclysm: a type Ia supernova.

Whether a white dwarf crosses that critical mass threshold because it collided with another white dwarf, gradually siphoned mass off of a less-dense companion star, or got swallowed by a full-fledged star is irrelevant. There are many ways to make a type Ia supernova, but the result is always the same: a spectacular supernova, where the progenitor white dwarf is completely destroyed. In April 2013, a type Ia supernova known as SN UDS10Wil was found at precisely this epoch in cosmic history: the most distant such supernova found to date.

At the centers of galaxy clusters, the largest galaxies with extremely massive black holes are found. As gas and other matter falls into that central region, those black holes can tear that material apart, forming an accretion disk and leading to rapid accelerations that produce jets of matter. These active galaxies are among the most energetic phenomena in the Universe.

DRAGONS FEAST AND SPIT FIRE

black holes attract and eject matter

UNIVERSE
Diameter: $1.29 \times 10^{13} \rightarrow 1.31 \times 10^{13}$ ly
  Observable: $3.22 \times 10^{10} \rightarrow 3.29 \times 10^{10}$ ly
Expansion: $191.44 \rightarrow 186.52$ km/s/Mpc
Density: $5.73 \rightarrow 5.41$ p/m³
Temperature: $7.79 \rightarrow 7.64$ K

ASTRONOMICAL OBJECTS
Galaxies: $2.32 \times 10^{13} \rightarrow 2.38 \times 10^{13}$
Stars: $6.39 \times 10^{20} \rightarrow 6.75 \times 10^{20}$
Black Holes: $5.45 \times 10^{18} \rightarrow 5.73 \times 10^{18}$
Planetary Systems: $8.63 \times 10^{20} \rightarrow 9.12 \times 10^{20}$
Worlds: $8.92 \times 10^{21} \rightarrow 9.51 \times 10^{21}$

WORLDS WITH LIFE
Simple Life: $4.62 \times 10^{17} - 1.05 \times 10^{19}$
Complex Life: 0
Intelligent Life: 0
Technological Life: 0
Interstellar Life: 0

ow that we're 3.5 billion years into the Universe's evolution, structures on very large cosmic scales are beginning to come together. The largest galaxy clusters now contain somewhere around 1,000 times the mass of the modern Milky Way, with gas and other matter continually falling into them from the intergalactic medium. But what's truly remarkable is what occurs in the central cores of these galaxy clusters: The most massive galaxies, including the ones with the highest-mass supermassive black holes, collect in the center, where they collide and merge. This leads to the activation of the most energetic engines in the Universe.

Even at this early stage of our cosmic history, the most massive black holes have already grown to masses that are tens of billions of times larger than the mass of our present-day Sun, with a few of them approaching the 100 billion solar masses mark. As galaxies interact, gas infalls, and mergers between galaxies within the cluster occur, matter can get funneled into the cluster center where the great supermassive black hole can feed on it. This results in an activation of the central galactic nucleus, creating jets, filaments, ultrahot gas, and much more.

As matter collects in the vicinity of the central black hole, it gets accelerated by the intense gravitational forces. The more matter there is, the greater the rate of collisions between matter particles becomes, which causes them to heat up and ionize. The closer the matter gets to the black hole's event horizon, the greater the speed, acceleration, and kinetic energy of that matter, causing it to heat up even further. And then, the fast-moving matter particles, being ionized and electrically charged, generate extraordinarily strong magnetic fields, which funnel the matter—now moving close to the speed of light—into accretion flows that head towards the black hole's center.

Some of this matter will be devoured by the central black hole, causing it to increase in mass and leading to the further growth of the size of the event horizon. But most of the matter driven into the center, up to 90–95% by many estimates, will get funneled away from the black hole itself. While the most common shape for this funneled matter is two jets, perpendicular to a black hole's accretion disk and opposed to one another, the central cores of galaxies are often more complex in their structure. Instead of having matter orbiting the central black hole in a single disk, the shape can be quite chaotic, creating a network of ultrahot, ultramassive filaments of matter.

The reason these features appear at the cores of galaxy clusters isn't only because that's where the most massive black holes reside but also because that's where the highest velocities of gas relative to a galaxy's central black hole can be found. As the nucleus of the galaxy becomes active, the hot, fast-moving matter generates bubbles of ionized plasma. Since there's already hot, ionized gas surrounding the central galaxies in galaxy clusters—many of which emit X-rays at temperatures of millions of degrees—these two sources of ionized particles interact.

The combination of magnetic fields and rapidly expelled gas creates networks of massive, elongated filaments, which can contain millions of solar masses' worth of material, and despite being only a few hundred light-years wide, can extend for up to tens of thousands of light-years. These active features can be found in a selection of galaxies throughout cosmic history, from the nearby NGC 1275 in the Perseus Cluster to the distant Spiderweb galaxy from more than 11 billion years ago.

After billions of years with only simple life existing in the Universe, the first complex life-forms start to emerge. With differentiated and specialized components to them, these organisms are larger and longer-lived than their simpler counterparts, arising first in the life-friendly waters of the world shown here.

IMAGE BY MARK A. GARLICK

THE FIRST COMPLEX LIFE

—— from simple life, complex life ——

UNIVERSE
Diameter: $1.31 \times 10^{13} \rightarrow 1.34 \times 10^{13}$ ly
  Observable: $3.29 \times 10^{10} \rightarrow 3.35 \times 10^{10}$ ly
Expansion: $186.52 \rightarrow 181.74$ km/s/Mpc
Density: $5.41 \rightarrow 5.11$ p/m³
Temperature: $7.64 \rightarrow 7.50$ K

ASTRONOMICAL OBJECTS
Galaxies: $2.38 \times 10^{13} \rightarrow 2.45 \times 10^{13}$
Stars: $6.75 \times 10^{20} \rightarrow 7.12 \times 10^{20}$
Black Holes: $5.73 \times 10^{18} \rightarrow 6.03 \times 10^{18}$
Planetary Systems: $9.12 \times 10^{20} \rightarrow 9.62 \times 10^{20}$
Worlds: $9.51 \times 10^{21} \rightarrow 1.01 \times 10^{22}$

WORLDS WITH LIFE
Simple Life: $4.88 \times 10^{17} - 1.13 \times 10^{19}$
Complex Life: $8.48 \times 10^{5} - 1.62 \times 10^{9}$
Intelligent Life: 0
Technological Life: 0
Interstellar Life: 0

By the present epoch, life has likely emerged on gargantuan numbers of worlds all throughout the Universe. More than a quintillion (10^{18}) worlds have, at some point, seen life emerge. Although life is doubtlessly short-lived on the majority of these planets and moons, there ought to be many cases where it survives and thrives for billions of years. On biologically successful worlds such as these, life not only has the opportunity to affect and transform the atmospheres, oceans, and continents, but could—dependent on conditions particular to each world—become complex and differentiated in far less time than it took for life to do so on planet Earth.

On every world, there's a limited supply of resources available for life to extract energy from. Life requires an energy source, an environment in which it can sustainably live, the ability to produce offspring or otherwise reproduce, and needs to be able to survive from predators and other pressures for long enough to do so. When conditions are stable and unchanging, life tends to evolve slowly: largely driven by random mutations. But when conditions change rapidly, extinctions become more common. When once-dominant organisms no longer completely fill their ecological niche, an opportunity opens up for other organisms to rise to prominence.

The worlds where evolution proceeds at the fastest pace will likely be the worlds where extinction events are common enough to frequently open up previously occupied ecological niches, but where none of them are quite severe enough to completely extinguish the presence of life on those worlds. It's possible, perhaps even probable, that life would arise and evolve at an accelerated pace compared to Earth wherever more energy is available, wherever there are more ecological niches to occupy,

wherever potential metabolic nutrient sources are more abundant, and wherever significant extinction events are more common.

From the time that life first arose on Earth to the time that complex, differentiated organisms arose—including multicellular organisms with specialized functions, and including organisms that reproduce sexually, mixing their genetic materials with another member of their species before passing their genes onto their offspring—it took upwards of three billion years. On other worlds, where evolution could not only have taken myriads of different pathways, but where mass extinction events were far more common than they were in our own planet's history, complex and differentiated life-forms might have arisen far more quickly.

On Earth, our planet existed for approximately four billion years before the dawn of the Cambrian explosion, with the very first fungi, plants, and animals only appearing a very short amount of time prior to that event. But elsewhere in the Universe, with precisely the right conditions in place, a diversity of creatures that we can scarcely imagine could reasonably have arisen in less than half that time. Even if the odds of it doing so are low, with so many stars, planets, moons, and other opportunities for life in the Universe, it's only a matter of time before life succeeds in precisely this fashion.

The major downside, of course, is the risk and timescale of life being completely annihilated. If extinction events are common and severe, then planet-wide extinctions are legitimate possibilities. If a living world's parent star is significantly more massive than our Sun, stellar heating would quickly boil that world's oceans, even if complex life arose swiftly. Creating and sustaining life is a challenge, but amid the cosmic carnage, unprecedented biological successes are surely commencing.

What gives this world its fascinating colors? While some of them are due to chemical processes, such as the oxidation of iron to produce reddish colors and basaltic lavas producing grayish-brown rocks, life processes can also be transformative. From ocean acidification to the production of oxygen and the consumption of carbon dioxide, an inhabited world will give off biosignatures that allow life's presence to be detected even from afar.

IMAGE BY MARK A. GARLICK

TERRA PRIMA TERRAFORMED

— *life transforms first Earth-like planet* —

UNIVERSE
Diameter: $1.34 \times 10^{13} \rightarrow 1.37 \times 10^{13}$ ly
  Observable: $3.35 \times 10^{10} \rightarrow 3.41 \times 10^{10}$ ly
Expansion: $181.74 \rightarrow 177.11$ km/s/Mpc
Density: $5.11 \rightarrow 4.83$ p/m³
Temperature: $7.50 \rightarrow 7.36$ K

ASTRONOMICAL OBJECTS
Galaxies: $2.45 \times 10^{13} \rightarrow 2.50 \times 10^{13}$
Stars: $7.12 \times 10^{20} \rightarrow 7.49 \times 10^{20}$
Black Holes: $6.03 \times 10^{18} \rightarrow 6.32 \times 10^{18}$
Planetary Systems: $9.62 \times 10^{20} \rightarrow 1.01 \times 10^{21}$
Worlds: $1.01 \times 10^{22} \rightarrow 1.07 \times 10^{22}$

WORLDS WITH LIFE
Simple Life: $5.13 \times 10^{17} - 1.21 \times 10^{19}$
Complex Life: $5.36 \times 10^{8} - 1.18 \times 10^{12}$
Intelligent Life: 0
Technological Life: 0
Interstellar Life: 0

All across the Universe, Earth-like planets around Sun-like stars are constantly being created from the recycled ash and debris from the deaths of previous generations of stars. As new stars form, particularly with sufficiently high abundances of heavy elements, a disk of matter forms around them from which planetary systems can arise: a protoplanetary disk. If enough heavy elements are present, then gravitational instabilities in that disk will begin to grow, and a small core of rock and metal will begin to form from the surrounding material. If that core gets large enough, more and more matter will accrete around it, leading to the formation of planetary systems with objects ranging from below the mass/size of Mercury all the way up to super-Jupiter worlds.

Some of these worlds, particularly the rocky planets and rocky moons around giant planets that are located at the right distance from their parent stars to possess liquid water on their surfaces, will have the right conditions where life can potentially arise on them. Furthermore, on some fraction of those worlds, life actually will not only arise but will sustain itself, developing a way to metabolize the available resources and reproduce. Where life survives and thrives for long periods of time, a living world will ensue.

At this point in cosmic history, the "first Earth-like world" to form, where life arose, thrived, and survived ever since, reaches about 1.5 billion years of age. This is a critical time for the planet, because for the first time, the atmosphere is completely transformed from its primitive state, showing clear evidence for a sustained, transformative biological presence on that world. Whereas very primitive Earth analog worlds are initially extremely rich in ammonia, methane, water vapor, as well as hydrogen and helium, the presence of life changes that story over time. After 1.5 billion years have passed, the atmosphere is now largely composed of nitrogen and carbon dioxide, with trace amounts of water vapor and—if life proceeds similarly to how it did on Earth—oxygen present.

Volcanic activity is also part of the evolutionary story of most Earth-like planets, which adds substantial quantities of water vapor, carbon dioxide, and nitrogen to the atmosphere. Most of that water vapor, if the conditions are right for liquid water on the planet's surface, will eventually become surface water, creating either a complete ocean world or a world with a mix of landmasses and water features. Most of the nitrogen will remain in the atmosphere, while carbon dioxide, when mixed with water and energy, can lead to the production of sugars. The earliest forms of photosynthesis could have arisen on an Earth analog as early as this in our cosmic history: fewer than four billion years after the Big Bang.

Each and every time there's a change in either the resource availability on a living planet or a transformative leap in the ability of organisms to make use of a previously unutilized resource, it raises the possibility of a novel class of organisms arising, with the capability of outcompeting the prevailing set of organisms currently occupying various ecological niches. The most efficient, successful, adaptable organisms will be the ones that survive, where their metabolisms then determine the future evolution of the planetary biosphere. All across the Universe, even these primitive life-forms are fundamentally reshaping the habitable environments of their home planets.

More massive than gas giant planets but not quite massive enough to initiate nuclear fusion in their cores, brown dwarfs are the "failed stars" of our cosmos. Many brown dwarfs have binary companions, and whenever two brown dwarfs merge, so long as their combined mass is greater than about 7.5% the mass of our Sun, they may yet transition to become full-fledged stars.

IMAGE BY MARK A. GARLICK

ALL THAT'S BEST OF DARK AND BRIGHT

—— brown dwarfs are too small to become stars ——

UNIVERSE
Diameter: $1.37 \times 10^{13} \to 1.39 \times 10^{13}$ ly
 Observable: $3.41 \times 10^{10} \to 3.48 \times 10^{10}$ ly
Expansion: $177.11 \to 172.85$ km/s/Mpc
Density: $4.83 \to 4.57$ p/m³
Temperature: $7.36 \to 7.23$ K

ASTRONOMICAL OBJECTS
Galaxies: $2.50 \times 10^{13} \to 2.59 \times 10^{13}$
Stars: $7.49 \times 10^{20} \to 7.95 \times 10^{20}$
Black Holes: $6.32 \times 10^{18} \to 6.61 \times 10^{18}$
Planetary Systems: $1.01 \times 10^{21} \to 1.07 \times 10^{21}$
Worlds: $1.07 \times 10^{22} \to 1.15 \times 10^{22}$

WORLDS WITH LIFE
Simple Life: $5.36 \times 10^{17} - 1.29 \times 10^{19}$
Complex Life: $5.12 \times 10^{9} - 1.29 \times 10^{13}$
Intelligent Life: 0
Technological Life: 0
Interstellar Life: 0

Inside of every star-forming region in the Universe, an incredible race is taking place. The densest, largest clumps of matter inside of a contracting molecular cloud are inevitably the most successful at attracting the surrounding matter in their vicinities. This determines their fates: They're inevitably going to draw in the most matter, not only becoming full-fledged stars but also achieving the highest masses of all stars that will compose the future star cluster that results from this star-formation event.

But for the clumps of matter that are less dense and smaller, their fates are much more uncertain. If there's enough available matter around them—and little enough ionizing external radiation from young, nearby stars—they'll have enough time to draw in sufficient amounts of mass to become full-fledged stars as well. If there isn't enough material available, as well as enough time for these clumps to grow large enough, it will be impossible for a star to form. Except for the gravitational "winners" in the race to form new stars, the remaining clumps of matter will simply lead to failed stars: without hot enough, dense enough cores to initiate the nuclear fusion of hydrogen into helium.

To become a true star, core temperatures need to cross a critical threshold of about 4 million K, which requires a total mass of at least 7.5% the mass of our Sun. If you get close, however—up to between 1.5% and 7.5% the Sun's mass—you can become a special but common type of "failed" star: a brown dwarf. With the core temperatures that these lower-mass objects can achieve, fusing hydrogen into helium isn't possible, but a heavy isotope of hydrogen with one proton and one neutron in its nucleus (deuterium)

can indeed fuse, illuminating them sufficiently so that clever observers can discover them.

Even though these objects are only about the physical size of Jupiter, despite being dozens of times as massive, they shine so brightly in infrared light that many of our space-based observatories have seen them in large numbers. They appear to be so numerous that there may be as many brown dwarfs throughout the Universe as there are true stars, of all types, combined. These objects have enough heat inside them that they're expected to continue to radiate in the infrared for trillions upon trillions of years.

But what's perhaps most remarkable about brown dwarfs is something we can understand by remembering this simple fact: Fully half of the stars in the Universe that form are members of multi-star systems. This should be true for brown dwarfs as well. In fact, many of the known brown dwarfs are found in binary systems, where two brown dwarfs orbit one another in a seemingly permanent gravitational dance. But gravitational dances aren't permanent, as given enough time, physical interactions and the emission of gravitational waves will cause orbits to decay. Many of these binary brown dwarfs, eventually, will inspiral and merge together. If the total mass of the new, post-merger object is suddenly greater than the "7.5% of the mass of the Sun" threshold, then all of a sudden, the new object will become a full-fledged star, after all.

Many brown dwarf pairs have already done this, while others won't do so for billions, trillions, or even quadrillions of years to come. Someday, the very last star to ever exist in the Universe will form in exactly this fashion: from the merger of two "failed stars" locked in an irresistible gravitational dance.

Just as optical lenses can bend and focus incoming light, large concentrations of matter can behave as gravitational lenses: bending, distorting, and magnifying the light from sources in the background. Here, a galaxy cluster at the upper right gravitationally lenses even more distant galaxies located behind it.

IMAGE BY MARK A. GARLICK

THROUGH THE LOOKING GLASS

—— galaxy clusters create gravitational lenses ——

UNIVERSE
Diameter: $1.39 \times 10^{13} \rightarrow 1.41 \times 10^{13}$ ly
  Observable: $3.48 \times 10^{10} \rightarrow 3.54 \times 10^{10}$ ly
Expansion: $172.85 \rightarrow 168.94$ km/s/Mpc
Density: $4.57 \rightarrow 4.35$ p/m³
Temperature: $7.23 \rightarrow 7.11$ K

ASTRONOMICAL OBJECTS
Galaxies: $2.59 \times 10^{13} \rightarrow 2.60 \times 10^{13}$
Stars: $7.95 \times 10^{20} \rightarrow 8.20 \times 10^{20}$
Black Holes: $6.61 \times 10^{18} \rightarrow 6.88 \times 10^{18}$
Planetary Systems: $1.07 \times 10^{21} \rightarrow 1.11 \times 10^{21}$
Worlds: $1.15 \times 10^{22} \rightarrow 1.19 \times 10^{22}$

WORLDS WITH LIFE
Simple Life: $5.57 \times 10^{17} - 1.37 \times 10^{19}$
Complex Life: $1.84 \times 10^{10} - 5.29 \times 10^{13}$
Intelligent Life: 0
Technological Life: 0
Interstellar Life: 0

Galaxy clusters—the largest, most massive gravitationally bound structures to form in the Universe—are finally appearing in great numbers. On cosmic scales ranging from tens to hundreds of millions of light-years, stupendous amounts of mass are accumulating in one central location. Containing anywhere from a few hundred to many thousands of galaxies, and with central galaxies that can contain close to a quadrillion stars inside, these collections of matter are continuously growing through two symbiotic processes: the accretion of matter from the intergalactic medium, and the attraction and eventual absorption of satellite systems, including smaller galaxies, galaxy groups, and even other galaxy clusters.

To an external observer who looks at one of these galaxy clusters, however, a spectacular sight awaits. Not only will they see the cumulative emitted light from all of the stars present within each of the galaxies inside of the cluster, but this will be augmented by a modified, magnified version of all of the stars and galaxies from behind that galaxy cluster. As the light from these background galaxies travels towards a massive galaxy cluster, the very presence of that cluster alters the properties of that light: stretching it, deflecting it, and in some cases, magnifying it for any properly aligned observer located on the opposite side of the cluster.

One of the consequences of living in a Universe where the gravitational force is described by Einstein's General Relativity is that wherever you have a large collection of mass in one place, the fabric of space becomes positively curved in that location. The more mass you have and the more densely packed that mass is, the greater the amount of curvature. Any light that passes through that region, therefore, gets bent, distorted, and magnified by that curvature, just like light rays that pass through an optical lens. The effect is known as gravitational lensing, and the most severe gravitational lensing effects in the Universe occur around the most massive, dense, and compact collections of mass: galaxy clusters.

Optically, these background galaxies will all have their shapes slightly distorted by the presence of the foreground galaxy cluster: compressed along the radial direction with respect to the cluster's center and stretched along the opposite direction, where an ellipse would be traced out around the cluster's center. Their positions will also be slightly offset from where they would appear in the cluster's absence. This effect is known as weak gravitational lensing and affects all objects.

But a few of these background galaxies with truly serendipitous alignments will be affected by strong gravitational lensing as well: a far more visually spectacular effect. Under certain geometrical alignments, background objects can be stretched into arcs or even complete rings and can often be seen along multiple light paths, where the same object can be seen in multiple different locations simultaneously. Owing to the fact that space and time are linked together, it's often the case that these multiple images will allow us to see strongly lensed background galaxies at different times. Whenever a supernova goes off in a background galaxy, observers often see a time delay of many months between when the supernova appears in one image versus another of the same galaxy.

Gravitationally, the effects of lensing on background galaxies enable a reconstruction of the foreground galaxy cluster's mass distribution. While the individual galaxies each contribute a significant amount of mass to the cluster, the majority can be found in the spaces between the galaxies: further evidence for the ubiquitous presence of dark matter.

Planet formation is a game of creation and destruction, as large masses can not only attract one another and merge together but also violently smash one another to smithereens. Around young stars, this dance continues for many millions of years in a protoplanetary environment, with only a few massive survivors typically emerging at the end of the planet-forming phase.

IMAGE BY JON LOMBERG

ROCKY ROADS TO LIVING WORLDS

—— rocky planets facilitate life ——

UNIVERSE
Diameter: $1.41 \times 10^{13} \rightarrow 1.44 \times 10^{13}$ ly
  Observable: $3.54 \times 10^{10} \rightarrow 3.60 \times 10^{10}$ ly
Expansion: $168.94 \rightarrow 165.15$ km/s/Mpc
Density: $4.35 \rightarrow 4.13$ p/m³
Temperature: $7.11 \rightarrow 6.99$ K

ASTRONOMICAL OBJECTS
Galaxies: $2.60 \times 10^{13} \rightarrow 2.64 \times 10^{13}$
Stars: $8.20 \times 10^{20} \rightarrow 8.54 \times 10^{20}$
Black Holes: $6.88 \times 10^{18} \rightarrow 7.16 \times 10^{18}$
Planetary Systems: $1.11 \times 10^{21} \rightarrow 1.15 \times 10^{21}$
Worlds: $1.19 \times 10^{22} \rightarrow 1.24 \times 10^{22}$

WORLDS WITH LIFE
Simple Life: $5.78 \times 10^{17} - 1.45 \times 10^{19}$
Complex Life: $4.60 \times 10^{10} - 1.49 \times 10^{14}$
Intelligent Life: 0
Technological Life: 0
Interstellar Life: 0

Rocky planets, perhaps the most fertile location in all the cosmos for life to arise, thrive, and persist for long periods of time, are continuously coming into existence wherever sufficiently enriched new stars form. In the inner regions of young stellar systems, heavy elements like carbon, oxygen, nitrogen, silicon, sulfur, and iron all collect. These raw atoms bind together, creating protoplanetesimals: fragments of solid rock and metal that cannot be blown away by their central star's (or protostar's) emitted particles and radiation. Over time periods ranging from hundreds of thousands to tens of millions of years, these protoplanetesimals interact, collide, merge, and—in the most successful of locations—grow to become gravitationally bound bodies with enough matter to pull themselves into spheroidal shapes.

These young, rock-and-metal-rich objects form the basis for planetary cores, and there are four main evolutionary pathways that they can take once they're formed. The first is gravitational ejection, where an interaction with a more massive body imparts so much kinetic energy and momentum to this object that it's kicked out of its stellar system entirely, roaming the Universe as a "rogue planet." The second is a collision and merger with a larger object, where it gets devoured by a larger, more successful protoplanetesimal, or even by a full-fledged planet or star in extreme cases.

The third evolutionary pathway, particularly interesting to Earthlings, is that these objects grow only to a point, remaining dominated by rock and metal: nonvolatile materials in the solid phase. Concluding their formation with only thin oceanic and atmospheric layers atop them, these terrestrial-like planets often have the potential to be Earth-like and may be the Universe's preferred locales for the development of biological life. And the fourth

scenario, leading to the most massive of planets, is where the rock-and-metal core grows to such magnitudes that even the lightest, most volatile gases stably collect around the planetary core, leading to a gas-rich, giant world where the atmospheric pressure can reach thousands, millions, or even billions of times the pressures found at Earth's surface. How large and massive, then, can a rocky planet get and still remain rocky, without transitioning into a gas giant world?

It turns out that for most planetary objects around most stars, the answer is surprisingly, perhaps even uncomfortably, close to the conditions found on Earth itself. Once a planetary core accumulates more than about two Earth masses—corresponding to a radius about 25–30% larger than planet Earth—it's going to successfully hold on to large quantities of even the lightest gases found in the Universe: hydrogen and helium. The solar wind particles and ultraviolet radiation emitted by a Sun-like star at a distance of Earth, Venus, or even Mercury would be unable to blow those light elements away, resulting in a thick envelope of gases, rendering such a world more like Uranus or Neptune than like any of the terrestrial planets presently known.

However, there's one way to make a larger "rocky" planet: bring it close enough to its parent star that winds and radiation combine to expose its planetary core. At very close distances, nearly all of a large, massive planet's atmospheric gases can be blown away by these effects, with each atom or molecule being "kicked" beyond the planet's escape velocity. In the most extreme cases, a planetary core up to double Earth's radius or more, with up to 10–12 times the mass of our planet, could remain rocky and relatively gas-free. The most massive of these rocky planets, although frequently at risk of being devoured by their parent stars, could easily survive for many billions of years: all the way to the present day.

This stellar system is rapidly moving away from the cluster, having been ejected via gravitational interactions. Isolated, hypervelocity stars and stellar systems are common in the Universe, created wherever multiple stars have too close of a gravitational interaction within a larger bound structure.

IMAGE BY JON LOMBERG

NOT ALL WHO WANDER ARE LOST

— *some stars are ejected and roam alone* —

UNIVERSE
Diameter: $1.44 \times 10^{13} \rightarrow 1.46 \times 10^{13}$ ly
  Observable: $3.60 \times 10^{10} \rightarrow 3.66 \times 10^{10}$ ly
Expansion: $165.15 \rightarrow 162.46$ km/s/Mpc
Density: $4.13 \rightarrow 3.92$ p/m³
Temperature: $6.99 \rightarrow 6.87$ K

ASTRONOMICAL OBJECTS
Galaxies: $2.64 \times 10^{13} \rightarrow 2.68 \times 10^{13}$
Stars: $8.54 \times 10^{20} \rightarrow 8.90 \times 10^{20}$
Black Holes: $7.16 \times 10^{18} \rightarrow 7.43 \times 10^{18}$
Planetary Systems: $1.15 \times 10^{21} \rightarrow 1.20 \times 10^{21}$
Worlds: $1.24 \times 10^{22} \rightarrow 1.30 \times 10^{22}$

WORLDS WITH LIFE
Simple Life: $5.97 \times 10^{17} - 1.53 \times 10^{19}$
Complex Life: $9.11 \times 10^{10} - 3.36 \times 10^{14}$
Intelligent Life: 0
Technological Life: 0
Interstellar Life: 0

verwhelmingly, whenever new stars form, they're created in very large numbers all in one location: in significant star-forming regions where lots of gas exists. Depending on how massive these gaseous clouds are, you're likely to wind up with a combination of open star clusters and globular clusters, with comparatively smaller numbers of stars that form in relative isolation. If there are enough heavy elements present, these stellar systems will tend to develop planets, many of which may be conducive to the emergence of life.

Over long periods of time—hundreds of millions of years for open star clusters, and billions of years or more for globular clusters—the gravitational interactions of the stars inside these objects, combined with interactions that take place with the external galactic environment, will gravitationally unbind the internal stars composing these clusters, causing them to fall apart. This typically occurs in three sequential stages.

In the first stage, internal gravitational interactions between the member stars of the cluster cause the heavier-mass stars to sink to the cluster's center, while the lighter-mass stars are kicked to higher-energy, less tightly bound objects. Many of the lightest stars will be expelled from the cluster entirely. In the second stage, external interactions with the larger galactic environment exert tidal forces on these collections of stars, stripping the lightest, most loosely held stars away from the cluster core, leaving only a central collection of more massive stars behind, on average. And in the third and final stage, there simply isn't enough mass left for the remaining stars to hold together given the chaotic gravitational forces acting on them, and the entire cluster dissipates, with the member stars strewn throughout their host galaxies.

But that's only the fate of most of these stars. A few of the stars on the low-mass side—luckily or unluckily, depending on your perspective—receive hypervelocity kicks from these gravitational interactions: a process astrophysicists call "violent relaxation." It applies equally as well to large concentrations of black holes as it does to dense stellar environments, in which three or more masses are bound together, with the two more massive ones tending to wind up more tightly bound and the third, lowest-mass one getting violently ejected.

On average, ejected stars receive "kicks" of a few tens of kilometers per second: enough to disperse them throughout their host galaxy. But whenever large numbers of stars form within a cluster, a few percent of the lowest-mass stars will receive much greater kicks: of hundreds or even a few thousand kilometers per second. At these breakneck speeds, they rocket through the interstellar environment with enough kinetic energy to escape from their home galaxy entirely, as well as larger-scale structures like galactic groups and clusters. As they speed off into the space that separates the galaxies, they become completely isolated from their sibling and cousin stars, destined to wander the Universe alone. After only a few hundred million years, they can find themselves isolated from any galaxies by millions of light-years.

Hypervelocity stars ejected from already-isolated galaxies can eventually wind up in locations that are hundreds of millions of light-years away from any other galaxies: in the deepest depths of intergalactic space. Perhaps surprisingly, these hypervelocity stars might still be able to maintain their planetary systems, and some of the worlds orbiting them might even possess life. Even the loneliest stars in the Universe might be home to creatures that, despite their isolation, can keep each other company amid the vast abyss of space.

The phenomena of "shooting stars" are simply tiny dust grains burning up in a planet's atmosphere, producing a brief streak of light that's comparable in brightness to a star. But in galaxies all across the Universe, a few stars rapidly streak through space, interacting with the interstellar medium and producing a bow shock and tail that can extend for more than 10 light-years.

IMAGE BY MARK A. GARLICK

GREYHOUNDS OF THE SKIES

—— *some stars move at very high velocities* ——

UNIVERSE
Diameter: $1.46 \times 10^{13} \rightarrow 1.49 \times 10^{13}$ ly
  Observable: $3.66 \times 10^{10} \rightarrow 3.72 \times 10^{10}$ ly
Expansion: $161.46 \rightarrow 158.08$ km/s/Mpc
Density: $3.92 \rightarrow 3.74$ p/m³
Temperature: $6.87 \rightarrow 6.76$ K

ASTRONOMICAL OBJECTS
Galaxies: $2.68 \times 10^{13} \rightarrow 2.72 \times 10^{13}$
Stars: $8.90 \times 10^{20} \rightarrow 9.23 \times 10^{20}$
Black Holes: $7.43 \times 10^{18} \rightarrow 7.69 \times 10^{18}$
Planetary Systems: $1.20 \times 10^{21} \rightarrow 1.25 \times 10^{21}$
Worlds: $1.30 \times 10^{22} \rightarrow 1.35 \times 10^{22}$

WORLDS WITH LIFE
Simple Life: $6.15 \times 10^{17} - 1.60 \times 10^{19}$
Complex Life: $1.52 \times 10^{11} - 6.33 \times 10^{14}$
Intelligent Life: 0
Technological Life: 0
Interstellar Life: 0

tars, located within any particular volume within practically any galaxy, tend to move at roughly the same speeds as one another while orbiting around the galactic center. In spirals and similar galaxies, where most of the stars are distributed within a central disk, stars at the same radius from the galactic center might only differ in speed from one another by about 5–10% of their total speed. From the perspective of any particular stellar system, the next-nearest star will remain as the closest system for tens of thousands of years; that's approximately how long it takes stars to drift significant distances relative to one another throughout interstellar space.

At the Sun's distance of 27,000 light-years from the center of a Milky Way–like galaxy, stars orbit at speeds of around 220 kilometers per second, while typically moving relative to one another at speeds between 10 and 20 kilometers per second. But relative to our Sun, the second-closest star system—Barnard's Star, at a distance of around 6 light-years away—moves at a whopping 142 kilometers per second. At these speeds and its present distance, its position changes in the sky by more than 0.3 full degrees every millennium. Over the span of a human lifetime, it will appear to shift by half of the full Moon's apparent size: the most of any known star visible from Earth.

Although our Sun does not yet exist when the Universe is just 4.2–4.3 billion years old, Barnard's Star already does and is likely moving at a speed similar to the stars around it. From the perspective of any star moving with mundane, run-of-the-mill speeds within its home galaxy, there will inevitably be nearby, fast-moving stars that rapidly shift their positions in the sky. Barnard's Star is colloquially

known as the "Greyhound of the Skies" due to its fast motion and is likely not an outlier: Most stars in the Universe will have at least one star close by that appears to move just as quickly as Barnard's Star.

But it gets better: At speeds exceeding ~100 kilometers per second relative to their surroundings, stars begin to substantially interact with the gas, dust, and plasma found in the interstellar medium. When such a fast-moving star enters its red giant phase—swelling to hundreds of times its original diameter—its energetic collisions with these interstellar particles will strip the outermost material of this giant star away. As the star rockets through the interstellar medium, it develops a tail that can extend for over 10 light-years in length. Even though what we commonly call a "shooting star" is simply a dust grain that burns up in a planet's atmosphere, there really are stars that streak through their home galaxies, leaving trails of stellar debris in their wakes. In our own Milky Way, the star Mira does this, with a tail 13 light-years long: corresponding to the distance it's traversed over the past 30,000 years.

In practically every location within each galaxy, while most stars move at approximately the same speed, a small but non-negligible fraction of them will speed through the interstellar medium at unusually high velocities. When these stars transition into the red giant phase, their outer layers will be stripped away by interactions with the matter in interstellar space, making them true shooting stars. Any observer with the right capabilities—either biological or technological—would be able to see "tails" emerging from these stars that span multiple degrees in the sky. These true shooting stars are found everywhere, all throughout our cosmic history.

Within a spiral galaxy, the most common location for new stars to form is along the dust-rich lanes of their spiral arms. As density waves move through the disk, they cause the compression and expansion of material inside, with compressed regions leading to new stars, the emission of ultraviolet light, and the pink glow of ionized hydrogen.

IMAGE BY JON LOMBERG

THE PIROUETTES OF GALAXIES

— most galaxies become spirals —

UNIVERSE
Diameter: $1.49 \times 10^{13} \rightarrow 1.51 \times 10^{13}$ ly
  Observable: $3.72 \times 10^{10} \rightarrow 3.78 \times 10^{10}$ ly
Expansion: $158.08 \rightarrow 154.79$ km/s/Mpc
Density: $3.74 \rightarrow 3.56$ p/m³
Temperature: $6.76 \rightarrow 6.65$ K

ASTRONOMICAL OBJECTS
Galaxies: $2.72 \times 10^{13} \rightarrow 2.75 \times 10^{13}$
Stars: $9.23 \times 10^{20} \rightarrow 9.56 \times 10^{20}$
Black Holes: $7.69 \times 10^{18} \rightarrow 7.95 \times 10^{18}$
Planetary Systems: $1.25 \times 10^{21} \rightarrow 1.29 \times 10^{21}$
Worlds: $1.35 \times 10^{22} \rightarrow 1.41 \times 10^{22}$

WORLDS WITH LIFE
Simple Life: $6.32 \times 10^{17} - 1.68 \times 10^{19}$
Complex Life: $2.30 \times 10^{11} - 1.09 \times 10^{15}$
Intelligent Life: 0
Technological Life: 0
Interstellar Life: 0

Even though the official peak of star formation occurred a few hundred million years prior to this moment, it's actually more of a plateau than a true peak. From when the Universe was 2.8 billion years old until it's 4.4 billion years of age, as measured from the start of the hot Big Bang, the star-formation rate remains at 90% or more of that peak value. While most of the stars that are created during this time arise from violent events—the mergers of galaxies, the infall of extragalactic matter, and starburst events that trigger the collapse of gas on a galaxy-wide scale—the second most common way stars are created remains important throughout the Universe's history: from the density waves within the arms of spiral galaxies.

The majority of galaxies we find in the Universe have a disk-like shape, as gravity causes matter to collapse along whichever of its three axes was initially the shortest. When gravitational collapse occurs, this structure will pancake (which is the astronomical term for going "splat") along that short axis, creating a disk. Whatever angular momentum the entire system possesses will cause that disk to rotate, with gravitation dragging the gas and stars inside into elliptical orbits about the galactic center.

However, galaxies don't generally possess uniform disks; most of the galaxies we see have concentrations of matter that are partitioned into high-density regions, along spiral arms, and low-density regions between those arms. While astronomers once thought that these arms were solid, long-lasting physical structures, we now know that they're not. Instead, as the material in a galaxy orbits around the center, both gas and stars move in and out of regions of various density. In the high-density regions, a pileup occurs, creating a galactic traffic jam. As a result, the apparent motion of the high-density spiral arms winds up being slower than any of the stars or gas that move within them.

These spiral structures are stable and self-sustaining owing to their gravitational interactions: The material that's farther out from the galactic center, at a greater radius, takes longer to complete a full revolution around the entire galaxy, while the material that's closer in, at a smaller radius, completes a revolution more quickly. The relative motions of the stars and other clumps of matter, combined with the gravitational forces between the matter at different radii, prevent the spiral arms from "winding up" into denser and tighter spiral patterns. Instead, they maintain a stable structure with only a few spiral arms.

Spiral arms universally possess two common properties. One is that the arms are always littered with bright, blue, hot, young stars, while the second is that the gas in those arms always possesses a pink-colored hue. The same underlying reason is the cause of both phenomena, as clumps of gas pile up in great densities lining the spiral arms themselves. That high-density gas gravitationally collapses, forming new stars, including the brightest, bluest, most energetic and also the shortest-lived stars of all. The radiation from these stars ionizes the gas around them, kicking electrons off of (mostly) hydrogen atoms. When those electrons fall back onto their nuclei, they cascade down the various energy levels, with the transition from hydrogen's third to second energy level producing that characteristic reddish-pink color.

When it comes to the Universe, what we often notice are the brightest, most spectacular examples of any phenomenon. But equally important are the quiet but relentless phenomena that occur alongside the spectacular bursts. After all, if there's one lesson the Universe has to teach us, it's the importance of everything that accumulates within it over time.

While the heavens may seem alight with a limitless canopy of stars, there are also many dark, nonluminous objects traveling through each galaxy. From ejected planets to clumps of mass that never quite grew to achieve stellar status, these "rogue planets" likely outnumber the stars in our Universe.

IMAGE BY MARK A. GARLICK

NO STARS, NO MASTERS

most planets exist outside stellar systems

UNIVERSE
Diameter: $1.51 \times 10^{13} \rightarrow 1.54 \times 10^{13}$ ly
   Observable: $3.78 \times 10^{10} \rightarrow 3.84 \times 10^{10}$ ly
Expansion: $154.79 \rightarrow 151.59$ km/s/Mpc
Density: $3.56 \rightarrow 3.40$ p/m³
Temperature: $6.65 \rightarrow 6.55$ K

ASTRONOMICAL OBJECTS
Galaxies: $2.75 \times 10^{13} \rightarrow 2.78 \times 10^{13}$
Stars: $9.56 \times 10^{20} \rightarrow 9.90 \times 10^{20}$
Black Holes: $7.95 \times 10^{18} \rightarrow 8.21 \times 10^{18}$
Planetary Systems: $1.29 \times 10^{21} \rightarrow 1.34 \times 10^{21}$
Worlds: $1.41 \times 10^{22} \rightarrow 1.46 \times 10^{22}$

WORLDS WITH LIFE
Simple Life: $6.47 \times 10^{17} - 1.75 \times 10^{19}$
Complex Life: $3.25 \times 10^{11} - 1.74 \times 10^{15}$
Intelligent Life: 0
Technological Life: 0
Interstellar Life: 0

Ever since the very first cloud of hydrogen and helium collapsed under its own gravity, initiating nuclear fusion reactions within its interior, the Universe has never been without stars. Once a sufficient number of heavy elements had been created in the interiors of massive stars— and returned to the Universe through phenomena like supernovae, neutron star mergers, and planetary nebulae— the new, enriched stars that were born would develop planets around them. Even though most of the early stars didn't have any planets, and despite the fact that many stars forming by this point in even very enriched galaxies will still lack them, the number of planets in the Universe, by now, ought to significantly exceed the number of stars.

But, as it turns out, there might be many more planets out there than even all of the stellar systems combined can account for. In fact, there might be up to as many as 100,000 planets for every star in the Universe, formed in the same regions where the stars themselves are born. In order to create stars, a massive cloud of molecular gas needs to contract under its own gravity, cooling in the process. As gas clouds contract, they begin to fragment, as regions that happen to initially have just a slightly greater than average density begin to preferentially attract successively more and more of the matter in that cloud.

The most initially massive regions grow the fastest, accumulating the greatest fractions of matter present in the initial cloud. These will be the locations where new stars arise and develop protoplanetary disks that eventually form a variety of planets of all sizes, masses, and distances from their parent stars. As these planets gravitationally interact with one another, their parent stars, and the dense interstellar environment, many will wind up colliding, getting absorbed by their parent star, or being ejected out of their home planetary system entirely. The ejected ones will be destined to wander throughout their galaxy as rogue—or orphan—planets that no longer possess a parent star.

But there's an even richer source for rogue planets than the ones ejected from planetary systems around stars: the clumps of matter that were growing within star-forming regions that simply couldn't attract enough matter quickly enough to become full-fledged stars. Just as there are a great many brown dwarfs—also known as "failed stars"— that arise within these regions, there should be even larger numbers of planetary-mass clumps of matter.

Some of these planets will be gas giants with a rock-and-metal core surrounded by large amounts of gas. Others will be lower in mass, largely composed of a mix of rock and metal, where the intense radiation from the surrounding stars strips off the majority of the light gases that helped form them. But it's the small, icy worlds created from planet-forming material that represent the overwhelming majority of these rogue planets. Regardless of composition or size, all of these worlds represent examples of rogue planets, with many of them likely possessing rich lunar systems of their own.

Cumulatively, there may be anywhere from a few hundred to over 100,000 of these rogue planets—most of which would have never had a parent star at all—for every star that's come into existence. Even this early on, where more than half of the Universe's stars have yet to form, the young Milky Way may already possess more than a quadrillion (10^{15}) planets, with most of them freely roaming through interstellar space.

Perhaps, upon reaching a certain level of technological advancement, one of the first tasks an intelligent civilization will perform is to search for and send messages to other potentially intelligent civilizations in the Universe. Although it took more than 13 billion years for humanity to arise within the Universe, other civilizations could have come into existence billions of years earlier.

IMAGE BY JON LOMBERG

RISE OF THE TECHNIUM

—— the first technological civilizations ——

UNIVERSE
Diameter: $1.54 \times 10^{13} \rightarrow 1.56 \times 10^{13}$ ly
  Observable: $3.84 \times 10^{10} \rightarrow 3.90 \times 10^{10}$ ly
Expansion: $151.59 \rightarrow 148.67$ km/s/Mpc
Density: $3.40 \rightarrow 3.24$ p/m^3
Temperature: $6.55 \rightarrow 6.45$ K

ASTRONOMICAL OBJECTS
Galaxies: $2.78 \times 10^{13} \rightarrow 2.80 \times 10^{13}$
Stars: $9.90 \times 10^{20} \rightarrow 1.02 \times 10^{21}$
Black Holes: $8.21 \times 10^{18} \rightarrow 8.46 \times 10^{18}$
Planetary Systems: $1.34 \times 10^{21} \rightarrow 1.38 \times 10^{21}$
Worlds: $1.46 \times 10^{22} \rightarrow 1.51 \times 10^{22}$

WORLDS WITH LIFE
Simple Life: $6.60 \times 10^{17} - 1.82 \times 10^{19}$
Complex Life: $4.26 \times 10^{11} - 2.57 \times 10^{15}$
Intelligent Life: $7.75 \times 10^{5} - 1.00 \times 10^{11}$
Technological Life: $1 - 6.22 \times 10^{7}$
Interstellar Life: $0 - 5.55 \times 10^{3}$

Once life becomes complex and differentiated, evolution proceeds at a different, accelerated rate compared to the prior times. Just as Moore's law predicts a constant "doubling time" for the number of transistors that can fit on an integrated circuit—which has led to a continuous exponential increase in computing power over more than half a century—a history of life on Earth shows there is a "doubling time" for genome complexity as well. If you were to examine all of the nonredundant genomes in all the organisms on Earth, you'd find that the amount of total information doubles every ~300–400 million years: a figure that has been constant since the dawn of life.

However, since the dawn of complex, differentiated organisms, the total genomic information contained within plants, animals, and fungi—including redundancies—has increased at an even greater rate. The earliest multicellular animals, likely some type of worm, arose less than one billion years ago, containing fewer than 100 million base pairs in its genome. Today, less than one billion years later, complex plants and animals contain genomes that are approximately 100 times more information-rich, enabling life's rapid evolution, diversification, and eventually, the emergence of intelligence.

Whenever there's a scarcity of resources, it's the fittest, most adaptable organism that finds a way to survive. That can be the fastest, strongest, or the most flexible organism. However, sometimes it's none of those but, rather, the most intelligent organism—the one that can most successfully solve the problem of acquiring the resources necessary for survival—that makes it. At some point, an intelligent enough creature arises and thrives for long enough that it begins to develop technology: where it harnesses the natural resources available in its environment to overcome its physical limitations. With sufficient technological advancement, even a weaker organism can come to dominate its environment.

With great power, however, comes great responsibility. Just as a yeast cell, unchecked, will consume all available resources until it poisons its environment, rendering it uninhabitable for all future yeast cells, a technologically advancing organism must restrain itself. If an intelligent species cannot live sustainably within its own environment, it will cause its own demise, leading to a mass extinction that may well include the intelligent species itself.

Still, the number of inhabited worlds is vast, and inevitably some species will be intelligent and successful enough to become the first technologically advanced civilization in the Universe—something that could reasonably happen even before the first five billion years have passed since the Big Bang. Such a civilization would be able to discover up to billions of inhabited planets even within their own galaxy, but not a single other technologically advanced civilization besides themselves. As they scoured the cosmos for others like them, they would discover a grim reality: They were the first, implying that they were truly alone.

Among the first civilizations in the Universe, some would no doubt self-destruct in various ways: through nuclear war, ecological collapse, or by poisoning their own environment. Others, however, would succeed at surviving their technological infancy, eventually exploring and perhaps even transforming and colonizing other worlds: first within their home star system, and then across the interstellar distances. With each new world where complex and differentiated life arises, the chance for an intelligent, technologically advanced civilization to emerge cannot be ignored. The Universe might have more such civilizations within it than most of us ever dared to imagine.

Acts of galactic cannibalism were more
common in the young Universe than they
are today, with this epoch marking the time
at which the Milky Way devours three of its
relatively massive neighbors. Bursts of star
formation and irregularly shaped features
are hallmarks of these cosmic mergers.

IMAGE BY JON LOMBERG

THE BIG GET BIGGER

—— *the Milky Way merges with Gaia-Enceladus* ——

UNIVERSE
Diameter: $1.56 \times 10^{13} \rightarrow 1.58 \times 10^{13}$ ly
  Observable: $3.90 \times 10^{10} \rightarrow 3.95 \times 10^{10}$ ly
Expansion: $148.67 \rightarrow 145.82$ km/s/Mpc
Density: $3.24 \rightarrow 3.10$ p/m³
Temperature: $6.45 \rightarrow 6.35$ K

ASTRONOMICAL OBJECTS
Galaxies: $2.80 \times 10^{13} \rightarrow 2.82 \times 10^{13}$
Stars: $1.02 \times 10^{21} \rightarrow 1.05 \times 10^{21}$
Black Holes: $8.46 \times 10^{18} \rightarrow 8.70 \times 10^{18}$
Planetary Systems: $1.38 \times 10^{21} \rightarrow 1.42 \times 10^{21}$
Worlds: $1.51 \times 10^{22} \rightarrow 1.56 \times 10^{22}$

WORLDS WITH LIFE
Simple Life: $6.73 \times 10^{17} - 1.89 \times 10^{19}$
Complex Life: $5.38 \times 10^{11} - 3.66 \times 10^{15}$
Intelligent Life: $5.36 \times 10^{6} - 8.70 \times 10^{11}$
Technological Life: $5.34 \times 10^{1} - 3.32 \times 10^{9}$
Interstellar Life: $0 - 1.10 \times 10^{6}$

Throughout the history of the Universe, galaxies don't simply grow by attracting matter into them from the space between the galaxies but through a phenomenon known as galactic cannibalism: where larger galaxies simply swallow up nearby, smaller ones. When we look to the distant past, we find that galaxies existed with lower masses but in greater numbers, and that as the Universe ages, it's the most massive galaxies that grow the fastest and devour the largest fraction of smaller ones. When we see a large galaxy today, 13.8 billion years after the Big Bang, we can confidently assert that it has very likely cannibalized many other smaller, lower-mass galaxies throughout its history.

This is true even for the Milky Way. Our most significant merger occurred at around 2.7 billion years after the hot Big Bang, when a galaxy known as the Kraken—up to ~20% of the Milky Way's mass at the time of the merger—was devoured by our home galaxy. But about two billion years later, three notable instances of galactic cannibalism occurred in rapid succession. First, an infalling dwarf galaxy was torn apart by the Milky Way's gravity, resulting in a large stream of stars—the Helmi stream— still visible today. Next, a galaxy known as the Sequoia galaxy was gobbled up, becoming incorporated into the Milky Way's disk. And finally, the most massive galaxy ever to be devoured by our own in all of history, the Gaia-Enceladus galaxy, was cannibalized.

These three instances of galactic cannibalism all occurred in very short succession: separated by less than a billion years total, and perhaps as little as just 200–300 million years. Each one of these three galaxies contained more than 100 million stars, with Gaia-Enceladus containing over one billion stars at the time of its merger. At the moment that the Milky Way consumed the last of these galaxies, it's plausible that Gaia-Enceladus was somewhere between the third and fifth most massive galaxy in our Local Group, behind only Andromeda, the Milky Way, and possibly Triangulum and/or the Large Magellanic Cloud. No other still-surviving galaxy could have matched it in size, mass, or number of stars.

Relative to the overall Milky Way, these galactic victims represent three of the five most massive galaxy mergers we've ever experienced. The way we've been able to reconstruct this piece of our past history doesn't come primarily from the Milky Way's stars but, rather, from the globular clusters found within our home galaxy. While the five most massive mergers, cumulatively, might have contributed only ~1% of the total stars now found in the Milky Way, they're responsible for approximately one-third (~33%) of the globular clusters we now possess.

Globular clusters are simply dense collections of very large numbers of stars that formed in one, or possibly a few, giant burst(s). Globular clusters have properties that are related to their host galaxies: when they form, where in the galaxy they form, and how much mass they form from. Based on the various properties of the ~150 globular clusters found within the Milky Way, many of them can be traced back to a common origin. At least five globulars were brought in by the merger that led to the Helmi stream, while at least three were brought in alongside the Sequoia galaxy, potentially including FSR 1758, a candidate for the Sequoia galaxy's remnant core. Remarkably, no fewer than 20 globular clusters were added during the Gaia-Enceladus merger. At no other point were so many massive galaxies devoured in such a short time, making this the greatest galactic feast in our Milky Way's history!

As asteroids, comets, and even more exotic interstellar objects speed throughout the galaxy, the combined effects of a planet's gravity and an unfavorable trajectory can lead to catastrophic impacts. Life's delicacy is on full display here, as such impacts can not only cause mass extinctions but also sterilize a living planet entirely.

IMAGE BY MARK A. GARLICK

CREATIVE DESTRUCTION

collisions destroy and create complexity

UNIVERSE
Diameter: $1.58 \times 10^{13} \rightarrow 1.61 \times 10^{13}$ ly
  Observable: $3.95 \times 10^{10} \rightarrow 4.01 \times 10^{10}$ ly
Expansion: $145.82 \rightarrow 143.04$ km/s/Mpc
Density: $3.10 \rightarrow 2.97$ p/m³
Temperature: $6.35 \rightarrow 6.26$ K

ASTRONOMICAL OBJECTS
Galaxies: $2.82 \times 10^{13} \rightarrow 2.84 \times 10^{13}$
Stars: $1.05 \times 10^{21} \rightarrow 1.08 \times 10^{21}$
Black Holes: $8.70 \times 10^{18} \rightarrow 8.95 \times 10^{18}$
Planetary Systems: $1.42 \times 10^{21} \rightarrow 1.46 \times 10^{21}$
Worlds: $1.56 \times 10^{22} \rightarrow 1.61 \times 10^{22}$

WORLDS WITH LIFE
Simple Life: $6.84 \times 10^{17} - 1.95 \times 10^{19}$
Complex Life: $6.56 \times 10^{11} - 5.04 \times 10^{15}$
Intelligent Life: $1.47 \times 10^{7} - 2.81 \times 10^{12}$
Technological Life: $3.31 \times 10^{2} - 2.50 \times 10^{10}$
Interstellar Life: $0 - 1.52 \times 10^{7}$

I n a Universe filled with cosmic violence, it's a wonder that life-friendly conditions continue to emerge in more and more places. As stars die and stellar remnants explode, all sorts of fireworks take place: supernovae, planetary nebulae, neutron star collisions, and tidal disruption events among them. With each one of these events, the interstellar medium—the gas-rich reservoirs that provide the material for new stars to form—becomes ever more enriched. For as long as sufficiently metal-rich gas remains in these galactic environments, whenever new stars are created, new planets, moons, and chances for life continue to emerge.

But enriched galaxies have a darker side to them: The more chances for life are created, the greater the number of hazards there will be flying throughout that galaxy as well. Close enough encounters or, in the worst-case scenario, a direct hit can transform a planet teeming with life into a barren, destroyed wasteland. These hazards aren't limited to stars and stellar remnants, including black holes, neutron stars, and white dwarfs, but also include the more common and less massive objects found throughout space: brown dwarfs, rogue planets, and ice-rich objects like those found in the outskirts of our Solar System.

On one hand, more massive objects can both cause more damage when they encounter a stellar system and disrupt the orbit of a potentially inhabited world from substantially farther away. On the other hand, there are greater numbers of lower-mass objects, increasing the opportunities for either gravitationally disrupting a living world's orbit or putting such a world at risk of a catastrophic collision. In either case, these unwelcome interstellar arrivals can bring about a swift and oftentimes unexpected end to life worldwide.

If ever a star were to pass within the approximate orbit of Saturn in our own Solar System, the orbits of Earth-like planets could be destabilized. Such an event would either hurl that world into the Sun, push it into a chaotic or highly eccentric orbit, or eject it entirely. In an environment similar to the Milky Way, this is uncommon, but not as uncommon as we might like: There's approximately a 1-in-10,000 chance, with each billion years that goes by, of such a catastrophic close encounter.

Smaller, lower-mass objects—like brown dwarfs and rogue planets—must get much closer to cause a catastrophe: closer than the Earth-Sun distance. A brown dwarf needs to pass within about 100 million kilometers of an inhabited world, while a Jupiter-mass object might require passing within a distance of 10 million kilometers to cause similar destruction. An Earth-mass object would need to get even closer: within about the orbit of Earth's moon. Although these are far more numerous, it's much rarer to have an encounter with such a low separation distance. All told, these two effects counteract one another, and the odds of an event like this causing a catastrophe are about the same as a living world being disrupted by a star: about 1 in 10,000 with each billion years that passes.

Finally, although collisions with comet-like or asteroid-like interstellar objects would be world-destroying, the greater danger comes from within one's own stellar system. Closer, high-mass encounters perturb a system's Oort cloud and Kuiper belt, leading to a greatly increased risk of a catastrophic collision. These likely cause the most common extinction events, about 100 times as frequent as the other effects combined. Regardless of method, unwelcome interstellar arrivals will lead to planet-wide extinctions at all times, all throughout the Universe.

BEWARE THE CALM SURFACE

— *life often lurks beneath icy crusts* —

UNIVERSE
Diameter: $1.61 \times 10^{13} \rightarrow 1.63 \times 10^{13}$ ly
  Observable: $4.01 \times 10^{10} \rightarrow 4.08 \times 10^{10}$ ly
Expansion: $143.04 \rightarrow 140.34$ km/s/Mpc
Density: $2.97 \rightarrow 2.83$ p/m³
Temperature: $6.26 \rightarrow 6.17$ K

ASTRONOMICAL OBJECTS
Galaxies: $2.84 \times 10^{13} \rightarrow 2.89 \times 10^{13}$
Stars: $1.08 \times 10^{21} \rightarrow 1.12 \times 10^{21}$
Black Holes: $8.95 \times 10^{18} \rightarrow 9.19 \times 10^{18}$
Planetary Systems: $1.46 \times 10^{21} \rightarrow 1.51 \times 10^{21}$
Worlds: $1.61 \times 10^{22} \rightarrow 1.67 \times 10^{22}$

WORLDS WITH LIFE
Simple Life: $6.94 \times 10^{17} - 2.02 \times 10^{19}$
Complex Life: $7.78 \times 10^{11} - 6.75 \times 10^{15}$
Intelligent Life: $2.89 \times 10^{7} - 6.32 \times 10^{12}$
Technological Life: $1.07 \times 10^{3} - 9.45 \times 10^{10}$
Interstellar Life: $0 - 8.5 \times 10^{7}$

ooking around you on the world you inhabit, it's easy to conclude that Earth is a water-rich world. Featuring lakes, rivers, seas, and oceans, 71% of Earth's surface is covered in water, with only the remaining 29% of the surface consisting of land. If you were to add all of Earth's water together, it would make up 0.02% of the entire mass of our planet. Despite comprising a thin layer just a few kilometers thick in even the deepest ocean depths, there's more than 500 times as much mass in the form of water on Earth as the entirety of Earth's atmosphere.

And yet, there are no doubt planets out there with far more water than Earth. Even in our own Solar System, Earth only possesses the fifth-greatest amount of water among rocky bodies, with Jupiter's moons Ganymede, Callisto, and Europa, as well as Saturn's moon Titan, all containing greater amounts of water than our home planet does. Ganymede, in particular, has the largest ocean in the Solar System. With 26 times the amount of water as Earth, a whopping 46% of Ganymede's volume is composed of water in both solid and liquid phases.

Hydrogen is the most abundant element in the Universe, but almost as soon as stars begin forming, oxygen quickly rises in abundance to become the third most abundant element. By this stage in cosmic history—a third of the way from the Big Bang to the present day—oxygen makes up nearly 1% of the atoms in the Universe, by mass. On a molecular level, the only chemical compound (i.e., a molecule or ion made of more than one type of atom) that's more abundant in the Universe than water (H_2O) is carbon monoxide (CO). While carbon monoxide can only exist as a liquid at very high pressures and relatively low temperatures, water is undoubtedly the most common liquid in the Universe.

A planet like Earth, where a significant fraction is covered in liquid water but substantial continents remain, may be a cosmic rarity compared with the alternatives: worlds without liquid water and worlds entirely covered in oceans. If liquid water is a necessary condition for life's emergence on a planet—a controversial but not necessarily incorrect stance—then the most common place to find life might not be on a planet that's similar to our own but, rather, on a world entirely covered in ocean: a water world.

Water worlds ought to come in two different varieties: worlds like Earth where water is simply more abundant, where ice only exists towards the colder, polar regions, and colder, ice-covered worlds with deep, thick, subsurface oceans. On Earth, life exists practically everywhere: on the surface, in the atmosphere, in lakes, rivers, tide pools, seas, and oceans, as well as around deep volcanic vents on the ocean floor. On a water world, even one covered in a layer of ice many kilometers thick, life could potentially arise and thrive even on a planet lacking a parent star.

Given the sheer number of planets with precisely these conditions, it's possible that water worlds—in particular, water worlds with icy crusts and subsurface oceans—may be the most abundant place in the Universe where life arises. If even primitive life-forms are found under the icy crusts of worlds like Europa or Enceladus within our own Solar System, it would revolutionize our thought process on just how common life is throughout our Universe. Wherever there's liquid water, life is certainly possible. The most common type of living world might be a continent-free water world, where parent stars are simply one option, not a requirement.

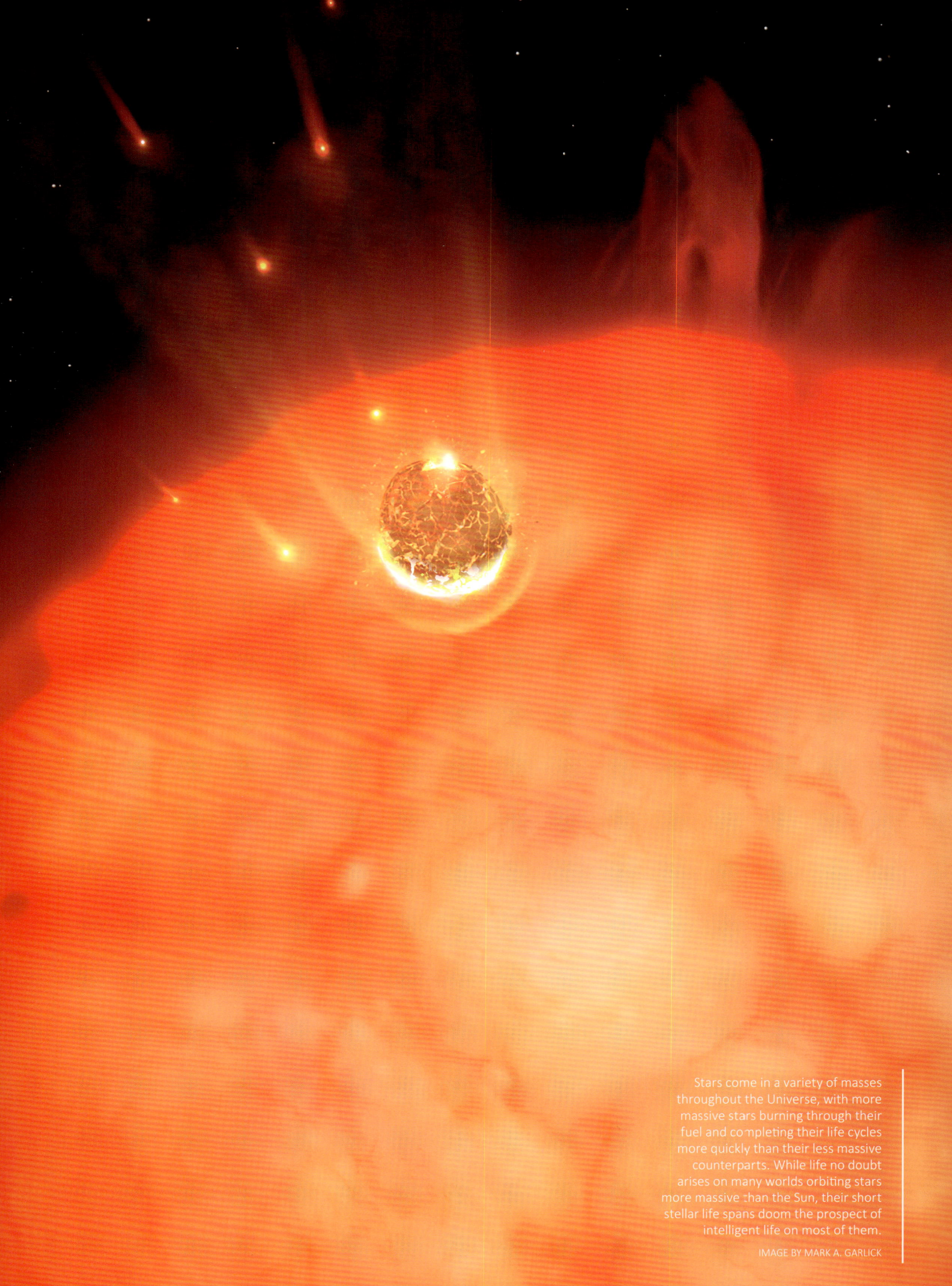

Stars come in a variety of masses throughout the Universe, with more massive stars burning through their fuel and completing their life cycles more quickly than their less massive counterparts. While life no doubt arises on many worlds orbiting stars more massive than the Sun, their short stellar life spans doom the prospect of intelligent life on most of them.

IMAGE BY MARK A. GARLICK

VIVE LA DIFFÉRENCE

— *energy gradients are the pathways to life* —

UNIVERSE
Diameter: $1.63 \times 10^{13} \rightarrow 1.65 \times 10^{13}$ ly
  Observable: $4.08 \times 10^{10} \rightarrow 4.13 \times 10^{10}$ ly
Expansion: $140.34 \rightarrow 137.88$ km/s/Mpc
Density: $2.83 \rightarrow 2.72$ p/m³
Temperature: $6.17 \rightarrow 6.08$ K

ASTRONOMICAL OBJECTS
Galaxies: $2.89 \times 10^{13} \rightarrow 2.88 \times 10^{13}$
Stars: $1.12 \times 10^{21} \rightarrow 1.14 \times 10^{21}$
Black Holes: $9.19 \times 10^{18} \rightarrow 9.42 \times 10^{18}$
Planetary Systems: $1.51 \times 10^{21} \rightarrow 1.54 \times 10^{21}$
Worlds: $1.67 \times 10^{22} \rightarrow 1.70 \times 10^{22}$

WORLDS WITH LIFE
Simple Life: $7.02 \times 10^{17} - 2.07 \times 10^{19}$
Complex Life: $8.92 \times 10^{11} - 8.69 \times 10^{15}$
Intelligent Life: $4.61 \times 10^{7} - 1.14 \times 10^{13}$
Technological Life: $2.39 \times 10^{3} - 2.39 \times 10^{11}$
Interstellar Life: $0 - 2.8 \times 10^{8}$

If there's one factor that appears to drive the diversity of life found in any particular location, it's an energy gradient between two different interfaces. The ocean/air and continent/air interfaces have a far greater abundance of life than the upper atmosphere, subterranean depths, or oceanic abysses do. Shallow bodies of water, like tide pools, lakes, ponds, and continental shelves, possess greater varieties of organisms than deeper or drier areas. Hydrothermal vents, where molten material pours from the lower layers of a world into a liquid ocean floating atop it, are hotbeds for life compared to their more stable surroundings.

It stands to reason—if this trend can be extrapolated to life beyond Earth—that hotter stars, which will bombard an orbiting planet with more high-energy radiation than a cooler star will, should compel life to evolve more quickly than a cooler, redder star will. With more energy available to stimulate biochemical reactions, it's possible that a greater variety of energy-extracting pathways will be available to organisms around such a star. Additionally, it's possible that some of those pathways will lead to greater overall efficiencies than are found on a world around a cooler, redder star.

This could lead to life evolving more quickly on worlds that orbit around more massive, hotter, bluer stars than our Sun. But even if that occurs, it doesn't necessarily stand to reason that complex, differentiated, and potentially intelligent life will be more common on these worlds than it is around stars that are either like our Sun or that are even less massive, cooler, and redder. The reason is simple but terrifying: The more massive a star is, the more luminous and bright it is, and hence the faster it burns through the nuclear fuel that allows stars to shine.

If a star twice as massive as another shone twice as brightly but contained twice as much fuel, the overall lifetime of a star wouldn't depend on its mass. But in reality, doubling a star's mass causes it to burn through its fuel eight times as quickly, which causes more massive stars to have considerably shorter lifetimes—and to experience stellar evolution far more rapidly—than their less massive counterparts. Stars rely on hydrogen fusion during the most stable portions of their life cycles, and once a star exhausts its core hydrogen, it evolves: first into a subgiant, then into a full-fledged red giant, becoming thousands of times as bright and swelling in size substantially. In short order, any living world will be roasted to the point of uninhabitability.

But the situation is even more dire than that. Stars, as they burn through the fuel in their core, actually heat up and become brighter over time. Our own Sun, over the past nearly four billion years, has increased in brightness and energy output by about 20–25%. An average star continually increases in brightness over billions of years, eventually getting hot enough to sterilize its inner planets. Around more massive stars, even though evolution on living worlds might proceed more rapidly, so will the life cycle of the parent star, leading to a rapid, catastrophic end to any biological activity on the surrounding worlds.

Life might emerge and evolve more rapidly around more massive stars, but the window it has to become complex, differentiated, intelligent, and technologically advanced is also far narrower. In the great cosmic race to evolve before a world becomes uninhabitable, stars like our Sun might hit the "sweet spot" for the emergence of intelligent life.

As stars form within the young Milky Way, the remnants of supernovae, kilonovae, and planetary nebulae all recycle heavy, processed elements back into the interstellar medium. Yesterday's stars contribute to today's molecular clouds of gas and will participate in tomorrow's episodes of star and planet formation. The fireworks happening within the Milky Way, at this time, will help lead to the later formation of Earth.

IMAGE BY JON LOMBERG

FRIGHTFUL BURSTS OF FIREWORKS

stars die and create new elements

UNIVERSE
Diameter: $1.65 \times 10^{13} \rightarrow 1.68 \times 10^{13}$ ly
  Observable: $4.13 \times 10^{10} \rightarrow 4.19 \times 10^{10}$ ly
Expansion: $137.88 \rightarrow 135.47$ km/s/Mpc
Density: $2.72 \rightarrow 2.61$ p/m³
Temperature: $6.08 \rightarrow 6.00$ K

ASTRONOMICAL OBJECTS
Galaxies: $2.88 \times 10^{13} \rightarrow 2.89 \times 10^{13}$
Stars: $1.14 \times 10^{21} \rightarrow 1.17 \times 10^{21}$
Black Holes: $9.42 \times 10^{18} \rightarrow 9.64 \times 10^{18}$
Planetary Systems: $1.54 \times 10^{21} \rightarrow 1.58 \times 10^{21}$
Worlds: $1.70 \times 10^{22} \rightarrow 1.75 \times 10^{22}$

WORLDS WITH LIFE
Simple Life: $7.09 \times 10^{17} - 2.13 \times 10^{19}$
Complex Life: $1.00 \times 10^{12} - 1.10 \times 10^{16}$
Intelligent Life: $6.68 \times 10^{7} - 1.85 \times 10^{13}$
Technological Life: $4.45 \times 10^{3} - 5.00 \times 10^{11}$
Interstellar Life: $0 - 7.24 \times 10^{8}$

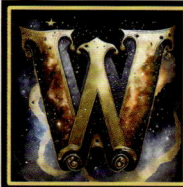

henever new stars form, cosmic fireworks are sure to ensue shortly thereafter. During large, significant star-forming episodes—responsible for creating most of the Universe's stars—gravitationally bound objects form with widely varying masses. Stars born with more than 8–10 solar masses' worth of material burn through their fuel the fastest, dying in core-collapse supernovae. Lower-mass stars still run out of fuel and die, returning their outer layers to the interstellar medium in a planetary nebula, leaving their centers to contract into white dwarfs. Those white dwarfs can then merge together or accrete material until they explode in a supernova of a different variety: "type Ia" supernovae. And neutron stars, themselves supernova remnants, can merge together, triggering a kilonova explosion that creates the heaviest elements of all.

All of these processes are examples of cosmic fireworks: responsible for enriching the gas in the interstellar medium with heavy elements essential for forming rocky planets, organic molecules, and life. The new generations of stars and planets that arise, however, will have different properties depending on how much time has passed between enrichment of the interstellar medium and the moment of new star formation. The reason is simple: While many of the elements that get created will be stable, some of them will radioactively decay. Whenever you examine a star or a planet, the elemental ratios of unstable-to-stable isotopes can not only tell you how old that stellar system is but also how much time has cumulatively passed from when those heavy elements were first created.

In the here and now, five billion years after the Big Bang, the Sun and Earth won't begin to form for another 4.2 billion years. But at this moment in cosmic history, something remarkable is happening: The cosmic fireworks that will wind up leading to the majority of the heavy elements that make up Earth and the Moon are occurring right now. Somewhere, tens of thousands of light-years from the center of the young Milky Way, hot on the heels of a series of galactic mergers, a large number of massive stars are reaching the end of their life cycles.

Many of the most massive stars are undergoing core-collapse supernova events, where heavy atomic nuclei are being compressed so tightly that they'll fuse into a massive ball of neutrons, releasing incredible amounts of energy and triggering the destruction of the star's outer layers. Very large amounts of neutrons get absorbed by the atomic nuclei in those outer layers, allowing heavy elements to be produced in great abundance: but only up to zirconium, the 40th element in the periodic table. Hot on the heels of those supernovae, however, many of the neutron star remnants produced in those explosions will then merge to form a different cataclysm: a kilonova.

Kilonovae occur when two neutron stars merge together, forming either a more massive neutron star or a black hole from the merger, but expelling about 5% of their total mass in the form of heavy elements, exclusively. The overwhelming majority of most of the heaviest elements in the Universe, including practically everything heavier than lead and bismuth, the 82nd and 83rd elements, respectively, originates from neutron star mergers. A single kilonova event can create about 10^{24} kilograms of gold: some three times the mass of a Mercury-like planet. By analyzing the various isotopes of any radioactive element that's present with a long enough half-life, we can determine how long ago those elements were created. Right now, 5.0 to 5.1 billion years after the Big Bang, the very elements that will someday make up the Earth are coming into existence.

What would it be like to have knowledge of the history and current status of every planet and star within a galaxy? With enough technological advances and the investment of enough resources over long enough time periods, even a single civilization could accomplish this task: creating the first *Encyclopaedia Galactica* in the Universe.

IMAGE BY JON LOMBERG

ENCYCLOPAEDIA GALACTICA

—— *the first galactic survey* ——

UNIVERSE
Diameter: $1.68 \times 10^{13} \rightarrow 1.70 \times 10^{13}$ ly
  Observable: $4.19 \times 10^{10} \rightarrow 4.25 \times 10^{10}$ ly
Expansion: $135.47 \rightarrow 133.13$ km/s/Mpc
Density: $2.61 \rightarrow 2.50$ p/m³
Temperature: $6.00 \rightarrow 5.91$ K

ASTRONOMICAL OBJECTS
Galaxies: $2.89 \times 10^{13} \rightarrow 2.91 \times 10^{13}$
Stars: $1.17 \times 10^{21} \rightarrow 1.20 \times 10^{21}$
Black Holes: $9.64 \times 10^{18} \rightarrow 9.87 \times 10^{18}$
Planetary Systems: $1.58 \times 10^{21} \rightarrow 1.62 \times 10^{21}$
Worlds: $1.75 \times 10^{22} \rightarrow 1.80 \times 10^{22}$

WORLDS WITH LIFE
Simple Life: $7.15 \times 10^{17} - 2.18 \times 10^{19}$
Complex Life: $1.11 \times 10^{12} - 1.36 \times 10^{16}$
Intelligent Life: $9.03 \times 10^{7} - 2.81 \times 10^{13}$
Technological Life: $7.34 \times 10^{3} - 9.26 \times 10^{11}$
Interstellar Life: $0 - 1.60 \times 10^{9}$

 omewhere, out there among the trillions of galaxies littering the observable Universe, some civilization arises to become the first to achieve one of the most remarkable milestones imaginable: creating an *Encyclopaedia Galactica,* an archive containing information about all of the planets and planetary systems within their galaxy. With life arising on so many worlds within the first few billion years after the Big Bang, some fraction of them have seen life survive, sustain itself, and thrive in an unbroken chain of successes. On a fraction of those, life became complex, differentiated, intelligent, and then technologically advanced. And then—much more recently—some of those civilizations survived their technological infancy, overcoming the many threats of catastrophe and extinction that a technologically advanced civilization faces. For the first time, one such civilization hasn't just survived but has achieved the feat of mapping their entire galaxy.

To someone whose life is confined to a single planet for all of their existence, the galaxy they're born into might seem so large that it might as well be endless. But in reality, galaxies are only ~100,000 light-years across, with rarer, more massive galaxies being larger and more common, less massive galaxies coming in on the smaller side. A light-year is certainly a long distance by some measures, but even 5.1 billion years after the Big Bang, objects can be seen as far away as 21 billion light-years. An astronomical object that's under a million light-years away, by comparison, is right in your own cosmic backyard. Neighboring civilizations, if they're lucky enough to arise within the same galaxy, could exchange many two-way speed-of-light messages during the lifetime of a galactic civilization. They could share culture, science, and possibly even visit one another.

With arrays of extremely large, sufficiently sensitive telescopes, the three-dimensional positions and properties of every star in a galaxy containing at least one sufficiently advanced civilization can be measured and cataloged. With sufficient knowledge of the stars, they can be further investigated for planets and planetary systems: moons, rings, and other satellites. Through a combination of techniques, the composition and contents of each such world can be revealed, including the presence and level of development of life.

Even if a distant planet or moon can only be imaged with single-pixel resolution, there's still an incredible amount of information that a sufficiently technologically advanced civilization can extract. By observing changes over time in color, brightness, and reflectivity at different wavelengths of light, they could learn how much of that world is oceans, ices, and land. They can see if the continents change color at different times of the year: like Earth's will change from green to brown and back again billions of years from now. They can see if clouds form on that planet, how severe the cloud cover is, and measure how fast the planet and its atmosphere are rotating.

Additionally, light from that world can be analyzed spectroscopically: by breaking it up into its constituent wavelengths. The presence and abundance of various gases can be determined, including how and whether there are any hourly, daily, seasonal, or annual variations. Over relatively short cosmic time intervals—just a few million years—the change in atmospheric contents could reveal what sort of feedback effects life is having on that planet. Extinct worlds, living worlds, saturated biospheres, intelligent life-forms, and even natural disasters can all be discovered directly. With a completed *Encyclopaedia Galactica,* an entire galaxy, so seemingly vast and endless, can finally be understood.

While the dynamics from within one's own stellar system can lead to catastrophic planetary impacts, the fact that stars routinely make close passes by one another in their galactic orbits can frequently lead to comet storms that pose existential threats to species living on rocky worlds. A single, massive impact from just one of them can easily wipe out most to all of the biodiversity on any inhabited planet.

IMAGE BY JON LOMBERG

LET SLEEPING COMETS LIE

—— *comets encircle stellar systems* ——

UNIVERSE
Diameter: $1.70 \times 10^{13} \rightarrow 1.72 \times 10^{13}$ ly
  Observable: $4.25 \times 10^{10} \rightarrow 4.31 \times 10^{10}$ ly
Expansion: $133.13 \rightarrow 131.00$ km/s/Mpc
Density: $2.50 \rightarrow 2.41$ p/m³
Temperature: $5.91 \rightarrow 5.84$ K

ASTRONOMICAL OBJECTS
Galaxies: $2.91 \times 10^{13} \rightarrow 2.92 \times 10^{13}$
Stars: $1.20 \times 10^{21} \rightarrow 1.23 \times 10^{21}$
Black Holes: $9.87 \times 10^{18} \rightarrow 1.01 \times 10^{19}$
Planetary Systems: $1.62 \times 10^{21} \rightarrow 1.66 \times 10^{21}$
Worlds: $1.80 \times 10^{22} \rightarrow 1.85 \times 10^{22}$

WORLDS WITH LIFE
Simple Life: $7.20 \times 10^{17} - 2.23 \times 10^{19}$
Complex Life: $1.20 \times 10^{12} - 1.65 \times 10^{16}$
Intelligent Life: $1.14 \times 10^{8} - 3.94 \times 10^{13}$
Technological Life: $1.08 \times 10^{4} - 1.51 \times 10^{12}$
Interstellar Life: $0 - 1.60 \times 10^{9}$

While stars move in their great gravitational dances throughout each galaxy, it's only a matter of time before two stars experience a close encounter. These encounters typically contain little risk of stars or planets between the two systems colliding but bring them within hundredths of a light-year of one another. On average, each stellar system within a galaxy experiences such an encounter every few billion years. For a few thousand years, an observer could witness a nearby star outshining everything else in the night sky, perhaps even including any moons around their planet. But the aftermath of such events is disastrous for inhabited planets, leading to extinction-level crises via comet showers.

On the outskirts of practically every stellar system, beyond the reaches of the final massive planet, large numbers of small but substantial objects abound. Far enough away from the parent star, all volatile chemical compounds—ammonia, water, methane, nitrogen, carbon monoxide and dioxide, etc.—persist in the solid phase: as ices. Beyond the final planet, these objects are distributed in a disk, similar to the Kuiper belt that will come to exist around the Sun. As you move farther out, however, these frozen objects become scattered and start to occupy a more spheroidal distribution, giving way to the inner and outer Oort clouds.

Under typically quiet circumstances, only small gravitational interactions occur between these objects, occasionally perturbing one of their orbits so that they travel towards the inner regions of their stellar systems. As these ice-rich bodies approach their parent stars, they heat up and start to get ionized, outgassing and producing tails, taking on familiar comet-like appearances. But whenever a large, massive interloper—a star, a rogue planet, or even a sufficiently massive gas or dust cloud—passes within the inner Oort cloud, it significantly perturbs enormous numbers of icy bodies all along its path. Wherever it traveled, a copious population of cometary bodies would be hurled into the inner regions of that stellar system over timescales of ~100,000 years.

These events are known as comet showers and could cause hundreds or even thousands of comets to be visible at once in a planet's night sky. While the odds of any particular comet colliding with any of the planets or moons in a stellar system is very low, a close gravitational pass from a star can cause literally millions of massive objects to become potential hazards. A strike from an object more than about 1 kilometer in diameter ought to be sufficient to drive most large land-dwelling species to extinction. The odds are strong that at least one intelligent species that made it past their technological infancy had to contend with such a natural disaster. With insufficient preparation, their civilization would be lost.

However, that doesn't necessarily translate into "game over" for intelligent life on that world. To completely drive life on a planet to extinction would require an impact from an object 20–50 kilometers in size, at least; such objects are extremely rare compared to the number of ~1 kilometer objects. A comet shower could cause a mass extinction, but not the total extinction of all life. Whenever a dominant species in an ecological niche goes extinct, the surviving species adapt to fill it. New organisms emerge from the survivors; evolution continues wherever life remains. Even around a Sun-like star where life emerged early on, more than a billion years of planetary habitability should still remain. If intelligent life arose and became technologically advanced once, it could yet happen again. After all, failure today paves the road to success tomorrow.

At long last, the first massive overdense objects, or MOOs, begin to form. At the centers of the most matter-rich galaxy clusters of all, enormous, giant elliptical galaxies arise, possessing hundreds of trillions of stars inside and weighing in at over a thousand modern Milky Ways. While other, more typical spirals and ellipticals can be found alongside them, these represent the most massive compact structures of all.

GIANTS AMONG GIANTS

—— *the first massive overdense objects* ——

UNIVERSE
Diameter: $1.72 \times 10^{13} \rightarrow 1.74 \times 10^{13}$ ly
  Observable: $4.31 \times 10^{10} \rightarrow 4.36 \times 10^{10}$ ly
Expansion: $131.00 \rightarrow 128.93$ km/s/Mpc
Density: $2.41 \rightarrow 2.31$ p/m³
Temperature: $5.84 \rightarrow 5.76$ K

ASTRONOMICAL OBJECTS
Galaxies: $2.92 \times 10^{13} \rightarrow 2.92 \times 10^{13}$
Stars: $1.23 \times 10^{21} \rightarrow 1.25 \times 10^{21}$
Black Holes: $1.01 \times 10^{19} \rightarrow 1.03 \times 10^{19}$
Planetary Systems: $1.66 \times 10^{21} \rightarrow 1.69 \times 10^{21}$
Worlds: $1.85 \times 10^{22} \rightarrow 1.88 \times 10^{22}$

WORLDS WITH LIFE
Simple Life: $7.24 \times 10^{17} - 2.28 \times 10^{19}$
Complex Life: $1.29 \times 10^{12} - 1.97 \times 10^{16}$
Intelligent Life: $1.39 \times 10^{8} - 5.34 \times 10^{13}$
Technological Life: $1.50 \times 10^{4} - 2.32 \times 10^{12}$
Interstellar Life: $0 - 2.99 \times 10^{9}$

Ever since the start of the hot Big Bang, when the Universe as we know it began from a hot, dense, almost perfectly uniform state, it's been expanding, cooling, and gravitating. Overdense regions—beginning with slightly more matter than their surroundings—preferentially attract more and more of the matter around them, causing them to grow larger and more massive. Eventually, they become gravitationally bound and collapse under their own mass, leading to the formation of cosmic structure.

Early on, only the smallest-scale structures form: individual star clusters and star-forming regions. But as time goes on, structure forms on larger and larger scales, leading to protogalaxies, full-fledged galaxies, and then galactic groups and eventually galaxy clusters. A few billion years ago, the first galaxy clusters formed, but by this point, the most massive ones aren't just growing bigger and bigger; they're causing other clusters and groups to fall into them. At last, for the first time, the largest galaxy clusters now contain more than a quadrillion (10^{15}) times the mass of a Sun-like star, forming a new class of cluster: massive overdense objects, or MOOs for short.

Galaxies and galaxy clusters normally show themselves through their emitted starlight, but three other components within these MOOs have more mass than all of their component stars combined. The first is gas: Among the thousands of significant-sized galaxies present within ultramassive galaxy clusters, gas within each galaxy and within the intracluster medium itself—the space between the galaxies—dwarfs the amount of mass present in stars. As individual galaxies speed through this intracluster medium, the gas inside each galaxy collides with the cluster gas, creating streams of star-forming matter ripped from the galaxies themselves. Over time, galaxies become

relatively gas-poor, sinking to the cluster's center and merging with one another. At the core of these MOOs, gas-poor elliptical galaxies well on their way to becoming "red and dead" become more common, while spirals become rarer.

The second component that outmasses the stars is very hot material: ionized plasma. As new stars form, waves of ultraviolet radiation kick bound electrons off of otherwise neutral atoms and molecules, creating a sea of ions and free electrons. Every time a photon passes through this sea of hot, charged particles, it interacts with them, as photons are just particles of light: electromagnetic radiation. This interaction kicks the passing photon to higher energies, an effect that imprints itself even on the leftover light from the Big Bang: the cosmic microwave background radiation. All this time after the Big Bang, 5.4 billion years on, this background radiation might have cooled down to only ~6 K above absolute zero, but photons passing through a MOO are kicked up to hotter temperatures and higher energies: an effect detectable from all throughout the visible Universe.

Meanwhile, the third and final factor is the most important one of all in terms of mass: dark matter. This invisible form of matter, which doesn't collide with normal matter and doesn't emit or absorb light, has been around since the earliest stages of the hot Big Bang and is the dominant driver of cosmic structure formation throughout all of our natural history. In each MOO that arises, only about 17% of the mass inside is due to all forms of normal matter combined, including planets, stars, gas, dust, and plasma alike. The remaining 83% of that mass is dark matter, whose presence is revealed by its gravitational effects. Dark matter forms the backbone of our cosmic web of structure, with MOOs appearing at the intersecting nexus of its most massive filaments.

As shown here, many star systems contain "hot Jupiters," or giant planets that orbit very close to their parent stars. The presence of a hot Jupiter appears to strongly disfavor the possibility of life as we know it arising within a stellar system, because rocky planets with life-friendly orbits will be destabilized by the central star-planet system.

IMAGE BY JON LOMBERG

STELLAR COMPANIONS

—— hot Jupiters stay close to their host star ——

UNIVERSE
Diameter: $1.74 \times 10^{13} \rightarrow 1.77 \times 10^{13}$ ly
  Observable: $4.36 \times 10^{10} \rightarrow 4.42 \times 10^{10}$ ly
Expansion: $128.93 \rightarrow 126.90$ km/s/Mpc
Density: $2.31 \rightarrow 2.23$ p/m³
Temperature: $5.76 \rightarrow 5.69$ K

ASTRONOMICAL OBJECTS
Galaxies: $2.92 \times 10^{13} \rightarrow 2.93 \times 10^{13}$
Stars: $1.25 \times 10^{21} \rightarrow 1.28 \times 10^{21}$
Black Holes: $1.03 \times 10^{19} \rightarrow 1.05 \times 10^{19}$
Planetary Systems: $1.69 \times 10^{21} \rightarrow 1.73 \times 10^{21}$
Worlds: $1.88 \times 10^{22} \rightarrow 1.93 \times 10^{22}$

WORLDS WITH LIFE
Simple Life: $7.27 \times 10^{17} - 2.32 \times 10^{19}$
Complex Life: $1.37 \times 10^{12} - 2.32 \times 10^{16}$
Intelligent Life: $1.64 \times 10^{8} - 7.02 \times 10^{13}$
Technological Life: $1.98 \times 10^{4} - 3.40 \times 10^{12}$
Interstellar Life: $0 - 8.45 \times 10^{9}$

Molecular clouds of gas, wherever enough heavy elements have built up for new stars to form with planets surrounding them, lead to planets appearing with different masses at wildly varying distances from the star. Rocky worlds with thin atmospheres as well as giant planets with thick hydrogen and helium envelopes form at various distances from their host stars, with gas giants just as likely to form in close, tight orbits around their parent stars as they are at any location. Right at this moment in history, within our Milky Way, one of these hot, massive, fast-orbiting planets—a "hot Jupiter"—forms around a newborn star: 51 Pegasi. And 8.3 billion years later, when human beings become technologically advanced here on Earth, this hot Jupiter planet, 51 Pegasi b, becomes the first planet we'll discover around a star other than our own.

Whipping around their parent stars at speeds in excess of 100 kilometers per second, these giant planets typically complete a full revolution in mere days. The temperatures on these worlds can rise into the thousands of degrees, and yet they can still contain volatile gases like methane, water vapor, and even hydrogen and helium. Hot Jupiters are relatively rare in the Universe, however, and there's a good reason for that. A delicate balance must occur during planet formation for a hot Jupiter to arise: a "Goldilocks" set of conditions, if you will.

Around practically all newly forming stars, a protoplanetary disk arises: primitive matter distributed around a central protostar. Close to the young star, radiation evaporates away ices and low-mass elements, leaving only rock-and-metal remnants. These components can add up to form a planetary core but won't have access to the volatile elements and compounds needed to make a gas giant. Far away from the young star, planetary cores can form, grow, and pull surrounding volatiles onto them to form gas giants. Gravitational interactions occur between objects within a protoplanetary disk, causing planets and planetesimals to migrate, collide, merge, and get flung into the parent star or ejected from the star system altogether.

The gas giants that form at intermediate distances from their parent star and migrate inward will "clean out" the inner regions of their systems, often getting swallowed by their parent star as they migrate too close to it: an Icarus-like fate. If they first form at larger distances, gas giants struggle to get close enough to their parent stars to clean out the system entirely, leaving a mix of terrestrial and lower-mass gas-rich planets interior to them. Only if a giant planet forms in the right location relative to its parent star—not too close and not too far, with neither too much nor too little mass interior to its orbit—can it migrate to a stable, hot, rapid orbit. These conditions are only achieved rarely, leading to a paucity of hot Jupiters.

Although it's possible, in theory, to form rocky, life-friendly planets exterior to these hot Jupiters, none have yet been found. The combined gravitational effects of both a parent star and a closely orbiting giant planet, when acting on a rocky planet that could potentially house liquid water on its surface with the right atmospheric conditions, serve to destabilize the less massive planet's orbit entirely. On timescales of hundreds of millions to a few billion years, such rocky planets—the best hope for intelligent life arising—will either be ejected or swallowed by one of the inner masses. These unstable conditions make complex life's emergence all but impossible. As exciting as these hot Jupiter–containing systems are, these instabilities render them unfavorable as far as life's chances are concerned.

At some point in the Universe's history, the moment of "first contact" will arise, where an intelligent civilization detects, for the first time, the presence of another intelligent life-form elsewhere in the Universe. Upon receiving such a signal, a response can immediately be crafted, where it will travel at the speed of light to initiate two-way communication for the first time.

IMAGE BY MARK A. GARLICK

FIRST CONTACT

—— *beings on different worlds make contact* ——

UNIVERSE
Diameter: $1.77 \times 10^{13} \rightarrow 1.79 \times 10^{13}$ ly
  Observable: $4.42 \times 10^{10} \rightarrow 4.48 \times 10^{10}$ ly
Expansion: $126.90 \rightarrow 124.91$ km/s/Mpc
Density: $2.23 \rightarrow 2.14$ p/m³
Temperature: $5.69 \rightarrow 5.61$ K

ASTRONOMICAL OBJECTS
Galaxies: $2.93 \times 10^{13} \rightarrow 2.94 \times 10^{13}$
Stars: $1.28 \times 10^{21} \rightarrow 1.31 \times 10^{21}$
Black Holes: $1.05 \times 10^{19} \rightarrow 1.07 \times 10^{19}$
Planetary Systems: $1.73 \times 10^{21} \rightarrow 1.77 \times 10^{21}$
Worlds: $1.93 \times 10^{22} \rightarrow 1.98 \times 10^{22}$

WORLDS WITH LIFE
Simple Life: $7.28 \times 10^{17} - 2.36 \times 10^{19}$
Complex Life: $1.43 \times 10^{12} - 2.72 \times 10^{16}$
Intelligent Life: $1.90 \times 10^{8} - 9.02 \times 10^{13}$
Technological Life: $2.51 \times 10^{4} - 4.78 \times 10^{12}$
Interstellar Life: $1 - 1.31 \times 10^{9}$

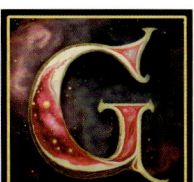

Given the large numbers of star systems with the right ingredients for life, there is no doubt very large numbers of planets, moons, and possibly other worlds where the right conditions for life's emergence persisted from very early times. Over the past 5.6 billion years, many of these worlds became inhabited, saw life survive and thrive on their surface, where it then evolved to become complex, differentiated, and even intelligent. On some unknown fraction of these worlds, a species would occasionally become technologically advanced, reaching for the stars, coming to understand the Universe in which they emerged, and discovering signs of life elsewhere all throughout their home galaxy.

But only now, long after the first *Encyclopaedia Galactica* has been created by the first species to reach that level of advancement, has an equally momentous achievement occurred: For the first time, two technologically advanced civilizations have come to exist at the same time in the same galaxy, and one receives a signal of the other's existence. This moment—known as "first contact"—would forever change the history of both civilizations.

But how? It's easy to imagine a wildly different set of possible outcomes, hugely dependent on the nature of each such species involved in this cultural exchange, as well as the distances between their civilizations. Are they curious and peaceful, having overcome their resource-hoarding natures to achieve a sustainable long-term presence on their world? Are they at significant risk of self-annihilation through war, disease, or ecological collapse? Have they both become technologically advanced within relatively short periods of one another, or is one many millions of years more advanced than the other? And have they made sense of the cosmos in fashions that are at all compatible with and communicable to one another?

This one event, of first contact between two advanced, intelligent civilizations, could arguably be the most significant event in all of cosmic history. Even from afar, with tens of thousands of light-years separating them in physical space, these interstellar "pen pals" could learn more about the Universe together than they ever could on their own. By comparing notes, they could share information about their respective corners of the cosmos, learning how the cosmos is both similar and different to what they've observed from their limited one-point perspectives and shedding any last prejudices they might have held that they occupied a privileged position within the Universe.

Even if they start out far apart from within the same galaxy, over time, the dynamics of stellar motion might bring them within ~100 light-years or less of one another, making the physical exchange of materials, artifacts, and even living beings potentially possible. But with timescales of tens to hundreds of millions of years required to bring two typical, random stars to the point of closest approach, the most likely outcome is far more sobering. All of a sudden, after an unknown length of time, one of the two civilizations will suddenly go silent. Whether from self-inflicted causes like nuclear war or the poisoning of their own environment or from external causes like an asteroid strike, no planet remains habitable to the same species forever. But with all the information they could transmit safely in the hands of another, surviving civilization, they might achieve a novel type of immortality: the knowledge, from across space and time, that their legacy to the Universe lives on.

On a world rich with complex life, two moons can be seen in the sky: evidence of an ancient, massive collision. If life was present on this world at the time the collision took place, it's possible that much of the material that was ejected was transported to other worlds or even other stellar systems, with life stowed away on board. The creatures seen in this image may yet have cousins they don't yet know about thriving on other worlds.

IMAGE BY MARK A. GARLICK

NATIVE LIFE, ALIEN ORIGINS

—— *life often descends from alien worlds* ——

UNIVERSE
Diameter: $1.79 \times 10^{13} \rightarrow 1.81 \times 10^{13}$ ly
  Observable: $4.48 \times 10^{10} \rightarrow 4.54 \times 10^{10}$ ly
Expansion: $124.91 \rightarrow 122.97$ km/s/Mpc
Density: $2.14 \rightarrow 2.06$ p/m³
Temperature: $5.61 \rightarrow 5.54$ K

ASTRONOMICAL OBJECTS
Galaxies: $2.94 \times 10^{13} \rightarrow 2.94 \times 10^{13}$
Stars: $1.31 \times 10^{21} \rightarrow 1.33 \times 10^{21}$
Black Holes: $1.07 \times 10^{19} \rightarrow 1.09 \times 10^{19}$
Planetary Systems: $1.77 \times 10^{21} \rightarrow 1.80 \times 10^{21}$
Worlds: $1.98 \times 10^{22} \rightarrow 2.01 \times 10^{22}$

WORLDS WITH LIFE
Simple Life: $7.29 \times 10^{17} - 2.40 \times 10^{19}$
Complex Life: $1.49 \times 10^{12} - 3.15 \times 10^{16}$
Intelligent Life: $2.15 \times 10^8 - 1.13 \times 10^{14}$
Technological Life: $3.09 \times 10^4 - 6.52 \times 10^{12}$
Interstellar Life: $1 - 1.95 \times 10^{10}$

Throughout the Universe, the cosmic story that leads to life continues to unfold. Stars burn through their fuel, building up heavy atomic nuclei through nuclear fusion, neutron capture, and stellar cataclysms. Those heavy atoms get recycled back into the interstellar medium, linking up to form chemical compounds such as complex molecules. As new stars and stellar systems form, these atoms and molecules bind together as components of protoplanetesimals; in the presence of heat, light, and other forms of energy, even greater complexity arises. After tens of millions of years, these proto-stellar systems evolve into full-fledged stars, each with its own set of planets, moons, and potential habitability.

In most cases, wherever life does arise, it does so because a rocky or icy world possesses the right conditions to make it possible and gets "lucky enough" in the unfolding of cosmic events. The surface, rich in organic molecules, eventually produces a combination that extracts energy from the resources in its environment, and makes copies of itself. Over time, that early form of proto-life doesn't die out but sustains a chain of survival and reproduction, eventually producing offspring that begins to diversify. On many of these worlds, life survives and thrives for long periods, perhaps becoming complex, differentiated, intelligent, or even technologically advanced.

This brings us back to the idea known as panspermia: where life originating from one world travels across either interplanetary or even interstellar distances to land on another. Practically all rocky planets experience bombardment from other objects—comets, asteroids, and interstellar interlopers alike—traveling through the Universe. Whenever an energetic-enough collision occurs, it won't simply cause devastation to the impacted world itself but will also kick up enormous amounts of debris. While most of that debris falls back onto that world, some direct hits will kick debris up with such energy that it can escape the atmosphere and gravitational pull of the world that was struck, sending material from a living world—including living or dormant life-forms—back into the Universe.

All of a sudden, this opens up the possibility of transporting living organisms from one world to another. If some of these life-forms wind up settling onto another world where the proper resources are in place to permit their metabolic activities, they could forever alter the biological history of that new world, whether within their original stellar system or after traveling across interstellar space to arrive at a new one.

If such transported life-forms happen to find a destination where the very processes that allow them to survive, thrive, and reproduce are permissible, it could potentially provide seeds for life on an otherwise uninhabitable world, or it could outcompete and replace any inferior life-forms that had arisen on a living world. Planets and moons that could have given rise to life, but which didn't for some reason or other, could get a second chance owing to life-forms of extraterrestrial origin that might be traveling throughout the Universe. If it turns out that it's quite difficult for life to arise from nonlife, this may even turn out to be a cosmically important process for creating living planets from nonliving ones altogether.

Perhaps there are living worlds where extraterrestrial life arrives and either one achieves dominance over the other or they interbreed: coexisting and coevolving. Perhaps some intelligent, spacefaring civilizations deliberately (or recklessly) bring life-forms from their own world to another. Life's frozen seeds may be traveling all throughout interplanetary space, just awaiting a new, permanent home.

What appears to be a technologically advanced civilization on a life-rich world might not necessarily be the first example of intelligent life to occur here. If the first species to rise to prominence dies off, even if their demise correlates with a mass extinction event, the surviving organisms will persist and evolve, perhaps giving rise to a more successful civilization down the line.

IMAGE BY JON LOMBERG

MY NAME IS OZYMANDIAS

—— civilizations can learn from past failures ——

UNIVERSE
Diameter: $1.81 \times 10^{13} \rightarrow 1.84 \times 10^{13}$ ly
  Observable: $4.54 \times 10^{10} \rightarrow 4.59 \times 10^{10}$ ly
Expansion: $122.97 \rightarrow 121.23$ km/s/Mpc
Density: $2.06 \rightarrow 1.99$ p/m³
Temperature: $5.54 \rightarrow 5.47$ K

ASTRONOMICAL OBJECTS
Galaxies: $2.94 \times 10^{13} \rightarrow 2.95 \times 10^{13}$
Stars: $1.33 \times 10^{21} \rightarrow 1.36 \times 10^{21}$
Black Holes: $1.09 \times 10^{19} \rightarrow 1.11 \times 10^{19}$
Planetary Systems: $1.80 \times 10^{21} \rightarrow 1.84 \times 10^{21}$
Worlds: $2.01 \times 10^{22} \rightarrow 2.06 \times 10^{22}$

WORLDS WITH LIFE
Simple Life: $7.30 \times 10^{17} - 2.43 \times 10^{19}$
Complex Life: $1.53 \times 10^{12} - 3.58 \times 10^{16}$
Intelligent Life: $2.37 \times 10^{8} - 1.38 \times 10^{14}$
Technological Life: $3.66 \times 10^{4} - 8.48 \times 10^{12}$
Interstellar Life: $1 - 2.73 \times 10^{10}$

 lthough individual forms of life might face particular challenges when it comes to long-term survival—for a species, eventual extinction is the norm, not the exception—inhabited worlds are incredibly resilient. Unless an event is able to extinguish 100% of the biological activity happening on an inhabited planet or moon, there's an excellent chance that every plausible ecological niche will once again find itself occupied in short order. Wherever there are resources to consume and life-forms capable of utilizing them, there are opportunities for new species to come into existence and rise to prominence.

Somewhere in the Universe, hundreds of millions of years ago, an intelligent, technologically advanced civilization arose and was driven to extinction by a natural event. Although that species is by now long gone, the planet recovered relatively quickly, and a new set of organisms began occupying those once-vacated niches. Those forms of life that are superior at gathering, hoarding, and protecting those necessary resources will be the most successful, outcompeting the others that would take their place if they could. Sometimes it's the fastest, strongest, or most adaptable organisms that achieve dominance within any given niche, but every once in a while, it's a different trait that leads to evolutionary prosperity: intelligence.

Intelligence is an empowering ability. Many species, if left solely to their own natural abilities—speed, strength, endurance, or resilience—would quickly succumb to the pressures of their environment. But through intelligence, the crafting and utilization of tools, and clever strategies, a weaker, slower, less physically fit specimen can succeed: surviving and thriving in a superior fashion to all those competing for the same resources.

And if intelligence can arise once on a planet, then so long as the planet remains conducive to life, there's no reason to think it won't be able to do so again. If life can become technologically advanced once on a planet, there's every reason to be optimistic—assuming that similar conditions still persist on that world—that multiple opportunities to reach for the Universe beyond their rocky home will arise down the line. When a second technologically advanced civilization does finally arise, they might even have remnants of the first such civilization's legacy to learn from.

Objects that were launched to geosynchronous orbit or beyond—including onto any lunar or planetary surfaces—might still persist even over timescales of hundreds of millions or even billions of years. Information-rich capsules, perhaps encased in hardened shells that could withstand the passage of time, could pose a mechanism for one-way communication from the original intelligent species towards any future evolutionary descendants or cousins. Even if such relics weren't left behind, evidence of mining activities to extract various elements or fuel sources would be identifiable even far into the future.

That means that the mistakes of the first intelligent, technologically advanced civilization to appear on a world—perhaps even mistakes that led to their own demise—would be mistakes that future intelligent inhabitants might be able to learn from. As long as life continues to survive and thrive on a world where complex, differentiated life has arisen, the potential for an intelligent, technologically advanced civilization to emerge cannot be discounted. Eight billion years from now, when human beings rise to prominence on Earth, perhaps some of them will wonder, "If our species doesn't make it, will there someday be another, here on Earth, that does?"

Behind a massive cluster of galaxies, some of the most distant objects of all can be seen: intrinsically faint and reddened by the expansion of the Universe but magnified by the effects of gravitational lensing. By measuring the recession of galaxies at all observed distances to high precision, a careful observer could finally deduce the presence of dark energy by noting a gradual change in the expansion rate.

IMAGE BY MARK A. GARLICK

THE FORCE DOTH AWAKEN

—— *expansion of the Universe accelerates* ——

UNIVERSE
Diameter: $1.84 \times 10^{13} \rightarrow 1.86 \times 10^{13}$ ly
 Observable: $4.59 \times 10^{10} \rightarrow 4.65 \times 10^{10}$ ly
Expansion: $121.23 \rightarrow 119.37$ km/s/Mpc
Density: $1.99 \rightarrow 1.91$ p/m³
Temperature: $5.47 \rightarrow 5.40$ K

ASTRONOMICAL OBJECTS
Galaxies: $2.95 \times 10^{13} \rightarrow 2.94 \times 10^{13}$
Stars: $1.36 \times 10^{21} \rightarrow 1.38 \times 10^{21}$
Black Holes: $1.11 \times 10^{19} \rightarrow 1.12 \times 10^{19}$
Planetary Systems: $1.84 \times 10^{21} \rightarrow 1.86 \times 10^{21}$
Worlds: $2.06 \times 10^{22} \rightarrow 2.09 \times 10^{22}$

WORLDS WITH LIFE
Simple Life: $7.29 \times 10^{17} - 2.46 \times 10^{19}$
Complex Life: $1.56 \times 10^{12} - 4.09 \times 10^{16}$
Intelligent Life: $2.59 \times 10^{8} - 1.68 \times 10^{14}$
Technological Life: $4.29 \times 10^{4} - 1.10 \times 10^{12}$
Interstellar Life: $1 - 3.83 \times 10^{10}$

Nearly 5.9 billion years after the hot Big Bang, the observable Universe has grown and evolved into a remarkably rich place. Trillions of galaxies now populate the observable Universe, with over a sextillion (10^{21}) stars among them. The leftover glow from the Big Bang—the cosmic microwave background—now sits at a mere 5.4 K above absolute zero. It's 23 billion light-years to the edge of the observable Universe, an ever-growing number as the Universe evolved from its initially hot, dense state.

Up until this moment in cosmic history, nothing seemed out of reach. If you sent a light signal to any distant object, then someday, you'd expect that light signal to eventually arrive at your intended destination. The expanding Universe has always behaved like a race: between the initial expansion, which works to drive everything apart, and gravitation, which works to pull everything back together. Up until now, the race looked like it was going to be a draw, with gravitation slowing the expansion—approaching a recession speed of 0—but never being able to reverse itself. Neither gravitation nor expansion would win in the end. If you sent out a light signal and waited long enough, eventually it would get to where you sent it.

But seemingly out of nowhere, distant galaxies are no longer slowing down as a matter-dominated Universe would imply. Instead, it's as though a foot has eased off the cosmic brakes while another presses the accelerator. With precision measurements, the conclusion is inescapable: A signal sent at the speed of light towards every distant galaxy presently observable can no longer reach all of them. The farthest galaxies—the ones visible at the farthest reaches of the visible Universe—will never receive the messages you are now transmitting.

It's as though there's more to the Universe than all the forms of matter and radiation combined. In addition to those forms of energy, a new one is emerging whose presence is only barely detectable. This energy, somehow, doesn't get less dense even as space expands and the volume of the Universe increases. It only seems to be appearing right now, after more than five billion years of cosmic evolution, after the matter and radiation densities have dropped to just 0.1% and 0.01%, respectively, of what they were when the first galaxies were forming some 200 million years after the hot Big Bang. Ever since that time, cosmic structure has been building up on larger and grander scales: galaxies, galactic groups, galaxy clusters, and most recently, massive overdense objects. But the end of that epoch is now in sight.

Gravitation is beginning to lose the great cosmic race. In very short order—predictably, about two billion years from now—the structures that haven't yet become gravitationally bound will suddenly be driven apart by the expanding Universe. It's as though a new, unseen form of energy has permeated all of space and is causing distant objects to accelerate away from one another faster than a Universe dominated by matter and radiation alone would permit. Whatever this new form of energy is, it doesn't appear to be clumping together like matter does, nor does it absorb or emit light. It simply causes the cosmic expansion to accelerate, driving distant objects apart. Someone discovering it for the first time might well call it what we do: dark energy. The great cosmic race—expansion versus gravitation—won't be a draw, after all. Expansion will win, and dark energy will be the reason why.

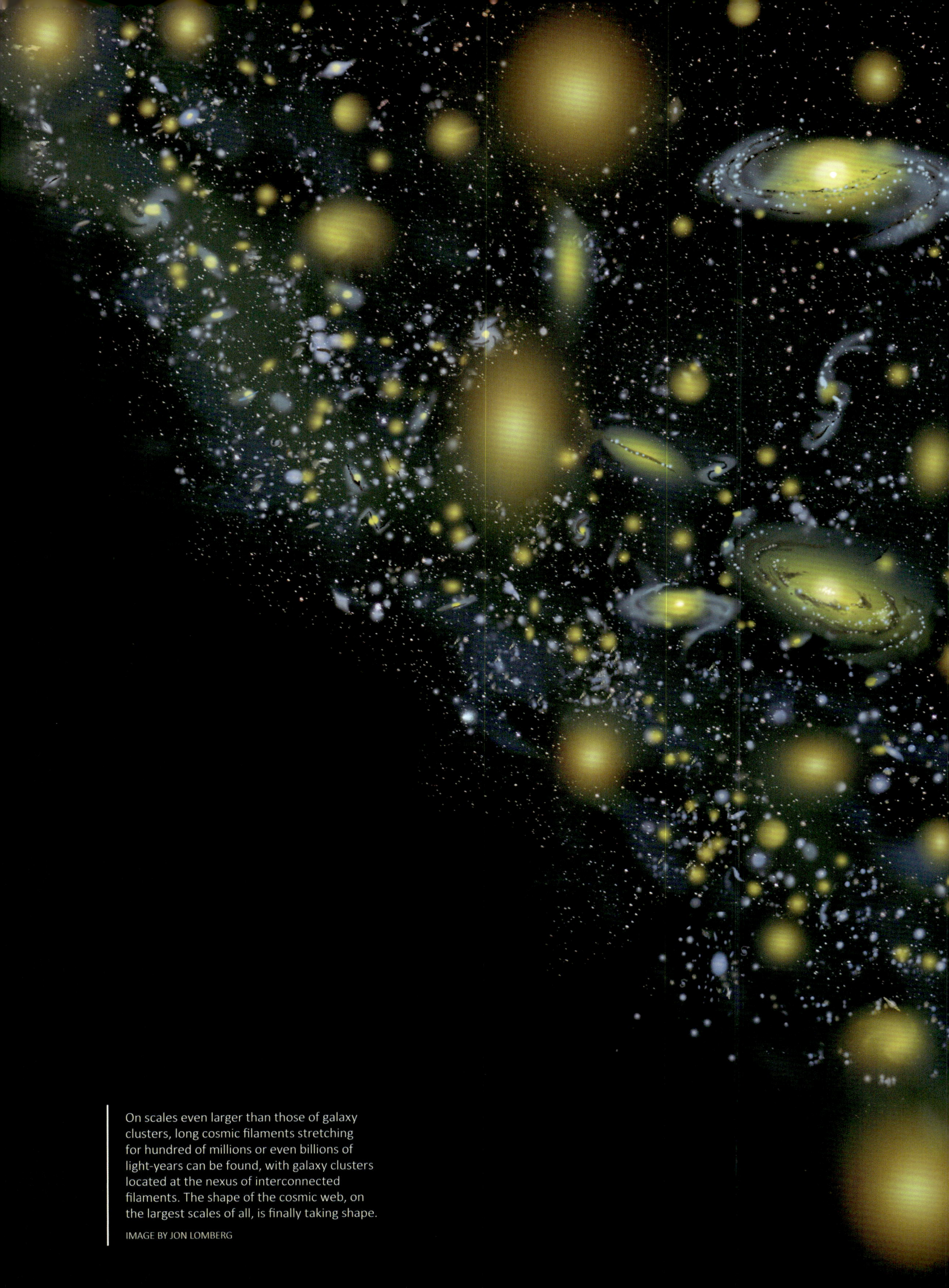

On scales even larger than those of galaxy clusters, long cosmic filaments stretching for hundred of millions or even billions of light-years can be found, with galaxy clusters located at the nexus of interconnected filaments. The shape of the cosmic web, on the largest scales of all, is finally taking shape.

IMAGE BY JON LOMBERG

GREAT WALLS OF GALAXIES

—— the largest objects in the Universe ——

UNIVERSE
Diameter: $1.86 \times 10^{13} \rightarrow 1.88 \times 10^{13}$ ly
  Observable: $4.65 \times 10^{10} \rightarrow 4.71 \times 10^{10}$ ly
Expansion: $119.37 \rightarrow 117.70$ km/s/Mpc
Density: $1.91 \rightarrow 1.84$ p/m³
Temperature: $5.40 \rightarrow 5.34$ K

ASTRONOMICAL OBJECTS
Galaxies: $2.94 \times 10^{13} \rightarrow 2.93 \times 10^{13}$
Stars: $1.38 \times 10^{21} \rightarrow 1.40 \times 10^{21}$
Black Holes: $1.12 \times 10^{19} \rightarrow 1.14 \times 10^{19}$
Planetary Systems: $1.86 \times 10^{21} \rightarrow 1.89 \times 10^{21}$
Worlds: $2.09 \times 10^{22} \rightarrow 2.13 \times 10^{22}$

WORLDS WITH LIFE
Simple Life: $7.28 \times 10^{17} - 2.49 \times 10^{19}$
Complex Life: $1.59 \times 10^{12} - 4.59 \times 10^{16}$
Intelligent Life: $2.78 \times 10^{8} - 1.99 \times 10^{14}$
Technological Life: $4.87 \times 10^{4} - 1.38 \times 10^{12}$
Interstellar Life: $1 - 5.10 \times 10^{10}$

ne of the great observational facts about the Universe is that the visible cosmic structures didn't all form simultaneously, but the smaller-scale structures formed early on while the larger-scale ones required much longer. This is true despite the fact that the same seeds of structure were present, initially, on all scales. Small, intermediate, and large cosmic scales alike all possessed imperfections—overdense and underdense regions—of roughly the same magnitude. But while individual galaxies began forming within the first few hundred million years of the hot Big Bang, it's only now, nearly six billion years on, that the largest cosmic structures begin to take shape. Instead of just star clusters, galaxies, and galaxy groups and clusters, the Universe now possesses great cosmic walls, the largest of which span over a billion light-years.

Why does it take so long for these ultra-large-scale features to come into existence? It's because gravity isn't an instantaneous force but only propagates from one massive object to another at a finite speed. That speed—the speed of gravity—turns out to be identical to the speed of light. Just as an observer in one location only sees a distant object as it was some time ago, when the light arriving now was emitted, it only experiences the gravitational effects of that same object as it was in the past: when the arriving gravitational signal was emitted.

Because massive objects gravitationally grow over time, the more distant an object is located from an observer, the less time it had to grow more massive compared to the relatively nearby objects. An object that's billions of light-years away will only exert a gravitational effect in proportion to how massive it was billions of years

ago. As a result of the finite speed of gravity, along with the ongoing effects of the Universe's expansion, cosmic structures on the largest scales require more time to form—and become gravitationally bound—than their smaller-scale counterparts.

Nevertheless, time is relentless in its passage. Given that the Universe still has most of its total energy in the form of matter—massive particles that clump together—gravitational growth is still progressing to larger and larger scales as time marches forth. While galaxies have existed for more than five billion years and galaxy clusters first came into existence around three billion years ago, it's only now that even grander structures are forming. The space between galaxy clusters is now lined with galaxy-rich filaments, and multiple galaxy clusters are beginning to collide and merge: being drawn in towards one another along those filamentary lines.

From the perspective of any observer who looks out at the cosmos and measures how galaxies clump and cluster, the richest nearby regions of the Universe will possess great galactic walls. These walls appear as Swiss cheese–like structures, where giant galaxy-free regions that gave up their matter to their surroundings long ago—cosmic voids—are surrounded by lines and sheets of galaxies. At their intersections, galaxy groups and clusters form, drawing in additional matter, including whole galaxies themselves, along those cosmic filaments.

For the first time, the largest objects in the Universe, great galaxy walls, are starting to take shape. For any observer located within one, they'd need to look at not just nearby galaxies but at galaxies billions of light-years away in order to discover the expanding Universe.

As intelligent civilizations arise in the Universe and become technologically advanced, they perhaps inevitably seek to answer the cosmic question of, "Are we alone?" A stable structure that wouldn't occur naturally, artificially assembled out of stars, is illustrated here: one possible way for an extraterrestrial intelligence to announce their presence to the rest of the Universe.

IMAGE BY MARK A. GARLICK

BEACONS OF LIGHT AND HOPE

—— some civilizations signal their existence ——

UNIVERSE
Diameter: $1.88 \times 10^{13} \rightarrow 1.91 \times 10^{13}$ ly
  Observable: $4.71 \times 10^{10} \rightarrow 4.76 \times 10^{10}$ ly
Expansion: $117.70 \rightarrow 116.07$ km/s/Mpc
Density: $1.84 \rightarrow 1.78$ p/m³
Temperature: $5.34 \rightarrow 5.28$ K

ASTRONOMICAL OBJECTS
Galaxies: $2.93 \times 10^{13} \rightarrow 2.95 \times 10^{13}$
Stars: $1.40 \times 10^{21} \rightarrow 1.43 \times 10^{21}$
Black Holes: $1.14 \times 10^{19} \rightarrow 1.16 \times 10^{19}$
Planetary Systems: $1.89 \times 10^{21} \rightarrow 1.93 \times 10^{21}$
Worlds: $2.13 \times 10^{22} \rightarrow 2.18 \times 10^{22}$

WORLDS WITH LIFE
Simple Life: $7.26 \times 10^{17} - 2.51 \times 10^{19}$
Complex Life: $1.60 \times 10^{12} - 5.13 \times 10^{16}$
Intelligent Life: $2.95 \times 10^{8} - 2.33 \times 10^{14}$
Technological Life: $5.45 \times 10^{4} - 1.69 \times 10^{12}$
Interstellar Life: $2 - 6.66 \times 10^{10}$

Despite the fact that the Universe is teeming with galaxies, stars, planets, and life, only very rarely does an intelligent, technologically advanced civilization arise that endures for long durations. In a significant fraction of galaxies, particularly those with smaller abundances of heavy elements, intelligent civilizations may be exceedingly rare. For those species that do become technologically advanced, their searches for other galactic intelligence may come up empty, leaving them wondering, "Are we truly alone?" Even though they're almost certainly not unique, in the absence of a positive detection of intelligent extraterrestrials, they would have no surefire evidence to the contrary.

But somewhere in the Universe, one lonesome civilization becomes so successful that they can announce their presence, sending out a beacon to distant stars and galaxies, more loudly and with greater reach than any other. Perhaps they've learned to leverage extraordinarily long-wavelength light, transmitting great amounts of information over very large distances, requiring only minuscule power outputs. Perhaps they've learned how to harness energy more effectively than other intelligent species, brute-forcing their way towards high signal-to-noise transmissions. Or perhaps they've learned to create bursts of information-containing signals from beyond the electromagnetic spectrum: by using neutrinos or gravitational waves. For as long as they remain technologically advanced and continue transmitting, they would be detectable across a larger volume of space than any other civilization in existence.

Even from millions of light-years away—perhaps in another galaxy or even another galaxy group or cluster—any technologically advanced civilization arising during this time would be capable of receiving that signal, concluding, "Yes, we are not alone!" Depending on what information was encoded, this less advanced civilization could benefit from unprecedented technological prowess. They might learn how to send a return message, survive numerous existential challenges, and discover the solution to many technological and resource-limited puzzles that they're encountering. And, perhaps most spectacularly, they'd become the first civilization to have two independent perspectives, from two different galaxies, that helped them to decode the great cosmic mysteries.

When such a signal at last arrives at and is decoded by a technologically emerging species, what would they think of it? "If another civilization can survive their technological infancy to achieve this level of sustained success in the cosmos," they might reason, "perhaps we can set aside our petty differences, our resource-hogging nature, and work together—as a unified planet—to solve our collective problems." Even if the initial message didn't come with a road map for how to survive and thrive with the newfound power and responsibility that comes with such technological power, the receipt of such a beacon could provide something invaluable to a burgeoning, intelligent species: hope.

The finite nature of the speed of light, however, provides a tremendous impediment to round-trip communications. Even a nearby extragalactic civilization would likely be located millions of light-years away; a separation distance of three million light-years would necessitate the passage of six million years for the transmitting civilization to receive a return message. Unless the long-term survival of advanced, intelligent species is commonplace, it's likely that the extinction of one or even both intelligent civilizations would occur before a true two-way exchange takes place. It is unknown whether contact from an intelligent extragalactic civilization would improve the odds of long-term survival or self-destruction throughout the cosmos. And now, 6.1 billion years after the Big Bang, one civilization gets to be the first to find out.

Within a massive galaxy cluster, starlight isn't simply concentrated within the individual component galaxies; large populations of stars also exist in the space between galaxies, creating a glow of intracluster light. Illustrated here in pink, that intracluster light is a tracer of dark matter, showcasing how there's more to the Universe than simply normal matter alone.

IMAGE BY MARK A. GARLICK

UNSEEN, BUT NOT UNFELT

—— dark energy and matter fill the Universe ——

UNIVERSE
Diameter: $1.91 \times 10^{13} \rightarrow 1.93 \times 10^{13}$ ly
  Observable: $4.76 \times 10^{10} \rightarrow 4.82 \times 10^{10}$ ly
Expansion: $116.07 \rightarrow 114.47$ km/s/Mpc
Density: $1.78 \rightarrow 1.71$ p/m³
Temperature: $5.28 \rightarrow 5.21$ K

ASTRONOMICAL OBJECTS
Galaxies: $2.95 \times 10^{13} \rightarrow 2.94 \times 10^{13}$
Stars: $1.43 \times 10^{21} \rightarrow 1.45 \times 10^{21}$
Black Holes: $1.16 \times 10^{19} \rightarrow 1.18 \times 10^{19}$
Planetary Systems: $1.93 \times 10^{21} \rightarrow 1.96 \times 10^{21}$
Worlds: $2.18 \times 10^{22} \rightarrow 2.21 \times 10^{22}$

WORLDS WITH LIFE
Simple Life: $7.23 \times 10^{17} - 2.53 \times 10^{19}$
Complex Life: $1.61 \times 10^{12} - 5.69 \times 10^{16}$
Intelligent Life: $3.11 \times 10^8 - 2.70 \times 10^{14}$
Technological Life: $6.01 \times 10^4 - 2.06 \times 10^{12}$
Interstellar Life: $2 - 8.55 \times 10^{10}$

 o matter where one looks, there seems to be more than just the laws and particles that govern stars, planets, and life. On these smaller scales, the physics of atoms and subatomic particles can explain every phenomenon that one can observe. Atomic nuclei—protons and neutrons— along with electrons, photons, and neutrinos, combine to produce every physical entity that can be detected and measured on these scales. From ionized plasmas to gases, liquids, solid rocks, metals, and ices, every material can be described by the known fundamental constituents of matter. They often assemble into larger structures: molecules, organisms, even planets, stars, and stellar remnants. From stellar and planetary systems all the way down to subatomic scales, every effect can be accounted for with a known cause.

But on galactic and larger scales—including galaxy groups and clusters, cosmic walls, and the entire observable Universe—there are two sets of novel effects that appear but whose causes are yet to be determined. On galactic and supergalactic scales, the amount of known matter that's present is insufficient to explain the observed gravitational effects. Stars move around within galaxies, particularly towards the outskirts, at speeds that exceed the known matter's predictions. When pairs of galaxies interact and merge, they accelerate more quickly than the observed matter predicts. When galaxies are measured within groups and clusters, they zip around at speeds that would cause them to escape from these structures, unless some additional source of mass is present. And when gravitational lensing occurs, background light bends and distorts by greater amounts than the known matter, alone, can explain.

In addition, when we examine the structures on the largest cosmic scales, from the cosmic web to distances between galaxies to the imperfections in the Big Bang's remnant light, they all require not only more mass than what we know normal matter can provide, but most of that mass—about 85% of it—can't be made of any of the particles we've discovered. Whatever is causing these excess gravitational effects neither absorbs nor emits light, nor does it collide with any identified species of matter. Perhaps cosmic discoverers of this effect would justifiably call it something akin to what Earthlings will call it billions of years from now: dark matter. It permeates the Universe, clumps together, and gravitates, but there's no foreseeable way to determine its cause.

Similarly, in addition to the matter and radiation in the Universe—including this mysterious source of additional gravitation—there's also a new phenomenon that's begun to appear over the past few hundred million years: an unexpected change in the cosmic expansion rate. Whereas the amount of matter and radiation present in any region of space becomes less as the Universe expands, the Universe now behaves as though there's a type of energy that maintains a constant density, even as the fabric of space itself expands. Unlike dark matter, this form of energy doesn't clump, cluster, or gravitate in the conventional way, but like dark matter, it doesn't appear to interact with light or matter in any sort of absorptive, emissive, or collisional way. It, too, is an effect with no identifiable cause, and might well be called elsewhere what Earthlings will call it billions of years hence: dark energy.

And so, 6.2 billion years after the Big Bang, only 11% of all the energy in the Universe is present in "known" forms of matter and radiation. The remainder is unexplained, leaving cosmic observers in the proverbial dark. Although the effects that go beyond the explanatory powers of known physics can be well described, the underlying causes behind these effects remain mysterious.

The most energetic stellar cataclysm of all is a gamma-ray burst. The highly collimated jets produced in these events can not only sterilize life on any planet they happen to strike but also outshine even the daytime Sun to an observer within the same host galaxy.

IMAGE BY MARK A. GARLICK

A CATACLYSM OF FIRE AND LIGHT

—— *21 quadrillion times brighter than the Sun* ——

UNIVERSE	ASTRONOMICAL OBJECTS	WORLDS WITH LIFE
Diameter: $1.93 \times 10^{13} \rightarrow 1.95 \times 10^{13}$ ly	Galaxies: $2.94 \times 10^{13} \rightarrow 2.93 \times 10^{13}$	Simple Life: $7.20 \times 10^{17} - 2.55 \times 10^{19}$
Observable: $4.82 \times 10^{10} \rightarrow 4.87 \times 10^{10}$ ly	Stars: $1.45 \times 10^{21} \rightarrow 1.47 \times 10^{21}$	Complex Life: $1.61 \times 10^{12} - 6.24 \times 10^{16}$
Expansion: $114.47 \rightarrow 113.03$ km/s/Mpc	Black Holes: $1.18 \times 10^{19} \rightarrow 1.19 \times 10^{19}$	Intelligent Life: $3.23 \times 10^{8} - 3.08 \times 10^{14}$
Density: $1.71 \rightarrow 1.66$ p/m³	Planetary Systems: $1.96 \times 10^{21} \rightarrow 1.99 \times 10^{21}$	Technological Life: $6.5 \times 10^{4} - 2.43 \times 10^{12}$
Temperature: $5.21 \rightarrow 5.16$ K	Worlds: $2.21 \times 10^{22} \rightarrow 2.24 \times 10^{22}$	Interstellar Life: $2 - 1.06 \times 10^{11}$

For most living beings, the brightest light they'll ever see comes from their parent star. This isn't because their star is intrinsically the brightest thing they'll encounter but, rather, because it's located extremely close by. Whereas a living world might be located only a few million kilometers from their life-giving star, the nearest objects that are intrinsically more luminous—other stars—are typically found multiple light-years away: hundreds of thousands of times as distant. Because the observed brightness drops off proportionally to the square of the distance between the source and the observer, the star that you orbit outshines the brightest stars in the night sky by a factor of billions.

And yet, under the right circumstances, a cosmic blast from well outside of one's home system can occasionally outshine even their planet's parent star. Very massive stars at the cores of the brightest, newest star clusters can shine as bright as several million "typical" stars combined. As they evolve, their colors and brightnesses vary, as the stars expand and contract over time. In the final moments of their lives, these stars can undergo spectacular cataclysms: events that destroy the star, leaving behind only a stellar remnant. These events, core-collapse supernovae, come in assorted flavors. The brightest ones of all can even, albeit briefly, outshine what a creature on a living world might recognize as their daytime Sun.

Inside a very massive star, light elements fuse into heavy ones at alarmingly rapid rates. Whereas most stars might take billions or even trillions of years to fuse all of their core's hydrogen into helium, an ultramassive star might get there in under two million years. Once the hydrogen is depleted, the nuclear fusion rate drops, becoming insufficient to hold it up against gravitational collapse, causing the star's core to contract further. With rapid contraction occurring in an environment where heat has no way to escape, the core heats up; when it reaches a sufficient temperature, the next element in line—helium—begins to fuse. Over a span of only a few thousand years, helium fusion gives way to carbon fusion, then neon-, oxygen-, and silicon-based fusion.

For most stars that make it this far, this represents the end of the line. These final fusion reactions liberate only small amounts of energy, and the entire core begins to collapse, heating up further. The central core forms a neutron star or, if it's too massive, a black hole, and triggers a runaway fusion reaction by injecting energy into the outer layers. In short order, the entire star is destroyed: a process that typically releases enough energy to shine as bright as tens of billions of typical stars all put together.

But under extreme conditions, some stars can create even brighter, more energetic cataclysms. The interior temperatures can rise so high that atomic nuclei enter higher-energy excited states, where they spit out subatomic particles: a process known as photodisintegration. At still-higher energies, individual photons spontaneously interact, producing matter-antimatter pairs of electrons and positrons: the pair-instability process. And when these stars collapse, if they're rotating rapidly, they can emit their energy not in a sphere but in collimated jets, creating catastrophes known as superluminous supernovae, hypernovae, or even the most energetic of all optical phenomena: gamma-ray bursts.

The highest-energy gamma-ray burst ever seen, GRB 080319B, occurs now, 6.3 billion years after the Beginning. It achieves a peak brightness equivalent to 21 quadrillion typical stars combined. To an observer within the same galaxy, it would outshine even their daytime Sun, an example of the most energetic burst of all time. The light from this event will arrive at the yet-to-be-formed Earth some 7.5 billion years from now, detectable by a sufficiently advanced civilization there.

All throughout the Universe, light-blocking dust is present in great amounts. But where does it come from? The most prolific sources of cosmic dust are the most massive, shortest-lived stars, which produce outflows of material to balance their growth. Intense stellar winds preferentially blow off the lighter outer layers, leading to the eventual creation of "dark nebulae" throughout space.

IMAGE BY JON LOMBERG

ALL COME FROM DUST

the building blocks of stars and worlds

UNIVERSE
Diameter: $1.95 \times 10^{13} \rightarrow 1.97 \times 10^{13}$ ly
  Observable: $4.87 \times 10^{10} \rightarrow 4.93 \times 10^{10}$ ly
Expansion: $113.03 \rightarrow 111.50$ km/s/Mpc
Density: $1.66 \rightarrow 1.60$ p/m³
Temperature: $5.16 \rightarrow 5.09$ K

ASTRONOMICAL OBJECTS
Galaxies: $2.93 \times 10^{13} \rightarrow 2.92 \times 10^{13}$
Stars: $1.47 \times 10^{21} \rightarrow 1.49 \times 10^{21}$
Black Holes: $1.19 \times 10^{19} \rightarrow 1.21 \times 10^{19}$
Planetary Systems: $1.99 \times 10^{21} \rightarrow 2.01 \times 10^{21}$
Worlds: $2.24 \times 10^{22} \rightarrow 2.27 \times 10^{22}$

WORLDS WITH LIFE
Simple Life: $7.16 \times 10^{17} - 2.57 \times 10^{19}$
Complex Life: $1.60 \times 10^{12} - 6.86 \times 10^{16}$
Intelligent Life: $3.35 \times 10^{8} - 3.52 \times 10^{14}$
Technological Life: $7.00 \times 10^{4} - 2.88 \times 10^{12}$
Interstellar Life: $2 - 1.32 \times 10^{11}$

By this point in the Universe's history, most of the stars that exist are found within large galaxies. The majority of these galaxies have substantial disks, spiral arms, and are steadily producing new stars. These massive spiral galaxies, including the young Milky Way, commonly possess an important, near-universal feature: They're rich in dust. In addition to the gases and plasmas that permeate interstellar space, copious amounts of dust grains and particles— collections of neutral molecules—are also particularly rich inside these galactic disks. Dust is excellent at absorbing optical starlight, radiating at long infrared wavelengths, and playing major roles in enabling the formation of new generations of stars. By contrast, in dust-depleted elliptical galaxies, no new stars are forming any longer.

But where does this cosmic dust come from in the first place? Perhaps surprisingly, the greatest source of cosmic dust comes not from the most numerous types of stars but from the most massive ones. A massive star, even early in its life, produces so much energy that it causes significant outflows of material: a method for young stars to balance their growth as new material accretes and falls onto them. The heaviest elements sink towards their centers, while the lightest ones preferentially remain on the surface: an example of buoyancy in action, even on stellar scales. These energetic outflows can easily blow off the hydrogen-rich outer layers.

As these stars evolve, their core material begins to change in composition, as hydrogen fuses into helium and the star initiates the burning of heavier elements, such as helium into carbon, carbon into neon, neon into oxygen, and oxygen into silicon. These fusion chains heat the star and increase its brightness but also increase the stellar wind strength. Because approximately 50% of all stars are born into multi-star systems, many pairs of massive binaries will have their winds affect one another. This is particularly effective for evolved, massive stellar pairs. In combination, a single pair of massive binary stars can produce a rocky planet's mass worth of dust every single year, with approximately 40% of that dust being carbon.

Although hydrogen and helium are the most abundant atoms in the Universe by far, it's carbon that's the most common atom found in the Universe's dust grains. These grains, in sparse environments, can remain as micron-sized or smaller grains, individually, while in denser environments—such as protoplanetary disks—they aggregate into dust particles, which can be much larger. Additional common elements found in dust include chromium, silicon, sodium, magnesium, aluminum, iron, nickel, sulfur, and oxygen. Even though plenty of dust particles are created by evolved, lower-mass stars that have entered the red giant phase, the greatest amounts of dust, overall, are created by the most massive stars.

This has always been the case, going all the way back to the very first stars that formed: the stars made of pristine material arising from the Big Bang itself. As the material from which stars form becomes more and more enriched with heavy elements, the dust helps new stars form by radiating heat away, enabling lower-mass stars to form beginning with the second (and subsequent) generations of star formation. Additionally, dust helps the interstellar medium cool down, helps form molecules, and can even help shield the dense gas clumps that form initial protostars. The presence of dust plays a major role in the present formation of stars, and where it's absent, conditions may no longer be conducive to the formation of new stars at all.

Whereas worlds with only simple, small, short-lived organisms provide the most common example of life in our Universe, the worlds that possess the greatest diversity and complexity of life-forms will dominate the biomass fraction of the Universe. Worlds such as the one illustrated here should be considered the most successful, at least from a biological point of view.

IMAGE BY MARK A. GARLICK

DIVERSITY IS NATURE'S WAY

life evolves in many shapes and sizes

UNIVERSE
Diameter: $1.97 \times 10^{13} \rightarrow 1.99 \times 10^{13}$ ly
  Observable: $4.93 \times 10^{10} \rightarrow 4.99 \times 10^{10}$ ly
Expansion: $111.50 \rightarrow 110$ 12 km/s/Mpc
Density: $1.60 \rightarrow 1.54$ p/m³
Temperature: $5.09 \rightarrow 5.04$ K

ASTRONOMICAL OBJECTS
Galaxies: $2.92 \times 10^{13} \rightarrow 2.92 \times 10^{13}$
Stars: $1.49 \times 10^{21} \rightarrow 1.51 \times 10^{21}$
Black Holes: $1.21 \times 10^{19} \rightarrow 1.22 \times 10^{19}$
Planetary Systems: $2.01 \times 10^{21} \rightarrow 2.04 \times 10^{21}$
Worlds: $2.27 \times 10^{22} \rightarrow 2.30 \times 10^{22}$

WORLDS WITH LIFE
Simple Life: $7.12 \times 10^{17} - 2.58 \times 10^{19}$
Complex Life: $1.59 \times 10^{12} - 7.46 \times 10^{16}$
Intelligent Life: $3.43 \times 10^{8} - 3.95 \times 10^{14}$
Technological Life: $7.42 \times 10^{4} - 3.34 \times 10^{12}$
Interstellar Life: $3 - 1.60 \times 10^{11}$

On any world where life arises, survives, and thrives, every resource-rich ecological niche that can be occupied by living organisms eventually will be. On some worlds, only a few select locations are potentially inhabitable: hydrothermal vents at the ocean bottoms of ice-covered worlds, the narrow region lining the day-night border along tidally locked worlds, or the cratered, shadowed polar regions of planets perilously close to their parent stars. Other worlds—much like planet Earth will someday become— overflow with a tremendous diversity of life: within the planetary crust, in the deep ocean, in shallow waters, on land masses, and even throughout the atmosphere. If one were to cumulatively consider all the forms of life that exist throughout the Universe, what life-forms would be the most abundant?

If one only counts the sheer number of organisms, the simplest, most primitive ones would vastly outnumber the more complex ones. Assuming that there are more worlds where simple life arises than where life becomes complex, differentiated, and possibly even intelligent, the most abundant form of life would likely be organisms with simple metabolisms that are capable of acquiring resources and hoarding them behind a protein-shrouded membrane. So long as these organisms can acquire the necessary resources to survive and live long enough to reproduce, they should be the most abundant forms of life of all.

On the other hand, one could simply choose to count the number of inhabited worlds. Even though life is certain to exist on rocky planets and moons orbiting their parent stars at the right distance for liquid water on their surfaces, perhaps those worlds aren't the most common locations for life. Perhaps it's the copious moons of outer gas giants,

driven by the internal heating from the tidal forces of the giant planets, where subsurface oceans frequently house simple life. Perhaps it's the even more numerous worlds in the Kuiper belts and Oort clouds of stellar systems, large mixes of ice and rock, where life arises most frequently, so long as liquid water persists beneath an icy crust. Or perhaps it's the worlds arising within failed stellar systems, which might yet be more numerous than all other planetary systems combined, that most commonly give rise to life. It's even conceivable that solidly frozen worlds, where life arose while they still possessed internal liquids, possess the most common forms of life of all. Perhaps they still house cryogenically frozen organisms, just waiting for the opportunity to reawaken.

But if one decides to measure abundance by biomass— or the total amount of mass present in the form of living organisms—then it's the most biologically successful worlds that will dominate. Wherever life is most successful, it will reach certain evolutionary milestones, enabling the rise of new forms of organisms that can simply harness greater sets of resources and become more massive than in locales where those milestones are not reached. The development of a nucleus enables more complex and more massive organisms than their simple prokaryotic precursors. The development of photosynthesis enables organisms to harness the power of light directly to create nutrients and the necessary energy for life. Multicellularity enables far larger, more massive, and longer-lived organisms than previous ones. And the ability to colonize the land can increase a planet's biomass by a factor of about 100 over a world possessing oceanic life alone. Life may be found on a wide variety of worlds, but if you measure abundance by biomass, the worlds with the most successful forms of life are certain to dominate.

In many stellar systems, there isn't going to be just one world present that's suitable for life but multiple ones. Around gas giant planets at the right distance from their parent star, multiple large, rocky moons could all be potential locations for life to survive and thrive. The possibility of biological exchange, or even deliberate colonization of one world by another, must be considered.

IMAGE BY MARK A. GARLICK

CLIMBING MOUNT IMPROBABLE

—— stellar systems can have multiple living worlds ——

UNIVERSE
Diameter: $1.99 \times 10^{13} \rightarrow 2.02 \times 10^{13}$ ly
 Observable: $4.99 \times 10^{10} \rightarrow 5.04 \times 10^{10}$ ly
Expansion: $110.12 \rightarrow 108.78$ km/s/Mpc
Density: $1.54 \rightarrow 1.50$ p/m³
Temperature: $5.04 \rightarrow 4.98$ K

ASTRONOMICAL OBJECTS
Galaxies: $2.92 \times 10^{13} \rightarrow 2.91 \times 10^{13}$
Stars: $1.51 \times 10^{21} \rightarrow 1.53 \times 10^{21}$
Black Holes: $1.22 \times 10^{19} \rightarrow 1.24 \times 10^{19}$
Planetary Systems: $2.04 \times 10^{21} \rightarrow 2.07 \times 10^{21}$
Worlds: $2.30 \times 10^{22} \rightarrow 2.34 \times 10^{22}$

WORLDS WITH LIFE
Simple Life: $7.08 \times 10^{17} - 2.59 \times 10^{19}$
Complex Life: $1.57 \times 10^{12} - 8.08 \times 10^{16}$
Intelligent Life: $3.50 \times 10^{8} - 4.40 \times 10^{14}$
Technological Life: $7.80 \times 10^{4} - 3.84 \times 10^{12}$
Interstellar Life: $3 - 1.92 \times 10^{11}$

In most stellar systems, there will inevitably be one world—a planet or moon—that's best suited for life to arise, survive, and thrive on. Even if there are multiple inhabited worlds within the same system, the difference in evolutionary complexity between the most successful world and the second most successful world should be extraordinary, where life's evolution beyond single-celled, asexually reproducing organisms should occur far earlier on one than another. But occasionally, two or more worlds within the same system will arise with similar environmental conditions and similar compositions, leading to a rare but fascinating scenario: life, possibly even intelligent life, arising on a world with a "planet B" well within reach.

The most common type of stellar system, where a series of major planets orbit a single, central star at well-separated radii, is relatively unlikely to develop two similarly inhabited planets. Even given similar masses, with different radii come different densities, compositions, amounts of energy received from their parent stars, and overall average temperatures. Even if multiple worlds within such a system give rise to life—perhaps even saturating and transforming their biospheres in the process—their differing properties are likely to lead that life down extremely divergent paths. Nevertheless, three separate sets of configurations can give rise to two similarly sized worlds with not only similar compositions but similar environments and external conditions as well, even over multibillion-year timescales.

The first such configuration is known as a double planet, where unlike a major planet orbited by a substantially smaller moon, two planets of very similar mass and size move around the same center of mass in orbit around their parent star. A sufficiently massive giant impact early on—one with enough energy to completely gravitationally unbind an initial planet—could lead to precisely this configuration, where ensuing gravitational interactions pull the debris into two comparably sized planets. A related configuration would be if two comparably massed worlds engage in orbit swapping: where they regularly alternate which one orbits closer to their parent star. The latter will exhibit a Solar System analog billions of years later, with Saturn's orbit-swapping moons Janus and Epimetheus.

A second way to obtain two worlds with similar properties and histories is to have a gas giant planet at the right distance from its parent star so that—were it rocky and water-rich, with a thin but substantial atmosphere—there could be liquid water on its surface. The gas giant itself is a lousy candidate for the development of complex life, but practically every gas giant that forms comes along with a circumplanetary disk: a disk of matter surrounding the planet capable of forming one or more large moons. If two or more massive, rocky moons form sufficiently far from the gas giant itself, they could be ideal candidates for coevolving similar forms of life. Billions of years from now, were our Solar System's Jupiter to exist at Earth's distance from the Sun, both Callisto and Ganymede would be incredible candidates for life.

And finally, in a binary star system, where one star is significantly less massive than another, there will be two points—the L4 and L5 Lagrange points—that are gravitationally stable, always maintaining the same, equal distances from both stars. If similarly massed planets form at these two locations, they'll have nearly identical conditions to one another over their entire histories. In any or all of these scenarios, there might not only be multiple inhabited planets but a true Planet B, where life from one inhabited world could survive and thrive on another one right in their own planetary neighborhood.

In a system where two stars are in close orbit near one another, mass transfer is a common occurrence. It isn't necessarily the most massive star that's the more successful stellar thief but, rather, the denser one: the one where more mass is packed inside a smaller volume. Where the stars touch, as in the case of a contact binary, they will eventually merge to form a single hot, blue star.

IMAGE BY MARK A. GARLICK

THIEVES IN THE NIGHT

stars can steal mass from other stars

UNIVERSE
Diameter: $2.02 \times 10^{13} \rightarrow 2.04 \times 10^{13}$ ly
  Observable: $5.04 \times 10^{10} \rightarrow 5.10 \times 10^{10}$ ly
Expansion: $108.78 \rightarrow 107.46$ km/s/Mpc
Density: $1.50 \rightarrow 1.45$ p/m³
Temperature: $4.98 \rightarrow 4.93$ K

ASTRONOMICAL OBJECTS
Galaxies: $2.91 \times 10^{13} \rightarrow 2.90 \times 10^{13}$
Stars: $1.53 \times 10^{21} \rightarrow 1.55 \times 10^{21}$
Black Holes: $1.24 \times 10^{19} \rightarrow 1.25 \times 10^{19}$
Planetary Systems: $2.07 \times 10^{21} \rightarrow 2.09 \times 10^{21}$
Worlds: $2.34 \times 10^{22} \rightarrow 2.37 \times 10^{22}$

WORLDS WITH LIFE
Simple Life: $7.03 \times 10^{17} - 2.60 \times 10^{19}$
Complex Life: $1.54 \times 10^{12} - 8.72 \times 10^{16}$
Intelligent Life: $3.55 \times 10^{8} - 4.89 \times 10^{14}$
Technological Life: $8.14 \times 10^{4} - 4.38 \times 10^{12}$
Interstellar Life: $3 - 2.29 \times 10^{11}$

ost stars in the Universe are found in singlet systems: with planets and moons in orbit around one primary, solitary parent star. However, the remainder exist within multi-star systems, as approximately 29% of all star systems form with multiple members inside: 23% are binaries, with two members; 5% are trinaries, with three stars; 1% are quaternary systems, possessing four stars; and although they're rare, systems with five or more stars inside represent about 0.2% of all star systems within a typical galaxy. Altogether, there are just as many stars present in multi-star systems as there are in common singlet systems.

While most multi-star systems have large separations, only influencing one another through their orbital, gravitational dances, there are a few major exceptions. Often, one member literally steals material from another: an example of stellar thievery. The simplest example arises when two stars are simply born very close to one another, in a very tight, rapid orbit. Stars all have large, extended gas-and-plasma envelopes that extend far beyond the internal, extremely hot region where light elements are fused into heavier ones. Remember, in order to initiate a nuclear fusion reaction, temperatures need to reach 4 million K, but at the edges of a star's photosphere, temperatures are much cooler: in the range of mere thousands of degrees.

During the main stage of their lives—while they're fusing hydrogen into helium in their cores—stars typically range in size from a few hundred thousand kilometers (for the coolest, reddest stars) to tens of millions of kilometers (for the hottest, bluest stars). Many stars that are found in binary systems have very rapid orbital periods, completing a full 360° revolution in mere days or even a fraction of a day. With the right combination of sizes and distances from one another, they can form a configuration known as a contact binary: where the fusion reactions inside each member continue independently, but their outer envelopes actually touch.

With a contact binary, the denser star can often siphon mass off of its companion in an act of stellar thievery. If enough mass is siphoned off, it can even change the internal temperature and the rate of fusion in both members of the system. Eventually, one of two fates emerge for these contact binaries: Either they merge together, becoming a hotter, brighter, bluer, but shorter-lived star, or one of them reaches the end of its life, dying in a stellar cataclysm while the other member lives on.

For multi-star systems with greater stellar separations, stellar thievery can still occur. After running out of hydrogen in their cores, stars typically swell into red giants, becoming diffuse, puffy, and up to hundreds of times as large as their earlier size. A denser companion star or stellar remnant can steal mass from the larger member, leading to a variety of fascinating fates. If the red giant was headed towards a supernova, the siphoning of sufficient amounts of mass can actually halt that process, leading instead to a white dwarf with unusually heavy elements inside. If the denser object is a white dwarf, the act of mass thievery can trigger a type Ia supernova; if it's a neutron star, it can lead to collapse into a black hole. And, in extremely rare cases, a red giant orbiting a neutron star or black hole can lead to the two fusing together, creating a hybrid star known as a Thorne-Żytkow object. Stellar thievery, governed by gravity, can truly be a fate-altering phenomenon for the stars involved.

For practically all stars born with at least half of the Sun's mass, they will someday evolve into the state shown here: the red giant phase of their lives. As the core runs out of hydrogen fuel, the outer layers expand while the core contracts, eventually heating up to initiate helium fusion. The star shown here is nearly a million times the Sun's volume and represents what Arcturus will look like 13.8 billion years after the Big Bang.

IMAGE BY MARK A. GARLICK

AWAKENING A RED GIANT

—— Arcturus becomes a red giant star ——

UNIVERSE
Diameter: $2.04 \times 10^{13} \rightarrow 2.06 \times 10^{13}$ ly
  Observable: $5.10 \times 10^{10} \rightarrow 5.16 \times 10^{10}$ ly
Expansion: $107.46 \rightarrow 106.17$ km/s/Mpc
Density: $1.45 \rightarrow 1.40$ p/m³
Temperature: $4.93 \rightarrow 4.87$ K

ASTRONOMICAL OBJECTS
Galaxies: $2.90 \times 10^{13} \rightarrow 2.90 \times 10^{13}$
Stars: $1.55 \times 10^{21} \rightarrow 1.57 \times 10^{21}$
Black Holes: $1.25 \times 10^{19} \rightarrow 1.27 \times 10^{19}$
Planetary Systems: $2.09 \times 10^{21} \rightarrow 2.12 \times 10^{21}$
Worlds: $2.37 \times 10^{22} \rightarrow 2.40 \times 10^{22}$

WORLDS WITH LIFE
Simple Life: $6.98 \times 10^{17} - 2.61 \times 10^{19}$
Complex Life: $1.52 \times 10^{12} - 9.38 \times 10^{16}$
Intelligent Life: $3.58 \times 10^8 - 5.39 \times 10^{14}$
Technological Life: $8.43 \times 10^4 - 4.96 \times 10^{12}$
Interstellar Life: $3 - 2.70 \times 10^{11}$

early seven billion years after the Big Bang, the Milky Way is transforming into an evolved, gas-rich spiral galaxy. Previous merger events—including Milky Way's devouring of the Kraken, Sequoia, Helmi stream, and Gaia-Enceladus— have all settled, resulting in a disk-shaped galaxy with a central bulge, spiral arms, and 100-plus globular clusters. Roughly 27,000 light-years away from the galactic center, within one of the spiral arms, a molecular cloud of gas collapses, triggering a new episode of star formation.

Thousands of new stars form from this enriched gas, ranging from hot, blue, massive, short-lived stars to cool, red, low-mass, long-lived stars. Containing about 30% the abundance of heavy elements that our Sun will form 2.5 billion years from now, the gas is sufficiently enriched to give rise to rocky planets, organic material, and possibly life.

In short order—over perhaps 10 million years—the most massive stars that form burn through their fuel, dying rapidly in core-collapse supernova events. When they were hot, highly luminous stars, their winds and radiation blew away the remaining star-forming material, returning it to the interstellar medium. With that gas now gone, only the newly formed stars, stellar systems, and failed stars survive in a dense swarm: an open star cluster. The supernova events that then occur can provide tremendous kicks to many of the stars in this cluster, particularly the lower-mass ones on the cluster's outskirts. Many such stars will be ejected early on, destined to roam the Milky Way for their remaining lifetimes.

Over the next few tens of millions of years, mutual gravitational interactions will begin to tear the star cluster apart: a process known as cluster dissociation. As the stars within the cluster begin to dissociate, they take diverging elliptical paths within the galaxy, gradually departing from the other members of their stellar nursery and only remaining bound to small numbers of stars apiece: forming singlet, binary, trinary, and other variants of multi-star systems. The hotter, bluer, and more massive a star is, the shorter its lifetime. While some open star clusters can persist for up to a billion years or even longer, most dissociate completely within tens or hundreds of millions of years.

Among the new stars birthed in this event is an F-class singlet star: slightly hotter, bluer, and more massive than the G-class star our Sun will someday be. Because of these properties, F-class stars shine more luminous than Sun-like stars do; they also burn through their core's nuclear fuel more quickly, resulting in a lifetime that's shorter than the 10- to 12-billion-year lifetime of a Sun-like star. This particular F-class star, after just under seven billion years, will run out of its core hydrogen. As the rate of radiation produced via nuclear fusion begins to drop, the core's outward radiation pressure will drop, too, and suddenly it no longer supports itself against gravitational collapse.

The core contracts, trapping the energy inside, and causes a rise in temperature. Eventually, the core's temperature exceeds about 26 million K, and the element helium—the main product of hydrogen fusion—begins fusing into carbon. All of a sudden, the star becomes a red giant, shining hundreds of times as bright as a Sun-like star. This red giant, which Earthlings will come to know as Arcturus, will be the closest red giant to human beings: the brightest star in Earth's northern celestial hemisphere.

As galaxies live for longer periods of time and undergo many bursts of star formation, their gas reservoirs get depleted over time. Once all of the gas is gone, no new stars can form, leaving behind elliptical galaxies in a state known to astronomers as "red and dead." As time marches onward, the star-formation rate drops, and the population of red and dead galaxies grows substantially, particularly inside galaxy clusters.

ELEGY FOR A STELLAR AGE

—— *star formation falls below half of its peak* ——

UNIVERSE
Diameter: $2.06 \times 10^{13} \rightarrow 2.09 \times 10^{13}$ ly
  Observable: $5.16 \times 10^{10} \rightarrow 5.21 \times 10^{10}$ ly
Expansion: $106.17 \rightarrow 104.90$ km/s/Mpc
Density: $1.40 \rightarrow 1.36$ p/m³
Temperature: $4.87 \rightarrow 4.82$ K

ASTRONOMICAL OBJECTS
Galaxies: $2.90 \times 10^{13} \rightarrow 2.89 \times 10^{13}$
Stars: $1.57 \times 10^{21} \rightarrow 1.59 \times 10^{21}$
Black Holes: $1.27 \times 10^{19} \rightarrow 1.28 \times 10^{19}$
Planetary Systems: $2.12 \times 10^{21} \rightarrow 2.15 \times 10^{21}$
Worlds: $2.40 \times 10^{22} \rightarrow 2.43 \times 10^{22}$

WORLDS WITH LIFE
Simple Life: $6.93 \times 10^{17} - 2.61 \times 10^{19}$
Complex Life: $1.49 \times 10^{12} - 1.01 \times 10^{17}$
Intelligent Life: $3.59 \times 10^{8} - 5.92 \times 10^{14}$
Technological Life: $8.66 \times 10^{4} - 5.59 \times 10^{12}$
Interstellar Life: $3 - 3.15 \times 10^{11}$

ver since the first clumps of matter gravitated, cooled, collapsed, and initiated nuclear fusion in their cores, the Universe has been relentlessly forming stars. The first stars, made of pristine matter, could only make large, massive stars, which lived and died very quickly, culminating in fantastic explosions and enriching the Universe with atoms forged inside those stellar furnaces. With those recycled insides participating in future generations of star formation, subsequent stars possess a wide variety of masses, including cool, red, very long-lived stars. Each time a large enough concentration of gas and dust collapses under its own gravity, new episodes of star formation commence, increasing the total number of stars in the Universe.

For a few billion years after the Big Bang, the star-formation rate increased dramatically. Regions of space that were born with slightly more matter than average grew, attracting the surrounding material into them, where the material self-gravitated, collapsed, and formed stars. Early, low-mass galaxies drew in gas from the intergalactic medium, leading to new episodes of star formation. On a larger scale, gravity attracted galaxies towards one another, where they merged together, triggering giant galaxy-wide waves of star formation known as starbursts. For approximately the first three billion years of cosmic history, the star-formation rate only rose and rose.

But now, nearly seven billion years after the Big Bang, that part of the story is far in the cosmic past. Even though galaxies bind together into groups and clusters as the cosmic web takes shape, the overall star-formation rate is much lower than it once was. In fact, right now marks an important milestone when the total average star-formation rate across the Universe is just half of what it was at its peak, a figure that continues to decline. Many factors contribute to why this is, and the ongoing decrease in new stars results from all of them combined.

Some galaxies have expelled all of their gas already, and without gas, there's no material for forming future generations of stars. When gas-rich galaxies merge together, starburst events ensue, creating large numbers of stars all at once. However, galaxy mergers are more common when the Universe is younger; by this point in time, even though they still occur—sometimes even between extremely large, evolved galaxies—they're far less frequent. Additionally, most stars are located within large, massive galaxies, which have violently energetic early histories but which settle down into quiet phases where only low levels of regular, ongoing star formation occur.

The underlying culprit behind the overall decrease is simply the relentless expansion of the Universe. As time marches on, the Universe gets less dense, and that affects every avenue of star formation. Isolated clumps of intergalactic matter are less likely to gravitationally grow and collapse. Galaxies that haven't already merged together with others or fallen into groups or clusters are now less likely to do so. Individual galaxies now draw smaller amounts of matter from their surroundings into them. And the biggest cosmic smashups of all, where galaxy clusters collide, are exceedingly rare.

For nearly the past four billion years, the average star-formation rate all across the Universe has been dropping, and it now drops below half of its ancient maximum. With fewer new stars come fewer new worlds with chances for life to arise on them. Although new stars will continue to form for trillions of years, the sobering fact is that most of the stars that will ever form have already done so. Our cosmic stellar peak is now a receding memory in the Universe's rearview mirror.

Mars-sized worlds have a problem: With such small sizes, their cores will cool and cease producing a protective magnetic field, which will lead their parent stars to strip away their atmospheres. Here, a civilization struggles to technologically save their dying planet from this fate. Once the atmosphere becomes too thin, liquid water will no longer stably exist on this world's surface.

IMAGE BY JON LOMBERG

A MARS BY ANY OTHER NAME

—— *Mars-like planets can support life* ——

UNIVERSE
Diameter: $2.09 \times 10^{13} \rightarrow 2.11 \times 10^{13}$ ly
  Observable: $5.21 \times 10^{10} \rightarrow 5.27 \times 10^{10}$ ly
Expansion: $104.90 \rightarrow 103.66$ km/s/Mpc
Density: $1.36 \rightarrow 1.31$ p/m³
Temperature: $4.82 \rightarrow 4.77$ K

ASTRONOMICAL OBJECTS
Galaxies: $2.89 \times 10^{13} \rightarrow 2.87 \times 10^{13}$
Stars: $1.59 \times 10^{21} \rightarrow 1.60 \times 10^{21}$
Black Holes: $1.28 \times 10^{19} \rightarrow 1.30 \times 10^{19}$
Planetary Systems: $2.15 \times 10^{21} \rightarrow 2.16 \times 10^{21}$
Worlds: $2.43 \times 10^{22} \rightarrow 2.45 \times 10^{22}$

WORLDS WITH LIFE
Simple Life: $6.87 \times 10^{17} - 2.62 \times 10^{19}$
Complex Life: $1.45 \times 10^{12} - 1.08 \times 10^{17}$
Intelligent Life: $3.59 \times 10^{8} - 6.48 \times 10^{14}$
Technological Life: $8.87 \times 10^{4} - 6.25 \times 10^{12}$
Interstellar Life: $3 - 3.65 \times 10^{11}$

By this point in the Universe's history, most of the stars that are now forming are rich enough in heavy elements that they'll arise along with planets and moons, including rocky worlds rich in organic molecules. Although the larger rocky worlds—between 10,000 and perhaps 16,000 kilometers in diameter—may make the best candidates for developing sustained, complex life, the smaller, more numerous worlds that form possess tremendous potential as well. With sufficiently thick atmospheres, substantial greenhouse effects, and protective magnetic fields, even a small, relatively remote world not so different from the planet Mars that will form in our own Solar System could become a thriving home to intelligent organisms.

In general, any planet or moon that forms will do so with a certain amount of initial heat inherent to it. Over time, it generates its own internal heat due to gravitational contraction and the internal decay of radioactive elements but radiates its heat out into the Universe over the entirety of its surface. The larger and more voluminous a world, the more total heat gets trapped inside; the smaller a world, the more quickly its heat gets radiated away.

Planets and moons also receive heat—the primary source of energy for powering biological activity—from any parent stars present within their systems. Although rocky worlds with thin atmospheres can possess liquid water on their surfaces at a specific distance from their parent stars, it's plausible for more distant worlds to develop and maintain liquid oceans simply by possessing thicker, heat-trapping atmospheres. Even a Mars-like world, small and distant, can maintain the conditions for life to thrive on its surface given a sufficiently thick atmosphere.

Somewhere in the Universe, a Mars-sized world came into existence with precisely these conditions: a small mass, a significant distance from its parent star, a substantial amount of internal heat with an enriched, metallic core, and a thick atmosphere rich in heat-trapping gases, with liquid water and organic materials on its surface. Just like it has on perhaps a quintillion worlds across the Universe, life arose on the surface of this world early on in its history. In a rarer event—but perhaps not so uncommon overall—life survived and thrived over long periods of time, adapting to the changing environment while evolving and developing greater complexity.

Inevitably, all Mars-sized worlds face a critical event, as particle-like streams from its parent star begin to strip away their atmospheres. Worlds possessing protective magnetic fields, where magnetic dynamos are active within their cores, can divert these charged particles away. Atmospheres can be maintained over multibillion-year timescales. But once the core sufficiently cools, the dynamo ceases, and only a remnant surface magnetic field, much smaller in magnitude, persists. Mars-sized worlds have only between one and two billion years before they lose enough of their atmospheres that liquid water becomes impossible on their surfaces.

Forming with a high fraction of radioactive, heavy elements may be able to extend this for a time, but there's another potential saving grace: the rise of an intelligent species. With sufficient knowledge and technology, an advanced civilization could generate its own artificial, planet-wide magnetic field, protecting its precious atmosphere and vastly extending its habitable lifetime. Life must evolve and intelligence must arise quickly for this to happen, but with so many opportunities for the improbable to occur, perhaps, somewhere, it actually did. Mars-like worlds might not be the most likely candidates for sustained, intelligent life, but in such a vast Universe, even unlikely events may be all but inevitable.

As one of the stars in this system begins its inevitable evolution into a red giant, its energy output increases. An intelligent civilization on one of its orbiting planets takes to the cloud tops, where conditions are cooler and more hospitable than on the now-superheated surface. This civilization will likely have only a short period of time before the increased energy output sterilizes life on their home planet.

IMAGE BY MARK A. GARLICK

SHELTER AMONG THE CLOUDS

—— *as stars die, so too do their worlds* ——

UNIVERSE
Diameter: $2.11 \times 10^{13} \rightarrow 2.13 \times 10^{13}$ ly
  Observable: $5.27 \times 10^{10} \rightarrow 5.32 \times 10^{10}$ ly
Expansion: $103.66 \rightarrow 102.56$ km/s/Mpc
Density: $1.31 \rightarrow 1.27$ p/m³
Temperature: $4.77 \rightarrow 4.72$ K

ASTRONOMICAL OBJECTS
Galaxies: $2.87 \times 10^{13} \rightarrow 2.86 \times 10^{13}$
Stars: $1.60 \times 10^{21} \rightarrow 1.62 \times 10^{21}$
Black Holes: $1.30 \times 10^{19} \rightarrow 1.31 \times 10^{19}$
Planetary Systems: $2.16 \times 10^{21} \rightarrow 2.19 \times 10^{21}$
Worlds: $2.45 \times 10^{22} \rightarrow 2.48 \times 10^{22}$

WORLDS WITH LIFE
Simple Life: $6.81 \times 10^{17} - 2.62 \times 10^{19}$
Complex Life: $1.42 \times 10^{12} - 1.14 \times 10^{17}$
Intelligent Life: $3.58 \times 10^{8} - 7.01 \times 10^{14}$
Technological Life: $9.00 \times 10^{4} - 6.88 \times 10^{12}$
Interstellar Life: $3 - 4.16 \times 10^{11}$

Every living world that comes into existence must someday face the inevitable: the moment when all biological activity ceases to occur. On many worlds, life will self-destruct; on others, climate change, volcanic catastrophes, violent impacts, gravitational interactions, or even stellar cataclysms will bring about its demise. But on inhabited planets where life persists in an unbroken chain, biological activity can survive all possible extinction events for billions of years. Eventually, however, all life succumbs to the inevitable. At some point, an inhabited world's parent star will evolve, heating up and roasting the planet. Beyond a certain threshold, extinction becomes certain.

Stars are born when nuclear fusion initiates in their cores, with hydrogen fusing into helium as the starting point. An extremely high-mass star might spend only one or two million years in this phase, but lower-mass stars can spend hundreds of millions, billions, or even trillions of years fusing hydrogen into helium. However, there's a catch: The more hydrogen gets converted into helium in a star's core, the greater the star's internal temperature becomes, increasing the size of its core, the overall rate of fusion, the total heat output, and the overall luminosity of the star.

Even after billions of years, these changes can be substantial. Over the hydrogen-burning life span of a star, it roughly doubles its overall energy output. A planet or moon that starts out receiving the proper amount of radiation for liquid water on its surface will eventually face the danger of its oceans boiling away. Once that occurs, even if life takes refuge in the clouds, the heat-trapping abilities of the various atmospheric layers will eventually render the planet's surface thoroughly uninhabitable.

There are mechanisms by which a planet could self-regulate for a time, prolonging the inevitable. Increased cloud cover could reflect additional sunlight away, cooling down the planet. A change in the color of a planet's surface could increase the efficiency at which it radiates energy into space. And the intervention of an intelligent species could even migrate the world farther away from its parent star, reducing the flux on the planet's surface even as the star's heat intensifies.

But at some point, the star's increasing energy output becomes overwhelming. As the star's core begins to run out of hydrogen, the rate of fusion in the core drops, causing the star to contract under its own gravity. Because stars are so thick and massive, the heat inside gets trapped, causing the core temperature to rise as the star contracts. As the temperature rises, the fusion region inside the star expands outward, and new hydrogen fuel begins to burn. Meanwhile, the increased core energy causes the outer layers of the star to expand, leading to a transition: from a "normal" star to a subgiant to, eventually, a full-fledged red giant star. Even though the temperature of the star's photosphere drops during this transition, its energy output will increase from a factor of a few dozen to several hundred, and over timescales of only a few tens or hundreds of millions of years.

Stars that have already lived for billions of years, orbited by worlds where life has survived and thrived since its inception, are undergoing this transition right now. All other stars that are still steadily burning their core fuel away are destined for this catastrophe in their future. With a finite amount of fuel in their cores, all stars will eventually die. As they go through their death throes, they'll inevitably take all living worlds in their vicinity down with them.

On planets with thick atmospheres that orbit close to their parent stars, novel and exotic types of precipitation often rain down from their clouds. Instead of liquid water, drops of liquid metal, rock, salt, and even gemstones fall from the skies, while electrical discharges are more common than even within Earth's most violent thunderstorms.

IMAGE BY MARK A. GARLICK

THE MANY SHADES OF RAIN

—— *rain on other worlds is not always water* ——

UNIVERSE
Diameter: $2.13 \times 10^{13} \rightarrow 2.15 \times 10^{13}$ ly
  Observable: $5.32 \times 10^{10} \rightarrow 5.38 \times 10^{10}$ ly
Expansion: $102.56 \rightarrow 101.37$ km/s/Mpc
Density: $1.27 \rightarrow 1.23$ p/m³
Temperature: $4.72 \rightarrow 4.67$ K

ASTRONOMICAL OBJECTS
Galaxies: $2.86 \times 10^{13} \rightarrow 2.85 \times 10^{13}$
Stars: $1.62 \times 10^{21} \rightarrow 1.64 \times 10^{21}$
Black Holes: $1.31 \times 10^{19} \rightarrow 1.32 \times 10^{19}$
Planetary Systems: $2.19 \times 10^{21} \rightarrow 2.22 \times 10^{21}$
Worlds: $2.48 \times 10^{22} \rightarrow 2.52 \times 10^{22}$

WORLDS WITH LIFE
Simple Life: $6.75 \times 10^{17} - 2.62 \times 10^{19}$
Complex Life: $1.38 \times 10^{12} - 1.22 \times 10^{17}$
Intelligent Life: $3.55 \times 10^{8} - 7.61 \times 10^{14}$
Technological Life: $9.11 \times 10^{4} - 7.61 \times 10^{12}$
Interstellar Life: $4 - 4.76 \times 10^{11}$

Rains and snows of a wide variety of types exist all throughout the Universe, with liquid water and water-based snows being only two representatives of thousands of possible outcomes that, no doubt, occur somewhere in the vast recesses of space. Of all the chemical compounds that could precipitate on a planet, moon, or other world found throughout the Universe, water might not even be the most commonplace. Most worlds are small, cold, and icy, yet still experience periods of relative hot and cold as they move through space. Many others have thick, gaseous envelopes surrounding them, with strong winds and large day/night temperature differences. Whenever a world possesses an atmosphere, gaseous compounds can precipitate out, entering either the liquid or solid phase, where they'll eventually fall towards the surface.

On icy worlds that orbit far from their parent stars, water isn't the only common compound that turns to ice. Carbon dioxide freezes, making dry ice. Methane can become solid as well, creating methane ice. Carbon monoxide, nitrogen, and even hydrogen molecules can freeze, creating ices of varying hardness and density. But, given enough external energy—which these worlds can receive as they experience bright enough direct sunlight—these volatile compounds can boil or sublimate away, entering a gaseous phase. As they rise through the various atmospheric layers, they precipitate out and form clouds. In liquid form, they produce rain; in solid form, they produce snow. Throughout the outer regions of stellar systems all across the Universe, and perhaps even in the depths of interstellar space, these exotic, ultracold rains and snows are the dominant form of precipitation in the Universe.

Many temperate and tropical worlds will experience the more familiar rains and snows made of liquid and solid water, respectively. But there are hotter worlds to consider as well: found in the inner reaches of stellar systems, coming in rocky and gas-rich varieties alike. On these worlds, where temperatures can rise into the thousands of degrees, even the solid surface can be vaporized by the intense conditions, leading to some of the most exotic forms of precipitation imaginable.

Gas giant worlds that orbit very close to their parent stars—known to astronomers as "hot Jupiters"—will actually have water molecules ripped apart into individual atoms, and those atoms will then become ionized: a consequence of experiencing their parent star's daytime light up close. When the planetary winds circulate that atmosphere back onto the cooler night side, water molecules re-form, creating a wild and unfamiliar water cycle. Heavier elements, including iron and corundum, can also be flung into the atmosphere on the daytime side. They also ionize but condense back into clouds when they drift over to the night side. They can form metal clouds, releasing metallic rain or, if the conditions are just right, even liquid gemstones.

Other hot worlds have been seen to have clouds and hazes made of hydrocarbons, salts, silicates, and even metallic features like aluminum oxide and titanium dioxide. The higher the most extreme temperatures on the world are, the heavier and more exotic the types of elements that can be present in their clouds. In the end, across all worlds, once a critical mass and density of material accumulates, precipitation occurs, with pressure, temperature, and the chemical compound in question determining whether it comes down in a solid or liquid phase. Some worlds will have snows of sands and salts; others of rubies and sapphires; still others of rocky material. The larger and hotter the world in question, the more wild and exotic its rains and snows can be.

One common type of planet found in the
Universe, but not in our own Solar System,
is a water world: a planet whose surface
is 100% covered by a liquid water ocean.
Despite the lack of continental landmasses,
biological organisms can weave together
to form networks of floating mats, allowing
fresh water to pool and for a variety of
complex life-forms to arise.

IMAGE BY MARK A. GARLICK

FLOATING ISLANDS OF LIFE

—— life can form landmasses on water worlds ——

UNIVERSE
Diameter: $2.15 \times 10^{13} \rightarrow 2.17 \times 10^{13}$ ly
  Observable: $5.38 \times 10^{10} \rightarrow 5.44 \times 10^{10}$ ly
Expansion: $101.37 \rightarrow 100.31$ km/s/Mpc
Density: $1.23 \rightarrow 1.20$ p/m³
Temperature: $4.67 \rightarrow 4.62$ K

ASTRONOMICAL OBJECTS
Galaxies: $2.85 \times 10^{13} \rightarrow 2.83 \times 10^{13}$
Stars: $1.64 \times 10^{21} \rightarrow 1.65 \times 10^{21}$
Black Holes: $1.32 \times 10^{19} \rightarrow 1.34 \times 10^{19}$
Planetary Systems: $2.22 \times 10^{21} \rightarrow 2.23 \times 10^{21}$
Worlds: $2.52 \times 10^{22} \rightarrow 2.53 \times 10^{22}$

WORLDS WITH LIFE
Simple Life: $6.69 \times 10^{17} - 2.61 \times 10^{19}$
Complex Life: $1.35 \times 10^{12} - 1.28 \times 10^{17}$
Intelligent Life: $3.51 \times 10^{8} - 8.17 \times 10^{14}$
Technological Life: $9.16 \times 10^{4} - 8.31 \times 10^{12}$
Interstellar Life: $4 - 5.36 \times 10^{11}$

From an Earth-centric perspective, we might presume that inhabited worlds will have a mix of landmasses and liquid water oceans. However, the Universe we inhabit might defy that particular expectation. Most of the worlds out there will either be dry, frozen, or so deluged with water that there won't be a single significant landmass that emerges above a planet-wide ocean. The completely dry worlds might be poor locations for living organisms to emerge; the frozen worlds might see their living organisms limited to the hydrothermal vents that exist on the seafloor. But worlds that are absolutely covered in water—where up to 70% of the world's entire composition is made of water—might not be doomed to the same fate.

Initially, deep ocean vents, where energy gets injected into an aqueous environment from a heated planetary core, might provide ideal conditions for the emergence of life. A combination of nutrients, the ability to metabolize them, and the capability of reproduction might be all it takes to give life its start. Beyond that, the development of a membrane and a few specialized functions might be all life needs to become widespread and move beyond the limited locations where it initially arose. After all, there's another energy source to be tapped: From above, some tiny remnants of whatever light strikes the top of the ocean can make it all the way down to those early, primitive life-forms.

Over hundreds of millions or perhaps even billions of years, organisms can develop the ability to take advantage of that light, using it to generate energy and perhaps even synthesize nutrients directly. With a virtually limitless source of energy coming from the light of a parent star, some sort of photosynthetic process is likely to arise on worlds all across the galaxy and Universe. Eventually, even on a continent-free world, life is likely to rise to the water's surface, where—at the interface of the liquid ocean and the gaseous atmosphere—it's receiving direct light from the star it orbits, with none of the various wavelengths reflected away by the water's surface.

Once life achieves the ability to survive and thrive in the direct light of its parent star, floating atop the watery ocean enshrouding the entire world, they might find benefits in becoming either unicellular or multicellular: where identical copies of the same organism weave themselves together to create an ocean-top mat. If the mat becomes thick and sturdy enough, it can accomplish something that would otherwise only be possible on a world with landmasses: to collect precipitation without having it mix with the salty ocean. These organic structures, for the first time, could create freshwater stores, even on a purely oceanic world.

These freshwater stores, all of a sudden, would create new ecological niches for life to inhabit and biomes that would have otherwise been impossible on a world without land. The larger and sturdier these ocean-top mats become, the larger and more diverse the types of living organisms that arise upon them can also become. In the most extreme cases imaginable, they might even create floating continents out of either the corpses of trillions of these organisms, all stitched together, or out of living organisms themselves. Even on an ocean-covered world that initially possessed fresh water, the presence and success of life can make it possible. Even on a world entirely shrouded in water, land-based life-forms can still emerge.

Red dwarf stars, the lowest-mass stars in the Universe, have the longest lifetimes and display the greatest amounts of stellar activity. While this may render most planets orbiting them inhospitable for tens or even hundreds of billions of years, perhaps in the far future these stars will stabilize, creating potentially habitable conditions long after the less massive stars have died away.

IMAGE BY MARK A. GARLICK

THE LIVES OF RED DWARFS

the most common and longest-lived stars

UNIVERSE
Diameter: $2.17 \times 10^{13} \rightarrow 2.20 \times 10^{13}$ ly
  Observable: $5.44 \times 10^{10} \rightarrow 5.49 \times 10^{10}$ ly
Expansion: $100.31 \rightarrow 99.27$ km/s/Mpc
Density: $1.20 \rightarrow 1.16$ p/m³
Temperature: $4.62 \rightarrow 4.58$ K

ASTRONOMICAL OBJECTS
Galaxies: $2.83 \times 10^{13} \rightarrow 2.83 \times 10^{13}$
Stars: $1.65 \times 10^{21} \rightarrow 1.67 \times 10^{21}$
Black Holes: $1.34 \times 10^{19} \rightarrow 1.35 \times 10^{19}$
Planetary Systems: $2.23 \times 10^{21} \rightarrow 2.26 \times 10^{21}$
Worlds: $2.53 \times 10^{22} \rightarrow 2.56 \times 10^{22}$

WORLDS WITH LIFE
Simple Life: $6.62 \times 10^{17} - 2.61 \times 10^{19}$
Complex Life: $1.31 \times 10^{12} - 1.35 \times 10^{17}$
Intelligent Life: $3.47 \times 10^{8} - 8.74 \times 10^{14}$
Technological Life: $9.18 \times 10^{4} - 9.03 \times 10^{12}$
Interstellar Life: $4 - 6.00 \times 10^{11}$

pon developing the capacity to perceive light, in any form, a biological organism's detection abilities will inherently be biased towards the brightest-shining, most easily visible objects. In any night sky, no matter which set of ultraviolet, visible, or infrared wavelengths an organism's sight is adapted to, the brightest, hottest, most luminous stars will shine most bright. Organisms that are more sensitive to bluer, shorter-wavelength light will preferentially see the youngest, highest-mass stars, as well as evolved blue supergiants. Organisms more sensitive to redder, longer-wavelength light will see large numbers of evolved red giant and supergiant stars: the ones entering the final stages of their lives. But the most common, enduring stars will remain invisible to all but the most sensitive infrared receptors: red dwarfs.

Somewhere between 70% and 82% of all stars that will ever be born in the Universe are red dwarf stars, which still cross that critical ~4 million K temperature in their cores, enabling them to fuse hydrogen into helium inside. However, the low masses of these stars—they can be no more than 40% as massive as a Sun-like star—ensure that only a relatively small volume of the star is undergoing nuclear fusion at any one time, leading to low luminosities and low temperatures but also very long lifetimes. Whereas the brightest stars might live only for millions or a few billion years at most, the longest-lived red dwarf stars can endure for tens of trillions of years.

Recall that the famed saying, "the candle that burns twice as bright burns half as long," is even worse for stars: A star possessing double the mass of another lives approximately just one-eighth as long. For the lowest-mass stars, however, this holds a fascinating implication: The slow rate of fusion, even with smaller amounts of fuel overall, ensures that long after those high-mass, bright stars evolve into stellar remnants, these red dwarfs will still endure, fusing hydrogen into helium in their cores for trillions of years to come.

The lowest-mass red dwarf stars might shine only 0.1% as brightly as a Sun-like star, which themselves are hundreds to millions of times fainter than giant or supergiant stars. But what nature fails to provide in brightness, it makes up for in sheer numbers and in duration. In those galaxies that have exhausted their supplies of gas and stopped forming stars several billion years ago—those "red and dead" elliptical galaxies—the more massive stars will simply run out of fuel and die, but the longer-lived, lower-mass stars will endure.

Inside red dwarf stars—the lowest-mass, longest-lived, most numerous class of star of all—the nuclear processes that cause stars to shine are so slow that a new process can occur inside of them: whole-star convection. In the higher-mass, shorter-lived stars, spent nuclear fuel simply builds up in the star's core, causing the star to contract, heat up, and evolve when the rate of nuclear reactions drops below a threshold. But in these low-mass red dwarf stars, the nuclear processes powering them are so slow that the spent nuclear fuel, in the form of helium, can actually get displaced towards the outer layers and replenished by new, unburned hydrogen fuel. This convective process includes the entire star's mass for red dwarfs, ensuring that when they do eventually die, practically the entire star will have converted into helium.

The longest-lived red dwarfs might endure for over 100 trillion years, or more than 10,000 times the present age of the Universe. With such long life spans, who knows what might arise on worlds around them?

This image illustrates the massive El Gordo galaxy cluster: one of the largest in the Universe at this time with a mass of more than two quadrillion Suns. Multiple galaxy clusters are colliding to merge and form this behemoth, with the characteristic emission of X-ray light revealing the heated gases caused by this cosmic smashup.

IMAGE BY MARK A. GARLICK

THE BIRTH OF EL GORDO

two galaxy clusters spawn a giant

UNIVERSE
Diameter: $2.20 \times 10^{13} \rightarrow 2.22 \times 10^{13}$ ly
  Observable: $5.49 \times 10^{10} \rightarrow 5.55 \times 10^{10}$ ly
Expansion: $99.27 \rightarrow 98.26$ km/s/Mpc
Density: $1.16 \rightarrow 1.13$ p/m³
Temperature: $4.58 \rightarrow 4.53$ K

ASTRONOMICAL OBJECTS
Galaxies: $2.83 \times 10^{13} \rightarrow 2.82 \times 10^{13}$
Stars: $1.67 \times 10^{21} \rightarrow 1.69 \times 10^{21}$
Black Holes: $1.35 \times 10^{19} \rightarrow 1.36 \times 10^{19}$
Planetary Systems: $2.26 \times 10^{21} \rightarrow 2.28 \times 10^{21}$
Worlds: $2.56 \times 10^{22} \rightarrow 2.60 \times 10^{22}$

WORLDS WITH LIFE
Simple Life: $6.56 \times 10^{17} - 2.60 \times 10^{19}$
Complex Life: $1.27 \times 10^{12} - 1.42 \times 10^{17}$
Intelligent Life: $3.41 \times 10^8 - 9.33 \times 10^{14}$
Technological Life: $9.16 \times 10^4 - 9.78 \times 10^{12}$
Interstellar Life: $4 - 6.68 \times 10^{11}$

ver since the first gravitationally bound clumps of mass began forming, there have been two main mechanisms by which cosmic structures grow. The first is gradually: by accreting matter from an initial structure's surroundings, as masses steadily grow by gravitational infall. But the second occurs suddenly: as nearby, massive structures get mutually pulled together, leading to the process of gravitational mergers. As massive structures grow and gravitate, they begin to attract one another; if they grow massive enough quickly enough, they can even overcome the expansion of the Universe.

These mergers, initially limited to clumps of mass on the scale of star clusters, eventually scale up to include protogalaxies, galaxies, and then groups and clusters of galaxies. For about the past three billion years, even fully formed galaxy clusters have occasionally attracted one another from tens of millions of light-years away, colliding together in a magnificent cosmic train wreck. And now, at this moment in cosmic history, the largest merger ever identified between two galaxy clusters—the colliding cluster simply known as El Gordo—finally takes place.

Composed of two smaller clumps, each of which is developed and massive enough to be considered a galaxy cluster all on its own, El Gordo has already grown to a mass of more than three quadrillion Suns at this moment in time. It's estimated that, across the entire Universe, there are only five clusters this large and massive by this point, with El Gordo being the only known example of such a massive, young cluster presently undergoing a major merger. The two smaller clumps are colliding together at several million kilometers per hour, drawn in towards one another until they reach nearly 1% the speed of light just prior to the merger's start.

In the moments just before the merger, the mutual gravitational pull of the galaxies on one another tidally disrupts the internal gas within each galaxy, leading to new waves of star formation inside them. When they begin to interact, three separate components within the two smaller clusters all behave in a unique fashion. The individual galaxies within each cluster are so small that they very rarely collide; they shoot right past one another the same way two handfuls of pebbles thrown at one another from across a field would. The gas contained inside each of the two clusters, however, being located mostly between the galaxies rather than within them, will interact, heat up, slow down, and "splat" together, emitting X-rays in the process. And finally, there's dark matter contained in these clusters, and it passes right through everything, only settling down in the aftermath of the merger to hold the newer, larger cluster together.

It's objects like these—colliding galaxy clusters—that provide some of the strongest astrophysical evidence for dark matter. When anyone observes the points of light that exist behind a galaxy cluster, the presence of all forms of mass bends the intervening space: a phenomenon known as gravitational lensing. When we see clusters in pre-collisional states, all of the mass is concentrated into two clumps: one corresponding to each of the clusters. When the clusters collide, the majority of the normal matter emits X-rays, located between the clusters, while the majority of the mass remains with the clusters. Then, when the mass settles into a single, post-collisional cluster, the mass begins returning towards the center: exactly what's seen with El Gordo. This separation between where the normal matter is and where the effects of gravity appear shows us that there must be more matter out there than normal matter can account for: hence the existence of dark matter. Given the time at which the El Gordo collision occurs, it's the largest, brightest, and most revealing galaxy cluster ever discovered in the Universe.

A compact, rapidly spinning stellar remnant, like a neutron star or black hole, can siphon mass off of a lower-mass, less dense companion star. Here, a red dwarf acts as the donor star, creating what astronomers call a low-mass X-ray binary (LMXB) system, identifiable from the copious amounts of X-rays emitted from the system.

IMAGE BY MARK A. GARLICK

STELLAR SYNCHROTRONS

—— *some binary star systems create X-rays* ——

UNIVERSE
Diameter: $2.22 \times 10^{13} \rightarrow 2.24 \times 10^{13}$ ly
  Observable: $5.55 \times 10^{10} \rightarrow 5.60 \times 10^{10}$ ly
Expansion: $98.26 \rightarrow 97.26$ km/s/Mpc
Density: $1.13 \rightarrow 1.09$ p/m³
Temperature: $4.53 \rightarrow 4.49$ K

ASTRONOMICAL OBJECTS
Galaxies: $2.82 \times 10^{13} \rightarrow 2.80 \times 10^{13}$
Stars: $1.69 \times 10^{21} \rightarrow 1.70 \times 10^{21}$
Black Holes: $1.36 \times 10^{19} \rightarrow 1.37 \times 10^{19}$
Planetary Systems: $2.28 \times 10^{21} \rightarrow 2.30 \times 10^{21}$
Worlds: $2.60 \times 10^{22} \rightarrow 2.61 \times 10^{22}$

WORLDS WITH LIFE
Simple Life: $6.49 \times 10^{17} - 2.59 \times 10^{19}$
Complex Life: $1.23 \times 10^{12} - 1.49 \times 10^{17}$
Intelligent Life: $3.35 \times 10^{8} - 9.93 \times 10^{14}$
Technological Life: $9.11 \times 10^{4} - 1.06 \times 10^{13}$
Interstellar Life: $4 - 7.41 \times 10^{11}$

inglet stars with planetary systems, like our own, aren't the cosmic norm. Just as frequently, stars are born as members of multi-star systems. Typically, they go about their lives—fusing hydrogen into helium in their cores—while simply orbiting one another as dictated by the laws of gravity. Eventually, one member of this stellar set burns through its core fuel the fastest, transforming into a red giant on its way to becoming either a white dwarf, a neutron star, or a black hole, as determined by its mass. When that occurs, the rules of the gravitational dance change, as the dying star loses mass as it evolves. What happens next can lead to a spouting stellar siphon—an X-ray binary—that occasionally becomes something even more spectacular.

The denser your massive object is, the stronger the gravitational field is at its surface. If an object of sufficiently low density passes close to a denser one, the denser object will exert tidal forces, pulling the nearer portion towards it with stronger gravitational forces than the remainder. Whenever a neutron star or black hole forms within a multi-star system, it can affect its companion, whether a normal star, a red giant, or even a white dwarf. If the conditions are right, the denser, more compact object begins stealing mass from the less dense object, siphoning off the more loosely held material.

As matter accumulates around the denser body, like a neutron star or black hole, a number of changes begin to occur. The stolen matter gets broken down into individual particles, as atoms are torn apart into nuclei and electrons, forming an accretion disk. These orbiting, rapidly swirling charged particles create strong magnetic fields, which in turn accelerate the charged particles. The accelerated particles form two jets—typically oriented perpendicular to the disk—that get ejected into the surrounding medium, emitting high-energy radiation in the form of X-rays. Finally, as matter is lost from the donor star, its nuclear reactions slow and cool, all while becoming less massive.

What happens next depends entirely on the donor star's nature. If the donor star is a white dwarf, it typically runs out of matter to donate after a certain point, putting an end to the X-ray emissions. It will still emit gravitational waves, inspiraling until it's close enough to its denser companion to trigger a cataclysmic tidal disruption event. If the donor star is a red giant, the dense companion continues accreting material until the donor star's entire outer envelope is stolen, putting an end to stellar evolution and leading to an exotic white dwarf rich with heavy elements. Alternatively, if the red giant ventures too close to its dense companion, the companion sinks to the giant star's center, creating a hybrid star with a neutron star or black hole at its core: a Thorne-Żytkow object.

However, the highest-mass X-ray binaries arise from pairs of massive blue supergiant stars, where one member's core collapses before the other. As the dense remnant—either a black hole or neutron star—begins siphoning mass off of its companion, it chaotically overflows, resulting in pulses and flares coming from the pair. Eventually, the lingering blue supergiant will reach the end of its life as well, typically dying in a core-collapse supernova event. Thereafter, the black hole/neutron star remnants that are left behind will inspiral due to emitted gravitational waves, eventually merging. If the remnants are both neutron stars, a kilonova event accompanied by a gamma-ray burst can occur as well. Many initial configurations create these stellar siphons, and while they all form X-ray binaries, the ultimate end result varies on a case-by-case basis.

While worlds in between the masses and size of Earth and Neptune are quite common in the Universe, any world that's more than about 30% larger in radius than Earth will have a thick atmosphere around it, rendering the surface uninhabitable. However, life-friendly conditions may still exist high up in the atmospheric clouds and hazes, permitting the possibility that wildly unfamiliar life-forms exist in these unfamiliar environments.

IMAGE BY MARK A. GARLICK

AN UNKINDLY DOPPELGANGER

— *a planet like Earth, but less hospitable* —

UNIVERSE
Diameter: $2.24 \times 10^{13} \rightarrow 2.26 \times 10^{13}$ ly
  Observable: $5.60 \times 10^{10} \rightarrow 5.66 \times 10^{10}$ ly
Expansion: $97.26 \rightarrow 96.29$ km/s/Mpc
Density: $1.09 \rightarrow 1.06$ p/m³
Temperature: $4.49 \rightarrow 4.44$ K

ASTRONOMICAL OBJECTS
Galaxies: $2.80 \times 10^{13} \rightarrow 2.79 \times 10^{13}$
Stars: $1.70 \times 10^{21} \rightarrow 1.72 \times 10^{21}$
Black Holes: $1.37 \times 10^{19} \rightarrow 1.38 \times 10^{19}$
Planetary Systems: $2.30 \times 10^{21} \rightarrow 2.32 \times 10^{21}$
Worlds: $2.61 \times 10^{22} \rightarrow 2.65 \times 10^{22}$

WORLDS WITH LIFE
Simple Life: $6.42 \times 10^{17} - 2.59 \times 10^{19}$
Complex Life: $1.19 \times 10^{12} - 1.57 \times 10^{17}$
Intelligent Life: $3.29 \times 10^{8} - 1.05 \times 10^{15}$
Technological Life: $9.04 \times 10^{4} - 1.14 \times 10^{13}$
Interstellar Life: $4 - 8.19 \times 10^{11}$

Within the disk of the Milky Way, individual stars orbit around the galactic center. As the stars move at different speeds relative to one another, they alternately bunch up and get farther apart, like a giant galactic traffic jam. The internal gas and dust does exhibit the same behavior, bunching up along the dense waves that compose the galaxy's spiral arms, periodically triggering new waves of star formation within them. All throughout the galactic disk, from several thousand light-years to perhaps 40,000 light-years from the galactic center, new stars form with rocky, potentially habitable planets. At this moment in cosmic history, one Sun-like star with an almost perfect planet forms: Kepler 452.

Our Sun and Solar System won't form for another 1.5 billion years, but Kepler 452 already has our Sun beat in a number of vital ways. First off, it possesses a greater fraction of heavy elements than our star system, with about 160% the abundance of elements such as iron compared to our own Solar System. Second, the star itself is 4% more massive, 11% larger, and about 20% more luminous than our Sun, while still possessing almost exactly the same temperature. And finally—and perhaps of most interest—it forms with its own system of planets, including one with an almost perfectly Earth-like orbit.

Known as Kepler-452b, this planet completes a full revolution around its parent star once every 385 days. Based on the distance from its parent star and the properties of the star itself, we know that if this planet were replaced by one identical to Earth, its average temperature would be merely 10°C (18°F) hotter than our own world. With a similar environment to our own planet and the potential for water in all three phases on it—water vapor in the atmosphere, liquid water on the surface, and solid ice at the poles and at high elevations—it raises the tantalizing possibility that this could evolve into a world teeming with life, perhaps with a ~1.5-billion-year head start on planet Earth.

There's only one problem: With a physical size that's about 60% larger than planet Earth—placing it into the category of planets known as super-Earths—the rocky core of this planet is almost certainly hanging on to a large, thick atmosphere rich in volatile chemicals. Even though its physical size is only 60% larger than planet Earth, Kepler-452b is about five times as massive, implying that its surface gravity is about double what ours is. These relatively small differences in physical properties translate into an enormous difference in surface conditions: Whereas planet Earth experiences an atmospheric pressure of 100,000 pascals (or 15.2 pounds per square inch), Kepler-452b should possess an atmosphere that's thousands of times as massive, leading to surface pressures that rival those found on Earth's ocean bottoms.

It turns out that a planet can only be about twice as massive as Earth (and about 25–30% larger in radius) before it loses the ability to have a thin, Earth-like atmosphere. With significantly thicker atmospheres, the majority of super-Earth planets are similar to miniature versions of Uranus and Neptune, not scaled-up versions of Earth, Mars, or Venus. It may still be possible for life to emerge and/or persist in the hazes, cloud tops, and high atmospheres of super-Earth worlds like Kepler-452b, but any thriving ecosystem of surface, ocean, and subterranean life would be very non-Earth-like on a planet such as this. Despite being the most Earth-like planet known to date around a star almost identical to our Sun, "almost perfect" doesn't quite add up in the quest for a second Earth.

Created long ago, an artifact roams the interstellar depths: launched into space by a now-extinct civilization, preserving both their cultural and cosmic history. Over time, it will inevitably pass in and out of planetary systems. Perhaps with the right engineering, it may survive long enough to be found by future explorers conducting galactic archaeology studies.

IMAGE BY MARK A. GARLICK

A DARKNESS UNFORGIVING

—— space is not empty, but filled with hazards ——

UNIVERSE
Diameter: $2.26 \times 10^{13} \rightarrow 2.29 \times 10^{13}$ ly
  Observable: $5.66 \times 10^{10} \rightarrow 5.72 \times 10^{10}$ ly
Expansion: $96.29 \rightarrow 95.32$ km/s/Mpc
Density: $1.06 \rightarrow 1.03$ p/m³
Temperature: $4.44 \rightarrow 4.40$ K

ASTRONOMICAL OBJECTS
Galaxies: $2.79 \times 10^{13} \rightarrow 2.78 \times 10^{13}$
Stars: $1.72 \times 10^{21} \rightarrow 1.73 \times 10^{21}$
Black Holes: $1.38 \times 10^{19} \rightarrow 1.40 \times 10^{19}$
Planetary Systems: $2.32 \times 10^{21} \rightarrow 2.34 \times 10^{21}$
Worlds: $2.65 \times 10^{22} \rightarrow 2.66 \times 10^{22}$

WORLDS WITH LIFE
Simple Life: $6.35 \times 10^{17} - 2.57 \times 10^{19}$
Complex Life: $1.15 \times 10^{12} - 1.64 \times 10^{17}$
Intelligent Life: $3.21 \times 10^8 - 1.12 \times 10^{15}$
Technological Life: $8.93 \times 10^4 - 1.22 \times 10^{13}$
Interstellar Life: $4 - 9.01 \times 10^{11}$

Throughout the abyss of interstellar space, where light-years typically separate individual star systems from one another, space is anything but empty. In addition to the light emitted from stars, dust, plasma, and the remnant light from the Big Bang itself, individual particles as well as aggregate structures roam freely through the Universe. Most of them were produced by inorganic processes: A combination of gravitation, electromagnetic binding, ionizing radiation, tiny collisions with energetic particles, and even massive collisions between giant astronomical bodies will produce everything from single particles to massive planetoids all roaming through space. But among them, a careful onlooker might very rarely encounter something simultaneously wondrous and frightening: a tattered messenger launched long ago from a onetime intelligent civilization.

Most of the material in the interstellar medium is made of hydrogen, either as a single, neutral atom—an ionized nucleus with a free electron—or as a diatomic molecule. Helium makes up most of the rest, with small amounts of heavier elements as well. Dust particles—very different from the type of dust you'd encounter on the surface of a planet—range in size from a few hundred atoms across up to the size of pebbles and rocks; they can exist in isolation or clumped together in a series of grains. Even though it's extremely dilute, with just one particle per cubic centimeter of space, it adds up quickly over the extreme distances separating star systems.

Additionally, remnants and ejecta from older stellar systems, including would-be comets and asteroids as well as rocky and icy debris kicked up from large collisions, travel all throughout each and every galaxy. For each star system that exists, there are likely trillions, and potentially even quadrillions, of boulder-sized and larger objects flying through the abyss of interstellar space. Occasionally, however, something else can be found among these cosmic pilgrims: metal-rich structures created by intelligent beings, from satellites to space probes to—just possibly—deliberate messages intended for any other intelligent extraterrestrials they might happen to encounter.

The problem with aiming for the stars is that, given enough time, even the emptiest galaxies of all will continuously bombard and erode objects that travel through the interstellar medium. Just as oceans will eventually erode mountains into rocks, pebbles, and eventually dust, a journey through interstellar space will do the same thing to matter. A typical comet-sized object might survive for a few hundred million or even—if it was particularly large initially—several billion years, but it will be ground into individual molecules, atoms, and subatomic particles, as will everything made of matter.

Unfortunately, this includes even intelligently created spacecraft intended to reach other advanced species. In the long run, there is no way to protect against the continuous onslaught of particles, dust grains, pebbles, and larger structures; energetic collisions will degrade, over time, the structural integrity of anything they contact. The most hardened, resilient structures might remain mostly intact for a time, but even an arbitrarily thick, solid lead shield won't be sufficient to protect whatever message is intended for extraterrestrial recipients for long. A galactic archaeologist might successfully decode an ancient, defunct spacecraft's origins and age, but several billion years in time could reasonably separate intelligent species from different star systems. Perhaps, before they go extinct, a fraction of advanced civilizations create some sort of cosmic time capsule, designed to endure the harsh environment of space as long as possible in a quest to make contact one last time, far outliving the species that first created them.

The multiple galaxy clusters shown here teeter on the precipice between becoming gravitationally bound to one another and getting caught up in the expansion of the Universe, where dark energy will accelerate them away from one another. The fate of these galaxies, and whether they'll merge into one cluster or forever remain separate, depends on their total mass, their separation distance, and their relative motions to each other.

IMAGE BY MARK A. GARLICK

DARK ENERGY DOMINATES

—— *the expansion of the Universe accelerates* ——

UNIVERSE
Diameter: $2.29 \times 10^{13} \rightarrow 2.31 \times 10^{13}$ ly
 Observable: $5.72 \times 10^{10} \rightarrow 5.77 \times 10^{10}$ ly
Expansion: $95.32 \rightarrow 94.38$ km/s/Mpc
Density: $1.03 \rightarrow 1.00$ p/m³
Temperature: $4.40 \rightarrow 4.35$ K

ASTRONOMICAL OBJECTS
Galaxies: $2.78 \times 10^{13} \rightarrow 2.77 \times 10^{13}$
Stars: $1.73 \times 10^{21} \rightarrow 1.75 \times 10^{21}$
Black Holes: $1.40 \times 10^{19} \rightarrow 1.41 \times 10^{19}$
Planetary Systems: $2.34 \times 10^{21} \rightarrow 2.36 \times 10^{21}$
Worlds: $2.66 \times 10^{22} \rightarrow 2.69 \times 10^{22}$

WORLDS WITH LIFE
Simple Life: $6.28 \times 10^{17} - 2.56 \times 10^{19}$
Complex Life: $1.11 \times 10^{12} - 1.71 \times 10^{17}$
Intelligent Life: $3.13 \times 10^8 - 1.18 \times 10^{15}$
Technological Life: $8.81 \times 10^4 - 1.30 \times 10^{13}$
Interstellar Life: $4 - 9.88 \times 10^{11}$

Following the Big Bang's first moments, structures have continuously formed, grown, and become gravitationally bound on increasingly larger cosmic scales. Because the Universe is expanding and gravity only propagates at the speed of light, it takes time for matter in different regions of space to feel one another's influence. As a result, clouds of gas form early on, star clusters form later, galaxies require longer timescales, galactic groups and clusters require even more time to form, and cosmic train wrecks, involving the collisions and mergers of multiple mature clusters, take even longer periods of time to occur. As the cosmic web takes shape, even larger apparent structures—galactic superclusters—begin to form as well, with some of them spanning billions of light-years in extent.

As the Universe crosses this critical threshold in time, something remarkable occurs: Every structure, on all scales, that isn't gravitationally bound at this moment will never become so. For the first time since the hot Big Bang, it's no longer matter and radiation that dominate the expanding Universe; a distant galaxy no longer appears to recede from each observer in a slower and slower fashion as time goes on. Instead, every object that isn't already gravitationally bound to you—as part of your stellar system, your galaxy, or your group or cluster of galaxies—will now never become so. As time progresses, these distant objects recede from you at a faster and faster rate.

Up until this time, it appeared that the Universe was on the path towards forming superclusters: networks of galaxy clusters connected by cosmic filaments made of a mix of normal matter and dark matter. Cosmic filaments and their points of intersection were becoming more dense and more well defined, not only drawing matter in from the underdense regions around them but also bringing various interconnected structures together. But as the matter and radiation throughout the expanding Universe drop below a critical density threshold, this new form of energy that appears to be inherent to space itself—dark energy—begins to drive these unbound structures apart. Superclusters, which are associations that will appear to persist for tens of billions of years to come, are phantasmal, unable to overcome dark energy's repulsive properties through mere gravitational attraction alone.

As we cross through this critical threshold, the stage is set for the ultimate fate of the largest-scale structures in the Universe. All of the individual bound structures that have already formed by this point—stellar systems, star clusters, galaxies, and galaxy groups and clusters—can remain gravitationally bound; they've overcome the expansion of the Universe. All of the still-forming structures that have drawn enough matter into them so that they're undergoing gravitational contraction can complete their gravitational collapse, eventually giving rise to a bound, large-scale structure. All of the galaxy groups and clusters that are being drawn towards one another will continue on their collision course, destined to eventually merge together.

But beyond the largest-scale bound structure that any object is a part of, whether a galaxy, galactic group, galaxy cluster, or set of infalling/merging galaxy clusters, every other bound structure not only recedes from it but also appears to recede at a faster and faster rate as time goes on. After just under eight billion years, the Universe isn't only expanding, but the expansion is also accelerating. The fate of the Universe is set: It will lead to a collection of individually bound structures forever speeding away from one another at ever-increasing speeds. The largest gravitational bindings, alas, have already occurred.

As the Sagittarius Dwarf Elliptical Galaxy passes through the Milky Way's plane, new episodes of star formation occur. Streams of gas and new stars temporarily flow between these two galaxies, most of which will wind up inside the larger Milky Way galaxy. Many new stellar cataclysms ensue, and those enriched elements will help form rocky planets and living worlds in future generations of stars.

IMAGE BY JON LOMBERG

THE DANSE MACABRE I
—— *the Milky Way merges with SagDEG* ——

UNIVERSE
Diameter: $2.31 \times 10^{13} \rightarrow 2.33 \times 10^{13}$ ly
  Observable: $5.77 \times 10^{10} \rightarrow 5.82 \times 10^{10}$ ly
Expansion: $94.38 \rightarrow 93.55$ km/s/Mpc
Density: $1.00 \rightarrow 0.97$ p/m³
Temperature: $4.35 \rightarrow 4.31$ K

ASTRONOMICAL OBJECTS
Galaxies: $2.77 \times 10^{13} \rightarrow 2.75 \times 10^{13}$
Stars: $1.75 \times 10^{21} \rightarrow 1.76 \times 10^{21}$
Black Holes: $1.41 \times 10^{19} \rightarrow 1.42 \times 10^{19}$
Planetary Systems: $2.36 \times 10^{21} \rightarrow 2.38 \times 10^{21}$
Worlds: $2.69 \times 10^{22} \rightarrow 2.71 \times 10^{22}$

WORLDS WITH LIFE
Simple Life: $6.22 \times 10^{17} – 2.55 \times 10^{19}$
Complex Life: $1.08 \times 10^{12} – 1.78 \times 10^{17}$
Intelligent Life: $3.06 \times 10^{8} – 1.24 \times 10^{15}$
Technological Life: $8.67 \times 10^{4} – 1.38 \times 10^{13}$
Interstellar Life: $4 – 1.07 \times 10^{12}$

Nothing lives forever in this Universe, not even the stars. Fueled by the nuclear fusion of light elements into heavier ones in their ultrahot cores, the lifetime and fate of a star is determined primarily by how much mass it's born with. The most massive stars will run out of their core fuel quickly—in only a few million years—dying in core-collapse supernovae and leaving either neutron stars or black holes behind. Less massive stars will live longer and die more gently, with their cores eventually contracting to form white dwarfs.

These stellar remnants can merge together to create their own cosmic fireworks, with neutron star–neutron star mergers creating explosive cataclysms known as kilonovae, where explosive ejecta create the heaviest elements found in nature. Similarly, white dwarf–white dwarf mergers create a different class of supernovae, where a runaway fusion reaction occurs and both stellar remnants are ripped apart and destroyed. What's not often appreciated is that these violent cataclysms—kilonovae and both classes of supernovae—can often lead to the formation of new stars: where cosmic death leads to stellar rebirth.

In gas-rich galaxies such as the Milky Way, molecular clouds of gas frequently teeter on the brink of gravitational collapse. A stellar cataclysm can provide a cosmic "nudge" as material slams into the cloud matter in an asymmetrical fashion. This imparts a kinetic motion to the gas, causing more and more of the mass to gather towards the central region. This creates a higher-density collection of gas, and if it's efficient enough at cooling, the cloud can either partially or wholly collapse under its gravity. The end result, if so, is the formation of new stars.

Approximately eight billion years after the hot Big Bang, a small galaxy—the Sagittarius Dwarf Elliptical Galaxy—passes through the Milky Way's galactic plane, perturbing the clumps of gas inside and triggering many new episodes of star formation. Simultaneously, an ongoing series of events are steadily occurring: Gas collapses to form new stars, the most massive ones die in supernovae, becoming neutron stars and black holes, where neutron stars merge together in kilonova events. White dwarfs also form later on from the less massive stars, with their mergers creating supernovae as well. These cataclysms inject large amounts of heavy elements, energy, and fast-moving material back into the Universe, where they can travel numerous light-years before slamming into any surrounding material.

In the aftermath of one such event—possibly a kilonova, possibly a supernova—a star almost as massive as the yet-to-form Sun comes into existence: Tau Ceti. With multiple super-Earth planets in its inner system and at least one planet larger and more massive than Jupiter beyond those, Tau Ceti has an enormous debris disk extending from beyond its Jovian planet all the way out into its Kuiper belt. Although its planets are all more massive than the rocky ones we have in our Solar System, some may yet be home to life, perhaps so foreign to what we understand life to be that we might not recognize it at first.

This star will achieve great importance ~5.8 billion years later, when humans arise on planet Earth. Tau Ceti, at that point, will become the closest singlet star system to our own with the same class of temperature, brightness, size, and longevity as our Sun. At just 12 light-years away, it's visible to the naked eye, and its eight candidate planets match our own Solar System. Born more than a billion years before we were, it may well have been a stellar cataclysm that made its formation possible.

The most accurate natural clocks in the Universe are millisecond pulsars: neutron stars that have accreted mass from a companion star to "spin up," with the most rapid rotators completing a full 360° spin in barely a single millisecond. Pulsars also contain the strongest magnetic fields known in the Universe: up to 100 billion times stronger than the strongest permanent magnets found on Earth.

IMAGE BY MARK A. GARLICK

NATURE'S TIMEKEEPERS

—— millisecond pulsars spin at a stable rate ——

UNIVERSE
Diameter: $2.33 \times 10^{13} \rightarrow 2.35 \times 10^{13}$ ly
  Observable: $5.82 \times 10^{10} \rightarrow 5.88 \times 10^{10}$ ly
Expansion: $93.55 \rightarrow 92.64$ km/s/Mpc
Density: $0.97 \rightarrow 0.94$ p/m³
Temperature: $4.31 \rightarrow 4.27$ K

ASTRONOMICAL OBJECTS
Galaxies: $2.75 \times 10^{13} \rightarrow 2.74 \times 10^{13}$
Stars: $1.76 \times 10^{21} \rightarrow 1.77 \times 10^{21}$
Black Holes: $1.42 \times 10^{19} \rightarrow 1.43 \times 10^{19}$
Planetary Systems: $2.38 \times 10^{21} \rightarrow 2.39 \times 10^{21}$
Worlds: $2.71 \times 10^{22} \rightarrow 2.73 \times 10^{22}$

WORLDS WITH LIFE
Simple Life: $6.14 \times 10^{17} - 2.54 \times 10^{19}$
Complex Life: $1.04 \times 10^{12} - 1.85 \times 10^{17}$
Intelligent Life: $2.97 \times 10^{8} - 1.30 \times 10^{15}$
Technological Life: $8.51 \times 10^{4} - 1.47 \times 10^{13}$
Interstellar Life: $3 - 1.17 \times 10^{12}$

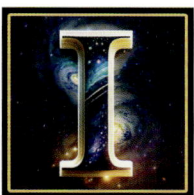

Individual objects are never stationary but, rather, spin about their axes and revolve around whatever gravitationally anchors them, whether a planet, star, or even an entire galaxy. Most of these rotational and orbital motions wind up changing substantially over time, as the gravitational forces acting on them tend to create behavior-altering effects that really add up over time. Planets orbiting their parent stars—especially planets with moons—will spin down over time, lengthening their days and causing them to slowly increase their orbital radii. Stellar evolution leads to stars changing size over time, altering their rotational periods and experiencing internal "starquakes," which rearranges their internal masses and causes them to spin faster. Even black holes (and other stellar remnants) in binary orbits experience gravitational radiation, causing their orbits to decay and changing their orbital periods over time, eventually leading to an inspiral and merger. No naturally occurring object, anywhere within the Universe, is capable of eternally keeping perfect time.

But of all the natural timekeeping devices that one could choose, one class of object comes closer than any other to being the most reliable clock of all: millisecond pulsars. Neutron stars are quite common by this point in our cosmic history, with a great many massive stars having died in core-collapse supernovae already. While the most extreme stars will see their cores form black holes, neutron stars are a more common fate. With a center that's only perhaps 10 kilometers across made exclusively of neutrons—surrounded by a thin shell of other heavy atomic nuclei containing protons and neutrons alike— these concentrated objects rotate about their axes, creating intense magnetic fields in their vicinities, arising from the charged particles within.

As the charged particles on the outskirts of the neutron star accelerate through the neutron star's intense magnetic field, they emit radiation that "pulses" in a particular direction once per rotation of the neutron star: a pulsar. In isolation, these neutron stars would simply slow their rotation, cool off, and settle into a quiet, nonpulsing state. However, these neutron stars very frequently don't exist in isolation but orbit alongside other massive objects: planets, other stellar remnants, or even active, living stars. If another source of mass is nearby enough to the dense neutron star, it can siphon material off of its less dense companion, creating a situation where the neutron star isn't slowing and cooling but, rather, spinning at higher and higher rotational speeds.

These neutron stars can spin so fast that they'll reach rotational speeds of up to about 60% the speed of light: the maximum speed achievable before either they break apart or before gravitational radiation would be emitted with such intensity it would slow the pulsar down at a greater rate than the accretion process could spin it up. The fastest such pulsars are known as millisecond pulsars, as they can complete nearly 1,000 full rotations each second. Once a millisecond pulsar's companion is ejected—a common occurrence in dense environments like a galaxy's center or a globular cluster—it remains stable for hundreds of millions or even billions of years, persisting with the same rotational period to within about one part in a quintillion (10^{18}).

Although pulsars, including millisecond pulsars, can "glitch" and abruptly change their rotational properties, they can also remain stable for as long as anyone's observed them. About two-thirds of all such millisecond pulsars have been found within globular clusters, hinting that perhaps the densest stellar habitats are the most common birthing grounds for these objects: the Universe's most perfect natural clocks.

While simply knowing about every star and planet within a galaxy might be enough to be considered an *Encyclopaedia Galactica,* a more ambitious goal would be to create a galaxy-wide network of intelligent, technologically advanced civilizations all communicating with one another at the speed of light. If one "seed" civilization colonizes the galaxy, they may even all be interrelated: as cosmic cousins.

IMAGE BY JON LOMBERG

COSMIC COLLABORATORS

—— advanced civilizations connect and share ——

UNIVERSE
Diameter: $2.35 \times 10^{13} \rightarrow 2.37 \times 10^{13}$ ly
  Observable: $5.88 \times 10^{10} \rightarrow 5.94 \times 10^{10}$ ly
Expansion: $92.64 \rightarrow 91.84$ km/s/Mpc
Density: $0.94 \rightarrow 0.92$ p/m³
Temperature: $4.27 \rightarrow 4.23$ K

ASTRONOMICAL OBJECTS
Galaxies: $2.74 \times 10^{13} \rightarrow 2.72 \times 10^{13}$
Stars: $1.77 \times 10^{21} \rightarrow 1.78 \times 10^{21}$
Black Holes: $1.43 \times 10^{19} \rightarrow 1.44 \times 10^{19}$
Planetary Systems: $2.39 \times 10^{21} \rightarrow 2.40 \times 10^{21}$
Worlds: $2.73 \times 10^{22} \rightarrow 2.74 \times 10^{22}$

WORLDS WITH LIFE
Simple Life: $6.07 \times 10^{17} - 2.52 \times 10^{19}$
Complex Life: $1.00 \times 10^{12} - 1.92 \times 10^{17}$
Intelligent Life: $2.89 \times 10^{8} - 1.36 \times 10^{15}$
Technological Life: $8.34 \times 10^{4} - 1.55 \times 10^{13}$
Interstellar Life: $3 - 1.26 \times 10^{12}$

It can't be said for certain that the Universe has unfolded according to some sort of grand plan. However, based on its initial conditions and the laws that govern it, certain phenomena seem all but inevitable. The initially hot, dense Universe expanded and cooled, giving way to the formation of atomic nuclei, neutral atoms, and eventually the first gravitationally bound structures, which formed stars within them. These structures grew through mergers and accretion, assembling to form galaxies, galaxy groups, and even large galaxy clusters, while the stars within them—living and dying and recycling the heavy elements formed inside into future generations—continued to form. When a sufficient abundance of heavy elements is reached, new stars form with rocky worlds around them, leading to the potential for life. Wherever life can survive and thrive over long periods of time, it can diversify, become complex and differentiated, and in some cases, intelligent enough to become technologically advanced.

These intelligent, technologically advanced civilizations are likely to come into existence at a variety of different times and places throughout our cosmic history, including in large, enriched galaxies all throughout the Universe. While many of these civilizations likely go extinct before either colonizing a world beyond the one they arose on or making "first contact" with another intelligent civilization, it's only a matter of time and a sufficient number of chances before both of these events occur somewhere. Perhaps it's right around now—a little over eight billion years after the hot Big Bang—that a significant number of intelligent civilizations within a single galaxy are all actively communicating with one another, creating the first civilized galactic network.

From any vantage point within a galaxy, a civilization is limited by the power and capabilities of their instruments, the laws of physics, and the speed of light. But a sufficiently large number of civilizations, distributed throughout a galaxy, could create a communication network that provides updates on the status of the various stars, planets, and living worlds with greater accuracy and comprehensiveness than any lone civilization could achieve. There would be many advantages to such a network, including facilitating the exchange of information and materials between civilizations, warnings about impending cataclysms or potential dangers, and perhaps even instructions for fledgling civilizations on how to survive their technological infancies.

For many intelligent civilizations, the greatest threat to their survival will come from environmental conditions, such as hazardous asteroids, a changing climate, or energetic events from their parent stars. For others, however, the great challenge will be overcoming their own primal nature: to consume resources past the point of sustainability, focusing on short-term, inconsequential problems while ignoring existential threats. Perhaps an instruction manual for long-term survival would focus on technologies to compel the rocky and icy bodies within a stellar system together into a single, gravitationally bound object. Perhaps it would include instructions on how a species can best survive disease, famine, and pestilence.

Perhaps, however, there's a greater immediacy to focus on how to overcome their most destructive impulses. Greed, plunder, and overconsumption may pose problems for all species. If there's no one else for an intelligent species to communicate with—to guide them, to share knowledge with them, or to give them hope that a future exists for them out there in the abyss of deep space—perhaps extinction is more probable. But perhaps in a Universe where numerous species are actively intercommunicating, kindness and compassion can bridge even vast interstellar distances.

Well after the Milky Way experienced its most recent major galactic merger, the Andromeda galaxy, the largest in our Local Group, merges with another relatively large galaxy. The intensified episode of star formation, as well as the eventual absorption of nearly all the stars and gas within the smaller galaxy, explains many of Andromeda's present-day features.

IMAGE BY MARK A. GARLICK

THE CANNIBAL NEXT DOOR

— *Andromeda merges with a major galaxy* —

UNIVERSE
Diameter: $2.37 \times 10^{13} \rightarrow 2.40 \times 10^{13}$ ly
 Observable: $5.94 \times 10^{10} \rightarrow 6.00 \times 10^{10}$ ly
Expansion: $91.84 \rightarrow 90.97$ km/s/Mpc
Density: $0.92 \rightarrow 0.89$ p/m³
Temperature: $4.23 \rightarrow 4.19$ K

ASTRONOMICAL OBJECTS
Galaxies: $2.72 \times 10^{13} \rightarrow 2.71 \times 10^{13}$
Stars: $1.78 \times 10^{21} \rightarrow 1.80 \times 10^{21}$
Black Holes: $1.44 \times 10^{19} \rightarrow 1.45 \times 10^{19}$
Planetary Systems: $2.40 \times 10^{21} \rightarrow 2.43 \times 10^{21}$
Worlds: $2.74 \times 10^{22} \rightarrow 2.77 \times 10^{22}$

WORLDS WITH LIFE
Simple Life: $6.00 \times 10^{17} - 2.51 \times 10^{19}$
Complex Life: $9.63 \times 10^{11} - 1.99 \times 10^{17}$
Intelligent Life: $2.80 \times 10^{8} - 1.43 \times 10^{15}$
Technological Life: $8.15 \times 10^{4} - 1.64 \times 10^{13}$
Interstellar Life: $3 - 1.36 \times 10^{12}$

very large galaxy that forms in the Universe grew from an initial seed of structure—a gravitationally overdense region—that contracted and became gravitationally bound. Over time, it grew to greater and greater mass via two methods: gravitational growth, where it draws matter from the surrounding spatial regions into it, and gravitational mergers, where it gobbles up smaller galaxies. Both of these methods add to the galaxy's mass, with the steady accumulation of matter ultimately responsible for the majority of a large galaxy's overall matter content.

However, it's the merger events, particularly the mergers of comparably large-mass galaxies that occur relatively late in the Universe's history, that can disrupt and cause novel, fascinating features to appear within a galaxy, even billions of years down the line. It's right at this time—about 8.3 billion years after the hot Big Bang—that the largest merger ever to occur in the Local Group occurs: to the Milky Way's galactic neighbor, the Andromeda galaxy.

At this time, Andromeda is only approximately the size and mass that the Milky Way will grow to in another 5.5 billion years, but another galaxy has been interacting with it for billions of years: one possessing about one-third of Andromeda's mass. The first time this smaller galaxy passed by Andromeda was about 3.5 billion years prior, when its gravitational influence triggered a tremendous episode of star formation within Andromeda. However, after a series of interactions, that companion galaxy now returns, portending the final fusion of these two Local Group members: the end stage of the largest galactic merger to occur in the history of our Local Group.

The earlier, first close gravitational pass likely ripped a large amount of interstellar gas out of the interacting galaxies, perhaps even giving rise to some of the more modest galactic members of the Local Group. By this point in cosmic history, those galaxies—which might include the Magellanic Clouds and/or the large satellite galaxies of Andromeda—now possess stars that are billions of years old. But this event, which saw the Local Group's largest and (at the time) third-largest galaxies completely merging together, determines many of the features that will come to define Andromeda.

As a result of this merger, Andromeda now contains the greatest proportion of atom-based matter in the entire Local Group. Andromeda and its satellites, in fact, contain more normal matter than the rest of the Local Group combined. The collision will become very violent, creating a massive, thick, dust-rich galactic disk within Andromeda. This galactic merger creates a huge central bulge in the post-merger galaxy, containing older, redder stars and up to 30% of the entire stellar mass of the galaxy. It may also help explain why, even billions of years into the future, Andromeda's core will continue to exhibit an unusual double structure: as though two independent components contributed to its formation.

This major merger, the largest in Local Group history, may not reach completion for several billion years. It also gives Andromeda a giant ring of gas and dust, is responsible for its enormous stream of old stars, and even helps explain why—unlike the Milky Way—about one-seventh of its stars are found strewn about throughout the diffuse outer halo of the galaxy. Additionally, Andromeda contains many more globular clusters than any other Local Group galaxy: about 460 total, or three times the total possessed by the Milky Way. Although late-time major mergers such as this are rare, they're also exceedingly important. After all, as cosmic detectives, the clues to reconstructing the Universe's violent history are written on the faces of its survivors.

Once dark energy has become important in the expansion of the Universe, all galaxies beyond a certain distance from one another will not be reachable or able to communicate with one another into the far future. While this advanced civilization can receive light from all the stars and galaxies in its sky, the signals it emits today will never reach the most distant, earliest galaxies that are visible in its skies.

IMAGE BY MARK A. GARLICK

WITHIN SIGHT, OUT OF REACH

the expanding Universe limits our reach

UNIVERSE
Diameter: $2.40 \times 10^{13} \rightarrow 2.42 \times 10^{13}$ ly
  Observable: $6.00 \times 10^{10} \rightarrow 6.05 \times 10^{10}$ ly
Expansion: $90.97 \rightarrow 90.20$ km/s/Mpc
Density: $0.89 \rightarrow 0.87$ p/m³
Temperature: $4.19 \rightarrow 4.15$ K

ASTRONOMICAL OBJECTS
Galaxies: $2.71 \times 10^{13} \rightarrow 2.70 \times 10^{13}$
Stars: $1.80 \times 10^{21} \rightarrow 1.81 \times 10^{21}$
Black Holes: $1.45 \times 10^{19} \rightarrow 1.46 \times 10^{19}$
Planetary Systems: $2.43 \times 10^{21} \rightarrow 2.45 \times 10^{21}$
Worlds: $2.77 \times 10^{22} \rightarrow 2.79 \times 10^{22}$

WORLDS WITH LIFE
Simple Life: $5.93 \times 10^{17} - 2.49 \times 10^{19}$
Complex Life: $9.28 \times 10^{11} - 2.06 \times 10^{17}$
Intelligent Life: $2.72 \times 10^{8} - 1.49 \times 10^{15}$
Technological Life: $7.96 \times 10^{4} - 1.72 \times 10^{13}$
Interstellar Life: $3 - 1.46 \times 10^{12}$

For billions of years, ever since the start of the hot Big Bang, seeing a distant object in the Universe implied the ability to both communicate with it and reach it. Communication means that light signals can be exchanged between two objects, despite the finite speed of light and the expansion of the Universe. Reaching it means that, if you left immediately in a spacecraft that could approach (but not achieve) the speed of light, you'd eventually arrive at your destination. This communicability and reachability apply to every planet, star, and galaxy visible within the Universe, including the ones that are forming right now. The reason for this was simple: No matter how distant an object was, it would recede from us at slower and slower speeds as time went on. If we could see it, you'd reason that we could reach it, too, even if we left at a later time.

But about half a billion years ago, the situation changed, as dark energy became the dominant factor in determining the expansion of the Universe. Distant objects no longer recede from one another more slowly with time but, rather, speed away relative to each other at faster and faster speeds as time goes on. All of a sudden, objects that once were within reach—receding below the speed of light due to the expansion of the Universe—begin speeding away from one another at an accelerating rate. As time marches forward, those objects will be driven apart by the expanding Universe so quickly that, eventually, a spacecraft or even a light signal that leaves one object can never arrive at another. And perhaps most frighteningly, that's already occurred for most of the objects visible from anywhere within the Universe.

Right now, the Universe is expanding at a specific rate: 90 km/s/Mpc. That means, for every megaparsec (about 3.26 million light-years) one object is away from another, it's driven away at 90 kilometers per second by the expansion of the Universe. Every object, therefore, that's more than about 11 billion light-years away from another cannot be reached by a signal transmitted at the speed of light. Even though the expansion rate itself is continuing to drop, these distant objects will recede faster and faster from one another as time goes on. Light that was emitted long ago can be glimpsed from objects as distant as 30 billion light-years at the moment, but even at the speed of light, no signals emitted today can ever be exchanged between galaxies that are too far apart.

Civilizations that reach a technologically advanced stage over the next several billion years will gain a unique perspective on the Universe: They'll be able to know, simultaneously, both how the Universe began and also what its ultimate fate will be. A Universe rich in stars and galaxies, where one can reconstruct the earliest stages of the hot Big Bang, will only persist for about 20 billion years. Beyond that time period, dark energy will drive the various bound clumps of matter apart by such a significant amount that other than the galaxies in one's local group or cluster, there won't be any others detectable for billions of light-years. Similarly, the effects of dark energy only become detectable after more than five billion years have passed since the start of the hot Big Bang; prior to that, the fate of the Universe could not be determined. Only during this "Goldilocks" period can an observer know it all: how it all began, how it grew up, and how it's going to end. In the meantime, most of the galaxies that can be seen are forever beyond one another's reach.

High-energy astrophysical sources, such as neutron stars and black holes, emit not just light but also energetic particles. When those particles smash into heavy atomic nuclei, they can blast them apart, creating lighter elements in a process known as spallation. The third, fourth, and fifth elements in the periodic table—lithium, beryllium, and boron—are primarily produced via this method.

IMAGE BY MARK A. GARLICK

THE CRUCIBLE OF COSMIC RAYS

origins of lithium, beryllium, and boron

UNIVERSE
Diameter: $2.42 \times 10^{13} \rightarrow 2.44 \times 10^{13}$ ly
  Observable: $6.05 \times 10^{10} \rightarrow 6.11 \times 10^{10}$ ly
Expansion: $90.20 \rightarrow 89.45$ km/s/Mpc
Density: $0.87 \rightarrow 0.84$ p/m³
Temperature: $4.15 \rightarrow 4.11$ K

ASTRONOMICAL OBJECTS
Galaxies: $2.70 \times 10^{13} \rightarrow 2.68 \times 10^{13}$
Stars: $1.81 \times 10^{21} \rightarrow 1.82 \times 10^{21}$
Black Holes: $1.46 \times 10^{19} \rightarrow 1.47 \times 10^{19}$
Planetary Systems: $2.45 \times 10^{21} \rightarrow 2.46 \times 10^{21}$
Worlds: $2.79 \times 10^{22} \rightarrow 2.81 \times 10^{22}$

WORLDS WITH LIFE
Simple Life: $5.86 \times 10^{17} - 2.47 \times 10^{19}$
Complex Life: $8.95 \times 10^{11} - 2.13 \times 10^{17}$
Intelligent Life: $2.64 \times 10^{8} - 1.55 \times 10^{15}$
Technological Life: $7.74 \times 10^{4} - 1.81 \times 10^{13}$
Interstellar Life: $3 - 1.56 \times 10^{12}$

Prior to the formation of the first stars in the Universe, the only elements that had come into existence were hydrogen, helium, and a tiny, minuscule amount of lithium—all created in the hot Big Bang. Once stars begin to form, nuclear fusion takes place within those stellar interiors, where hydrogen and helium fuse into carbon, nitrogen, oxygen, and many heavier elements. When those stars evolve—into red giants, planetary nebulae, and even supernovae—they can create elements that work their way up the periodic table. These atoms are augmented by colliding stellar remnants, like white dwarfs and neutron stars, that help fill out the missing and less abundant elements and isotopes.

But three elements—lithium, beryllium, and boron—can't be made in stars or stellar remnants in any way. These three elements—heavier than helium but lighter than carbon—are indeed found all throughout the Universe but weren't created by stellar fusion nor by the bombardment of neutrons, which are the methods that create all other elements once the Universe has cooled off from the hot Big Bang. The majority of the Universe's lithium, and all of its beryllium and boron, are instead created in an entirely unique way: through a process known as cosmic ray spallation.

Once stars and stellar remnants come into existence, they readily generate their own magnetic fields by causing the motion of charged particles. These magnetic fields, particularly in the vicinity of neutron stars and black holes, can become extremely powerful, with the strongest ones accelerating particles very close to the speed of light. These accelerated particles get strewn throughout the Universe, traveling until they have the good fortune to strike another particle. Most of the time, that particle will either be a photon, an electron, or a hydrogen atom, but every once in a while, a heavier element—one created by a star or stellar remnant—will get caught in the cross fire. From the perspective of these light elements, that's the key interaction that leads to their creation.

Just as nuclear fusion can create new elements by building lighter atomic nuclei into heavier ones, the opposite process can occur: nuclear fission. Fission takes place whenever a heavy atomic nucleus gets split into lighter components, typically by absorbing a particle, forming a metastable state, and decaying into a series of smaller, lighter nuclei. But when a heavy atomic nucleus gets struck by a high-energy cosmic particle, it isn't fission that occurs but, rather, a process known as spallation. Instead of forming a new, unstable nucleus that decays, the incoming cosmic ray, upon striking a heavy nucleus, splits it apart into large numbers of smaller bound states: free protons and/or neutrons, intermediate-mass nuclei, and light elements. This process occurs not only in deep space but in the atmospheres and on the surfaces of planets, moons, asteroids, and comets.

Although cosmic ray spallation produces a small fraction of many of the elements found in our cosmos, it's the number one source of lithium, beryllium, and boron. In fact, it's the only source of beryllium and boron—elements that happen to be essential to many known biological processes. Compared to the abundances of the lighter and heavier elements—helium and hydrogen on the lighter side; carbon, nitrogen, and oxygen on the heavier side—these spallation-produced elements are outnumbered by factors of millions. Nevertheless, despite not being produced by stars or stellar remnants, these elements have come to be. It's only by cosmic rays striking heavy atomic nuclei that their existence is possible at all!

In the outskirts of a giant elliptical galaxy, a star about 50% the mass of our Sun can be seen forming, with a protoplanetary disk around it. Whereas stars like our Sun may support life on the planets around it for up to a few billion years, these lower-mass stars remain stable for much longer. With tens of billions of years of stable temperatures and conditions to look forward to, perhaps the planets that form in environments like these are the most life-friendly of all.

IMAGE BY JON LOMBERG

GOLDILOCKS STARS

K-class stars are conducive to life

UNIVERSE
Diameter: $2.44 \times 10^{13} \rightarrow 2.46 \times 10^{13}$ ly
  Observable: $6.11 \times 10^{10} \rightarrow 6.16 \times 10^{10}$ ly
Expansion: $89.45 \rightarrow 88.71$ km/s/Mpc
Density: $0.84 \rightarrow 0.82$ p/m³
Temperature: $4.11 \rightarrow 4.08$ K

ASTRONOMICAL OBJECTS
Galaxies: $2.68 \times 10^{13} \rightarrow 2.66 \times 10^{13}$
Stars: $1.82 \times 10^{21} \rightarrow 1.83 \times 10^{21}$
Black Holes: $1.47 \times 10^{19} \rightarrow 1.48 \times 10^{19}$
Planetary Systems: $2.46 \times 10^{21} \rightarrow 2.47 \times 10^{21}$
Worlds: $2.81 \times 10^{22} \rightarrow 2.82 \times 10^{22}$

WORLDS WITH LIFE
Simple Life: $5.79 \times 10^{17} - 2.45 \times 10^{19}$
Complex Life: $8.62 \times 10^{11} - 2.19 \times 10^{17}$
Intelligent Life: $2.55 \times 10^{8} - 1.61 \times 10^{15}$
Technological Life: $7.56 \times 10^{4} - 1.89 \times 10^{13}$
Interstellar Life: $3 - 1.66 \times 10^{12}$

Only 8.5 billion years after the hot Big Bang, somewhere around 80% of all the stars that will ever exist have already been born. The most massive stars are the shortest-lived ones and the least numerous, while the less massive a star is, the longer it lives before its nuclear fuel runs out. The trade-off for a long life, however, is significant: The longer a star lives, the redder, cooler, and fainter it will be, with a dearth of ultraviolet and visible light output. In order for life to arise and evolve into something complex, however, a parent star must live sufficiently long: billions of years at least, ruling out the most massive, shortest-lived stellar systems. Somewhere, in between the heaviest and lightest stars, are the ones most likely to give rise not only to life but to complex, differentiated, and even intelligent life.

Below a certain threshold—about 40% of our Sun's solar mass—a star will simply be a red dwarf, slowly but steadily fusing hydrogen into helium for 150 billion years or more. Red dwarfs make up at least 75% of all stars; are prone to flares, bursts, and X-ray emissions; and will easily lock any orbiting planets to them, where one side always faces the star and one side always faces away. The vast majority of their emitted starlight is infrared radiation; visible and ultraviolet light are hard to come by. Such worlds are in constant danger of receiving a sterilizing dose of radiation, of having their atmospheres stripped away, and of facing permanently scorching dayside and cryogenic nightside conditions. Although they're the most common star type in the Universe, they may be unfriendly to life.

Conversely, stars that form above a certain threshold—around 1 solar mass or greater—will only be able to support life for a few billion years, at most, before the star's interior heats up sufficiently to boil any liquid water present on an orbiting world's surface. Even if life arises, survives, and becomes complex and differentiated on a world around such a star, it only possesses a brief window of time before complete extinction becomes inevitable. Evolution must proceed rapidly and serendipitously at many stages for an intelligent civilization to arise, and even in those cases, the brief timescales for habitability may make second or third chances exceedingly rare.

Still, somewhere in between a short-lived star and a very low-mass star must lie a "sweet spot," where the star remains stable for long periods of time while minimizing the existential dangers to life on any orbiting worlds. The best candidates, astronomically, are known as K-class stars: stars between about 40% and 80% of the mass of our Sun. Such star systems should remain habitable for anywhere between 20 billion and 150 billion years, while still producing substantial amounts of ultraviolet and visible light radiation. They are no more prone to flares or outbursts than Sun-like stars but only increase in temperature and luminosity very slowly. Orbiting rocky worlds at the right distance for liquid water on their surfaces are unlikely to become tidally locked as well, allowing for day/night rotation that's separate from a world's annual revolution.

Worlds that orbit these yellow/orange-colored stars may have the best opportunity to develop not only life but intelligent, technologically advanced life over the entirety of the Universe. Approximately 15% of all stars fall into this category, while only 5% of stars possess the mass of the Sun or greater. With long lifetimes, stable systems, and large numbers, these K-class stars may be the most life-friendly in all the Universe.

Although the upper limit on black holes is set only by how much mass they can devour, the lowest-mass black holes are created through the mergers of neutron stars. If two neutron stars have a combined mass that's more than about 2.5 solar masses, then the post-merger product can become a black hole. Below that threshold, the remnant will remain a neutron star.

IMAGE BY MARK A. GARLICK

WHEN GODS DIVIDE BY ZERO

—— how black holes form ——

UNIVERSE
Diameter: $2.46 \times 10^{13} \rightarrow 2.49 \times 10^{13}$ ly
  Observable: $6.16 \times 10^{10} \rightarrow 6.22 \times 10^{10}$ ly
Expansion: $88.71 \rightarrow 87.98$ km/s/Mpc
Density: $0.82 \rightarrow 0.80$ p/m³
Temperature: $4.08 \rightarrow 4.04$ K

ASTRONOMICAL OBJECTS
Galaxies: $2.66 \times 10^{13} \rightarrow 2.66 \times 10^{13}$
Stars: $1.83 \times 10^{21} \rightarrow 1.85 \times 10^{21}$
Black Holes: $1.48 \times 10^{19} \rightarrow 1.49 \times 10^{19}$
Planetary Systems: $2.47 \times 10^{21} \rightarrow 2.50 \times 10^{21}$
Worlds: $2.82 \times 10^{22} \rightarrow 2.86 \times 10^{22}$

WORLDS WITH LIFE
Simple Life: $5.72 \times 10^{17} - 2.44 \times 10^{19}$
Complex Life: $8.29 \times 10^{11} - 2.26 \times 10^{17}$
Intelligent Life: $2.47 \times 10^{8} - 1.67 \times 10^{15}$
Technological Life: $7.35 \times 10^{4} - 1.97 \times 10^{13}$
Interstellar Life: $3 - 1.77 \times 10^{12}$

Whenever anyone looks out at the distant Universe, there are two sets of things to observe: the objects that reveal themselves directly by emitting or absorbing light, and the invisible objects whose presence can only be inferred from their indirectly imprinted signatures. As massive and compact as black holes are, they fall into that second category and are only detectable by the effects they have on the matter they accelerate and on the light that gets bent by their gravity. Nevertheless, there are nearly a quintillion black holes across the observable Universe by this point. While the most massive ones can rise into the hundreds of billions of solar masses, there's a strict lower limit to black hole masses: about 2.5 solar masses. Below this threshold, there are no black holes.

The most common way that black holes arise is from the remains of a once-massive star. Most stars born with about 8 solar masses or more are destined to die in a core-collapse supernova, which will leave behind a remnant neutron star or a black hole, depending on the specific details of stellar death. However, the lower-mass stars exclusively produce neutron stars; it's only the heavier ones that can create black holes. As a result, core-collapse supernovae produce black holes anywhere from ~10 solar masses on up but rarely create any lower masses than that.

While supernovae might be the most common mechanism for making black holes, core-collapse supernovae aren't the only way such objects are created. In general, a star generates radiation in its core, with that radiation exerting pressure that balances the gravitational force, holding the star up against collapse. Processes can occur, however, that cause the radiation pressure in that core to drop. Stars can run out of nuclear fuel in their cores; they can produce energetic photons that split into matter-antimatter pairs; they can produce large quantities of rapidly escaping, noninteracting neutrinos. If the pressure drops rapidly enough, the entire star can rapidly, directly collapse, forming a black hole of the mass of the entire star, with no supernova or other visible cataclysm forthcoming. Unfortunately, this creates heavier black holes, not lighter ones.

Outside of stars, it's also possible for enough gas-rich material to condense into a single region of space, forming a black hole directly. This likely happened early on in the Universe: around the time the first stars were forming. Many of the black holes created through this process would have masses of tens of thousands of solar masses, serving as the seeds for supermassive black holes that exist at the centers of galaxies. Although it's theoretically possible for lower-mass black holes to form at this time—a theoretical concept known as primordial black holes—that scenario lacks any observational evidence supporting it.

Instead, the Universe produces the lowest-mass black holes that actually exist from the merger of two neutron stars. Just as many of the star systems that form have two or more stars in them, many of the stars that die in a core-collapse supernova have companions with similar masses, frequently leading to a system where two neutron stars orbit one another. As they orbit each other, they emit gravitational waves, causing the orbits to decay. Eventually, they'll inspiral and merge, creating a kilonova event that results in a new, more massive neutron star if their combined mass is below 2.5 solar masses, but creating a black hole at higher masses. These are the lowest-mass black holes of all, where space is curved more strongly than anyplace else in the Universe.

There are many ways for an intelligent, technologically advanced civilization to drive itself to extinction, as the existential threat of a powerful enough set of weapons can ruin a planet's habitability for many of its home species. Here, a global thermonuclear war leads to the end of an intelligent civilization, as the last survivors flee in starships in search of a more hospitable environment.

IMAGE BY MARK A. GARLICK

NOW I AM BECOME DEATH

some technological life destroys itself

UNIVERSE
Diameter: $2.49 \times 10^{13} \rightarrow 2.51 \times 10^{13}$ ly
  Observable: $6.22 \times 10^{10} \rightarrow 6.27 \times 10^{10}$ ly
Expansion: $87.98 \rightarrow 87.27$ km/s/Mpc
Density: $0.80 \rightarrow 0.78$ p/m³
Temperature: $4.04 \rightarrow 4.00$ K

ASTRONOMICAL OBJECTS
Galaxies: $2.66 \times 10^{13} \rightarrow 2.64 \times 10^{13}$
Stars: $1.85 \times 10^{21} \rightarrow 1.86 \times 10^{21}$
Black Holes: $1.49 \times 10^{19} \rightarrow 1.49 \times 10^{19}$
Planetary Systems: $2.50 \times 10^{21} \rightarrow 2.51 \times 10^{21}$
Worlds: $2.86 \times 10^{22} \rightarrow 2.87 \times 10^{22}$

WORLDS WITH LIFE
Simple Life: $5.65 \times 10^{17} - 2.42 \times 10^{19}$
Complex Life: $7.97 \times 10^{11} - 2.32 \times 10^{17}$
Intelligent Life: $2.39 \times 10^{8} - 1.73 \times 10^{15}$
Technological Life: $7.14 \times 10^{4} - 2.06 \times 10^{13}$
Interstellar Life: $3 - 1.87 \times 10^{12}$

 omewhere in the Universe, civilizations networked together across their home galaxy, and possibly beyond, are lamenting what might have been. On a temperate, rocky world around a modest, unremarkable star, life arose and thrived in an unbroken chain for billions of years. Despite a large number of extinction events—some caused by pollutants generated by life itself, others caused by cosmic events such as asteroid strikes—life evolved to become complex, differentiated, intelligent, and tool-using. At some point, perhaps by random chance, a species on that planet became clever enough to develop advanced technology, receiving messages from the other actively communicating societies within their galaxy. They began transmitting their own signals in an attempt to exchange information, hoping to achieve two-way communication. Although their signals did indeed generate responses from the existing, more advanced societies, they were never received. Instead, only silence ensued, creating a mournful puzzle of a lost civilization.

To reconstruct what happened, civilizations within that galaxy would likely first turn to their own astronomical tools. With a powerful enough set of observatories, they could directly identify and image the home planet of this lost civilization: something they've perhaps been doing ever since the first message arrived. They would find a planet whose surface brightness and colors changed over time, providing evidence of planet-wide biological activity and an atmosphere with variable weather. They'd be able to measure the atomic and molecular contents of that atmosphere through the technique of spectroscopy, identifying the composition and ratios of nitrogen, oxygen, carbon dioxide, methane, water, and other contents. The lack of novel elements suddenly appearing could rule out asteroid strikes and volcanic activity.

They could, over time, collect enough photons to measure the temperature of the planet and note that there was a substantial, relatively abrupt drop that occurred on a planet-wide scale. Initially, the cause of such an event would be ambiguous. Was there some sort of environmental disaster that occurred from plundering resources that affected the biosphere, leading to an ecological disaster that froze the planet? Did natural processes push the planet over an environmental tipping point, triggering the onset of an ice age? Perhaps, in an effort to mitigate rising temperatures from the rapid release of fossil fuels, they attempted to geoengineer a climate solution, only to tip the scales too far in the opposite direction, leading to their extinction?

All of those would be legitimate possibilities that would create their own observable signatures, but none of them came to pass in this instance. Instead, the "smoking gun" for the civilization's demise came in the form of cosmic particles: neutrinos. The product of nuclear reactions, neutrinos carry with them, encoded in their flavor and energy spectrum, a signature of how they were created. With a sensitive enough neutrino detector, even from hundreds or thousands of light-years away, an advanced civilization could detect the rapid detonation of uncontrolled fusion reactions: evidence of planet-wide nuclear war. In one swift burst, a civilization could render their planet inhospitable for every living land creature larger than an iguana.

In the aftermath of a planet-wide nuclear war, the resulting fires would lead to an incredible amount of soot that would reflect sunlight from high in the atmosphere. This would lead not only to a widespread cooling known as nuclear winter but famine as well, causing a mass extinction event that could never occur through natural processes. From across their galaxy and perhaps even beyond, this lost civilization serves as a cosmic warning to all planets with the capacity for self-destruction.

When celestial objects pass too close to very dense masses, such as black holes, the forces on the "near part" and "far part" of those objects will be sufficient to rip them apart. These tidal forces can tear apart even planets and stars, creating a cataclysmic, destructive phenomenon known as a tidal disruption event. The long, thin streams of destroyed material are reminiscent of spaghetti.

IMAGE BY MARK A. GARLICK

IN THE MANNER OF SPAGHETTI

differential effects of gravity stretch things

UNIVERSE
Diameter: $2.51 \times 10^{13} \rightarrow 2.53 \times 10^{13}$ ly
  Observable: $6.27 \times 10^{10} \rightarrow 6.33 \times 10^{10}$ ly
Expansion: $87.27 \rightarrow 86.57$ km/s/Mpc
Density: $0.78 \rightarrow 0.76$ p/m³
Temperature: $4.00 \rightarrow 3.97$ K

ASTRONOMICAL OBJECTS
Galaxies: $2.64 \times 10^{13} \rightarrow 2.63 \times 10^{13}$
Stars: $1.86 \times 10^{21} \rightarrow 1.87 \times 10^{21}$
Black Holes: $1.49 \times 10^{19} \rightarrow 1.50 \times 10^{19}$
Planetary Systems: $2.51 \times 10^{21} \rightarrow 2.53 \times 10^{21}$
Worlds: $2.87 \times 10^{22} \rightarrow 2.89 \times 10^{22}$

WORLDS WITH LIFE
Simple Life: $5.58 \times 10^{17} - 2.40 \times 10^{19}$
Complex Life: $7.66 \times 10^{11} - 2.39 \times 10^{17}$
Intelligent Life: $2.30 \times 10^{8} - 1.79 \times 10^{15}$
Technological Life: $6.93 \times 10^{4} - 2.15 \times 10^{13}$
Interstellar Life: $3 - 1.99 \times 10^{12}$

atter and radiation, amid all else that occurs, continue to always gravitate. On large cosmic scales, even as the Universe expands, gravitation draws matter into overdense regions, leading to the formation of stars, galaxies, and cosmic collections of galaxies. On smaller scales, gravitation causes molecular gas clouds to collapse, leading to new star clusters. Each stellar system that's birthed has the opportunity to develop its own system of planets and moons, which remain in stable orbits due to gravity. And stars tug on one another as they pass by while revolving around the galactic center, perturbing their motions and leading to the possibility of gravitational ejections.

But gravitation isn't exerted equally over the entirety of an object; rather, the portions that are closer to an external mass will experience a greater attraction than the portions farther away. Similarly, the portions that are above, below, or left or right of the object's center will be attracted downward, upward, rightward, or leftward (respectively) relative to the center of the object. The effect of these differential forces—noting that the force all over the object differs depending on location—shows up as tides on most planets and moons but can lead to remarkable phenomena in more extreme circumstances.

When two galaxies approach one another, the tidal forces cause the gas within each galaxy to compress and stretch out along the imaginary line connecting those galaxies, which triggers gas collapse and the formation of new stars. When a smaller galaxy approaches a larger one, the small galaxy frequently undergoes a starburst due to these tidal effects: where the entire galaxy becomes a star-forming region.

When a rocky or icy moon orbits a massive planet, tidal forces have a tremendous effect. They heat the moon's interior, leading to moonquakes and thermal activity, such as geysers and possibly volcanoes. The most volcanically active worlds in all the Universe should be moons orbiting close to the outer extents of gas giants, which could lead to worldwide eruptions and resurfacing events on timescales of mere years. At even closer distances, a moon could be destroyed entirely by these tidal forces, shattered into a ring of debris that eventually rains back down onto the planet.

Whenever massive objects—asteroids, planets, stars, etc.—venture too close to a stellar remnant, they can be torn apart as well in a tidal disruption event. The secondary object can be even more massive than the remnant itself; it only needs to be less dense to be torn apart. White dwarfs can tear apart stars; neutron stars can tidally destroy white dwarfs; black holes can disrupt and destroy any object composed of matter at all. In fact, black holes are the most extreme object in this regard, and the tides they generate are so severe that they lead to a process known as spaghettification.

Any object that ventures too close to a black hole will experience the strongest differential forces found anywhere in the Universe. Solid objects that have chemical bonds between them will have those bonds broken, as the matter gets stretched along the direction towards the black hole and compressed in all other dimensions until a long, thin, spaghetti-like strand is created. As the matter approaches the black hole's central singularity, the tidal forces become even more severe. Molecules will be broken apart into individual atoms; atoms will be stripped of their electrons; atomic nuclei will be torn apart into protons and neutrons, and then further decomposed into quarks and gluons. In the end, any form of matter will be spaghettified by a black hole, as its gravitational effects are truly irresistible.

When stars begin to evolve into red giants, their energy output increases tremendously: several times over just a few tens of millions of years. Even if life was thriving on this world at the start of this period, by the end, it will be nothing more than a scorched, barren, atmosphere-free remnant of rock and metal.

IMAGE BY JON LOMBERG

SNAPSHOTS IN TIME

we see the past, never the present or future

UNIVERSE
Diameter: $2.53 \times 10^{13} \rightarrow 2.55 \times 10^{13}$ ly
  Observable: $6.33 \times 10^{10} \rightarrow 6.39 \times 10^{10}$ ly
Expansion: $86.57 \rightarrow 85.88$ km/s/Mpc
Density: $0.76 \rightarrow 0.74$ p/m³
Temperature: $3.97 \rightarrow 3.94$ K

ASTRONOMICAL OBJECTS
Galaxies: $2.63 \times 10^{13} \rightarrow 2.61 \times 10^{13}$
Stars: $1.87 \times 10^{21} \rightarrow 1.88 \times 10^{21}$
Black Holes: $1.50 \times 10^{19} \rightarrow 1.51 \times 10^{19}$
Planetary Systems: $2.53 \times 10^{21} \rightarrow 2.54 \times 10^{21}$
Worlds: $2.89 \times 10^{22} \rightarrow 2.90 \times 10^{22}$

WORLDS WITH LIFE
Simple Life: $5.51 \times 10^{17} - 2.38 \times 10^{19}$
Complex Life: $7.36 \times 10^{11} - 2.45 \times 10^{17}$
Intelligent Life: $2.22 \times 10^{8} - 1.85 \times 10^{15}$
Technological Life: $6.71 \times 10^{4} - 2.23 \times 10^{13}$
Interstellar Life: $3 - 2.10 \times 10^{12}$

napshots are usually the only views anyone gets of the Universe. At any moment, observers looking out at the Universe will see things exactly as they are at that particular time: right as the light from each observable object is reaching them. The farther away an object is, the longer the light must travel to reach its destination, implying that looking back to greater distances means that one sees the Universe as it was back in time: when the light that's arriving now was first emitted. But rather than a cosmic snapshot, an intelligent, technologically advanced civilization that endures for long periods of time— thousands, millions, or even 100 million years—would witness an evolving Universe. Not only are there overall cosmic changes, but individual stellar and planetary systems metamorphose as well. In many spectacular instances, these changes aren't merely smooth and gradual but can radically transform their local environments.

On a cosmic scale, the changes that occur over ~100 million years are small and slight: on the order of 1–2% changes in things like the temperature, radius, and star-formation rate in the Universe. Individual stars that aren't involved in some sort of cataclysm or that don't run out of the nuclear fuel they're presently burning in their cores might become slightly larger, hotter, and more luminous, but these changes are gradual and smooth as well. However, the Universe is a messy place, filled with planets, moons, stars, galaxies, and trillions of times as many hazards—comets, asteroids, and interstellar mixes of rock, metal, and ice—as there are stars. With sextillions of stars and stellar systems sprinkled across the visible Universe, the impact that routine cosmic collisions have cannot be underestimated.

When a rocky world gets struck by a massive, energetic impactor, an enormous suite of debris gets kicked up. In under a million years—a cosmic blink of an eye—a system of moons will form, with more massive moons generally forming close to the main object and smaller, less massive ones farther away. Similarly, the innermost moons that form can get torn apart by gravitational interactions into a ringed system, which often falls back onto its parent world, raining debris until the ring is entirely gone.

The lunar systems around gas giant worlds are fragile and precarious. Asteroids and comets can strike them, and because their masses can be so low, it takes relatively little energy from a random collision to blast them apart. While many of these large moons acquire massive craters, some of them are blown to smithereens, transforming what was once a moon into a diffuse, thin-ringed system around the giant planet. While moons can be destroyed in an instant, the ringed systems don't last long either. Over the span of anywhere from 50 million to 200 million years, a system of rings as impressive as Saturn's can evaporate away entirely.

On inhabited worlds, however, the changes can be the most dramatic of all. The contents of a life-bearing world's atmosphere can change tremendously, as the relative balance between gases such as nitrogen, oxygen, carbon dioxide, methane, and water vapor can change by 10 percent or more in a 100-million-year time span. New kingdoms of life can emerge and rise to dominate a world's biomass, while others can go extinct entirely. Intelligent civilizations can rise and fall; entire planet-wide ecosystems can emerge or collapse. No matter how granularly you examine the Universe—including over relatively brief periods of time—there's no guarantee that drastic changes won't have occurred since the last time you checked in.

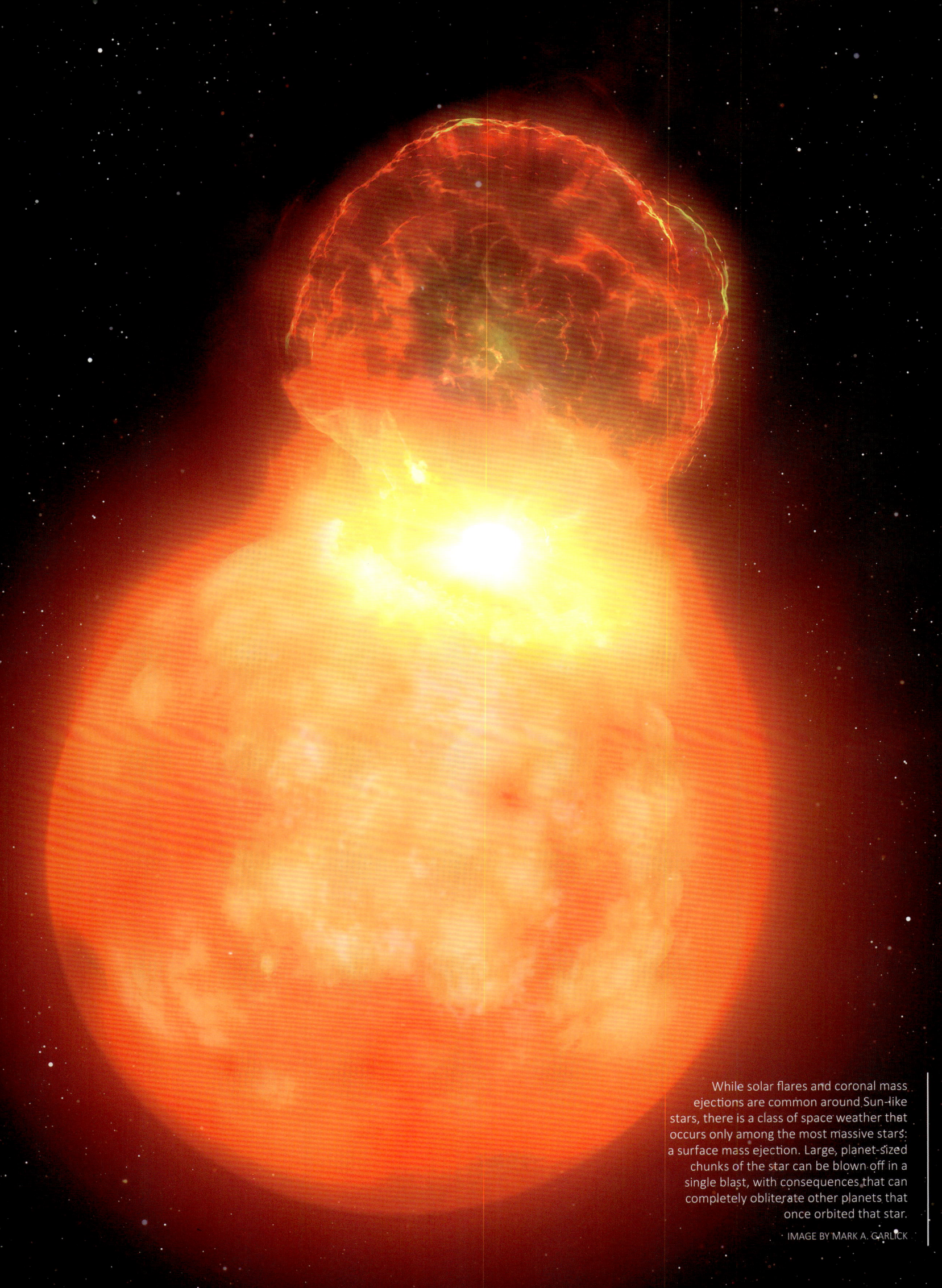

While solar flares and coronal mass ejections are common around Sun-like stars, there is a class of space weather that occurs only among the most massive stars: a surface mass ejection. Large, planet-sized chunks of the star can be blown off in a single blast, with consequences that can completely obliterate other planets that once orbited that star.

IMAGE BY MARK A. GARLICK

COOKING UP SOLAR STORMS

solar magnetic fields kink and expel plasma

UNIVERSE
Diameter: $2.55 \times 10^{13} \to 2.58 \times 10^{13}$ ly
  Observable: $6.39 \times 10^{10} \to 6.44 \times 10^{10}$ ly
Expansion: $85.88 \to 85.21$ km/s/Mpc
Density: $0.74 \to 0.72$ p/m³
Temperature: $3.94 \to 3.90$ K

ASTRONOMICAL OBJECTS
Galaxies: $2.61 \times 10^{13} \to 2.60 \times 10^{13}$
Stars: $1.88 \times 10^{21} \to 1.89 \times 10^{21}$
Black Holes: $1.51 \times 10^{19} \to 1.52 \times 10^{19}$
Planetary Systems: $2.54 \times 10^{21} \to 2.55 \times 10^{21}$
Worlds: $2.90 \times 10^{22} \to 2.92 \times 10^{22}$

WORLDS WITH LIFE
Simple Life: $5.44 \times 10^{17} - 2.35 \times 10^{19}$
Complex Life: $7.07 \times 10^{11} - 2.52 \times 10^{17}$
Intelligent Life: $2.14 \times 10^8 - 1.91 \times 10^{15}$
Technological Life: $6.49 \times 10^4 - 2.32 \times 10^{13}$
Interstellar Life: $3 - 2.22 \times 10^{12}$

Most stars appear to radiate a constant amount of energy out into the Universe, but that's not strictly true. Sure, it's true on average, over long periods of time, and for most stars. However, all stars vary in how much energy they output: most by only about 0.1%, but some by very large amounts of 99% or more. Some stars brighten and fade very regularly, as their outer layers heat up, expand, cool down, condense to become partly opaque, and then contract, falling back onto the star, where they become ionized and transparent, and the cycle repeats all over again. But there's another phenomenon common to practically all stars that occurs far less regularly: flares and other forms of space weather. Although most such events are relatively benign, the more intense ones are capable of not only endangering life on orbiting worlds but also imposing total planetary destruction.

Within a star, the temperatures and energies are so great that neutral atoms—made of light, negatively charged electrons bound to heavy, positively charged atomic nuclei—can no longer stably exist. Instead, one or more electrons are routinely ionized by the energy within the star, leading to a roiling sea of hot, charged, rapidly moving particles. These charges in motion create electric currents, which in turn create magnetic fields. These magnetic fields can build up between different regions in the star and also between the star and the surrounding hot, rarefied corona. As these disconnected regions move around, they can cause magnetic fields to disconnect and reconnect, with magnetic reconnection events releasing tremendous amounts of energy. This can result in the emission of space weather events, including solar flares, coronal mass ejections, or, in the most extreme cases, catastrophic events known as surface mass ejections.

Solar flares and coronal mass ejections send streams of charged particles outward from the star that created them, usually with relatively low energies and speeds of only a few hundred kilometers per second. In those instances, organisms on an orbiting world's surface would be safe; the atmosphere will absorb those particles, creating beautiful but harmless auroral displays. However, the most energetic classes of flares and coronal mass ejection events can strip the atmospheres from otherwise habitable planets, irradiate their surfaces, and wreck any possibilities for complex life. Solar flares are so common on red dwarf stars—the lowest-mass, most common type of star—that it's not certain that worlds in orbit around them can maintain the existence of an atmosphere for very long at all. On technologically advanced worlds with electrical and electronic infrastructure, even routine space weather events can overload and destroy power grids and everything connected to them. The resulting damage can be both colossal and catastrophic, potentially significant enough to cause the full-scale collapse of a civilization.

Even the worst of these events, however, cannot compare to the most destructive form of space weather of all: a surface mass ejection. The most massive stars will reach a point in their evolution when they begin fusing carbon (and heavier elements) in their cores, becoming cool, diffuse, and enormous: with diameters in excess of one billion kilometers. Large chunks of the star itself—as massive as a Mars-sized planet—can be blown off all at once. If an orbiting planet is caught within the blast, it could be either drawn into the parent star or blown apart entirely: an act of cosmic infanticide. While it's notable that stars often give rise to life on the worlds that revolve around them, it's a sobering truth that they can snatch it away in a mere instant, too.

What appears to be an absolutely unremarkable moment in cosmic history, the fragmentation of a collapsing molecular cloud, is arguably the most important event in cosmic history from a human-centric perspective: the first stages in the formation of the Sun and Solar System. Currently just a protostar, this larger, cooler, and nonuniform ball of gas is still contracting, on the cusp of initiating nuclear fusion in its core.

IMAGE BY MARK A. GARLICK

HERE COMES THE SUN

—— *the Sun begins to form* ——

UNIVERSE	ASTRONOMICAL OBJECTS	WORLDS WITH LIFE
Diameter: $2.58 \times 10^{13} \rightarrow 2.60 \times 10^{13}$ ly	Galaxies: $2.60 \times 10^{13} \rightarrow 2.58 \times 10^{13}$	Simple Life: $5.37 \times 10^{17} - 2.33 \times 10^{19}$
Observable: $6.44 \times 10^{10} \rightarrow 6.50 \times 10^{10}$ ly	Stars: $1.89 \times 10^{21} \rightarrow 1.90 \times 10^{21}$	Complex Life: $6.81 \times 10^{11} - 2.57 \times 10^{17}$
Expansion: $85.21 \rightarrow 84.62$ km/s/Mpc	Black Holes: $1.52 \times 10^{19} \rightarrow 1.53 \times 10^{19}$	Intelligent Life: $2.07 \times 10^{8} - 1.96 \times 10^{15}$
Density: $0.72 \rightarrow 0.70$ p/m³	Planetary Systems: $2.55 \times 10^{21} \rightarrow 2.57 \times 10^{21}$	Technological Life: $6.29 \times 10^{4} - 2.39 \times 10^{13}$
Temperature: $3.90 \rightarrow 3.87$ K	Worlds: $2.92 \times 10^{22} \rightarrow 2.94 \times 10^{22}$	Interstellar Life: $3 - 2.32 \times 10^{12}$

 herever gas-rich galaxies persist, stars are born via two different mechanisms. The more spectacular method arises when two galaxies tug on each other or collide and merge, triggering a galaxy-wide collapse of gas and forming millions of stars (or more) in short order. But most of the stars that form—particularly this late in the cosmic game—come from a second, more mundane source: the gradual and periodic collapse of molecular gas clouds lining the arms of spiral galaxies. Stars, gas, and all forms of matter move in and out of a galaxy's spiral arms as it rotates, bunching up along the arms themselves and spreading out along the space between them. When gas bunches up within those arms, cooling and becoming sufficiently dense, it can trigger new bursts of star formation, typically leading to hundreds or thousands of new stars. In one fairly typical galaxy—the Milky Way—such an event is occurring right at this critical moment.

Most star-forming regions lining the arms of spiral galaxies are small: no more than a couple dozen light-years across. Most contain only a few thousand solar masses' worth of material: mostly hydrogen and helium gas. And most undergo gravitational collapse unevenly, with numerous regions of star formation occurring throughout at various times, frequently separated by millions or even tens of millions of years. As a result, these nebulae wind up forming a few collections of stars known as open star clusters, dominated by a few handfuls of bright, blue, massive stars alongside hundreds or even thousands of smaller, fainter, redder, low-mass stars. About 25,000 light-years from the center of the Milky Way, the gas within one of our galaxy's spiral arms forms stars through exactly this process.

It's the rarest of stars—the bluest, most brilliant ones—that ultimately bring an end to the formation of new stars, as the ultraviolet radiation they emit is responsible for ionizing and evaporating away the remaining star-forming material. The cosmic race that occurs within these nebulae is a race between gravity, which works to attract and collapse matter from these overdense clumps of gas into full-fledged stars, while the radiation from already-birthed stars works to blow that matter away. About half of the stars that form are singlet systems, with a lone central star, while the other half wind up as members of multi-star systems: binaries, trinaries, or even greater numbers of stars all mutually bound together. For hundreds of millions of years, these newborn stellar systems remain in a large collection where hundreds or thousands of them coevolve together: as part of an open star cluster.

At this point, the Milky Way's gas-rich disk is sufficiently enriched so that practically all of the stars that form within such a nebula should form with planets around them. About 75–80% of the stars that form will be low-mass, red dwarf stars; fewer than 1% will be of the massive, bright, blue, hot variety that are destined to die in core-collapse supernova events. But in between is a potential sweet spot: a set of stars that will burn through their fuel at a stable, relatively steady rate for billions of years but that won't tidally lock their orbiting planets to them, won't flare enough to strip away their atmospheres, and won't regularly create X-ray bursts that might sterilize any early forms of life that might emerge.

Somewhere, within one of the open star clusters that emerges as the last vestiges of gas gets evaporated away by the short-lived massive stars inside, the protostar that we'll someday call our Sun is forming. As the first nuclear fusion reactions ignite within its core, the moment arrives that we can finally declare, "Our star is born."

The early environment around the young Sun is incredibly violent, as protoplanetesimals gravitate, collide, and blast one another apart countless times before settling down into a stable system of orbiting, gravitationally bound bodies. Many of the young, massive protoplanets that form are either swallowed by the Sun or ejected into interstellar space.

IMAGE BY MARK A. GARLICK

DANCE OF THE PLANETS

the Solar System forms

UNIVERSE
Diameter: $2.60 \times 10^{13} \rightarrow 2.62 \times 10^{13}$ ly
  Observable: $6.50 \times 10^{10} \rightarrow 6.55 \times 10^{10}$ ly
Expansion: $84.62 \rightarrow 83.97$ km/s/Mpc
Density: $0.70 \rightarrow 0.68$ p/m³
Temperature: $3.87 \rightarrow 3.83$ K

ASTRONOMICAL OBJECTS
Galaxies: $2.58 \times 10^{13} \rightarrow 2.57 \times 10^{13}$
Stars: $1.90 \times 10^{21} \rightarrow 1.91 \times 10^{21}$
Black Holes: $1.53 \times 10^{19} \rightarrow 1.54 \times 10^{19}$
Planetary Systems: $2.57 \times 10^{21} \rightarrow 2.58 \times 10^{21}$
Worlds: $2.94 \times 10^{22} \rightarrow 2.95 \times 10^{22}$

WORLDS WITH LIFE
Simple Life: $5.30 \times 10^{17} - 2.31 \times 10^{19}$
Complex Life: $6.53 \times 10^{11} - 2.63 \times 10^{17}$
Intelligent Life: $1.99 \times 10^{8} - 2.02 \times 10^{15}$
Technological Life: $6.08 \times 10^{4} - 2.48 \times 10^{13}$
Interstellar Life: $2 - 2.44 \times 10^{12}$

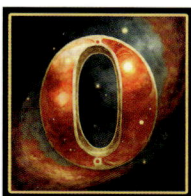 ur newborn Sun fuses light elements into heavier ones in its core—just like thousands of other stars currently within the same stellar nursery—causing winds and radiation emanating from its surface to impact the surrounding matter. Light, tenuous chemical compounds are ionized and blown away, particularly close to the star itself. Material in the same plane as the rotating Sun collapses, forming a thin disk containing planet-forming material. And instabilities and imperfections in this disk lead to collisions and gravitational growth, forming protoplanets and planetesimals rapidly. The early, massive protoplanets grow the most quickly, carving gaps into the protoplanetary disk and accumulating significant rock-and-metal cores, while the central, young Sun's energy output serves to evaporate the surrounding material. This race to form planets and moons is a vitally important struggle in every stellar system's earliest days.

Our most massive planet—Jupiter—grows up the most quickly. Accumulating a massive core more rapidly than any other imperfection, Jupiter quickly becomes capable of holding on to even the lightest elements—hydrogen and helium—owing to its immense gravity. Four other massive planetary cores also form early on, with three growing into the Saturnian, Uranian, and Neptunian system, while still another planet is ejected during these early stages. Interior to all of them, it's likely that early protoplanets and protoplanetesimals don't survive, being swallowed by the Sun or ejected via gravitational instabilities.

But the Solar System takes several tens of millions of years to finish forming planets and moons, and that means planetary material is still available to coalesce into smaller worlds interior to Jupiter. While the circumplanetary disks surrounding Jupiter, Saturn, Uranus, and Neptune

fragment into a rich system of moons—and as the asteroid belt and Kuiper belt form from a mix of metals, rocks, and volatile ices—at least five inner worlds form less than 250 million kilometers from the Sun. Along with Mercury, Venus, Earth, and Mars, a fifth rock-rich world known as Theia—about the size of Mars—also forms.

About 50 million years after nuclear fusion ignites in the Sun's core, a remarkable collision occurs between Theia and Earth. This collision is by far the most energetic event in our planet's history, kicking up over 100 quintillion tons of material into a puffy, donut-shaped ring that surrounds Earth, which grows another 5–10% more massive from this giant impact. The young Earth is extremely hot during this time: about 2,700 K, or sufficient to glow red-hot all across its surface. Some of the material in the donut-shaped ring—known as a synestia—falls back onto Earth, but the remainder forms a large satellite: Earth's Moon. Of all the satellites in the Solar System, our Moon is the largest to form from an impact event.

Because of the temperature of early Earth, many elements on the Earth-facing side of the Moon, such as calcium and aluminum, become depleted. On the far side of the Moon, however, these elements can easily accumulate, explaining why the lunar far side has a thicker crust with a different chemical composition than the lunar near side. In less than 100 million years, from the first ignition of nuclear fusion inside the young Sun, the Solar System has already begun to look very similar to its present configuration: with four inner, rocky worlds, including Earth with its large Moon, a rich asteroid belt, four giant planets each with its own lunar system, and an ice-rich Kuiper belt beyond Neptune. With an atmosphere rich in carbon dioxide, methane, hydrogen sulfide, and other volcanic gases, our home planet is finally born.

After a massive collision kicked up an enormous cloud of debris around the proto-Earth, it led to the formation of our Moon, which was much hotter and closer to Earth in these early stages of the Solar System's history. The young Earth possessed a very different atmosphere to the one we know today and completed a full 360° rotation in just six to eight hours: a much shorter day than at present.

IMAGE BY MARK A. GARLICK

BENEATH A STAR-FILLED SKY

—— *new stars fill the sky of proto-Earth* ——

UNIVERSE
Diameter: $2.62 \times 10^{13} \rightarrow 2.65 \times 10^{13}$ ly
  Observable: $6.55 \times 10^{10} \rightarrow 6.61 \times 10^{10}$ ly
Expansion: $83.97 \rightarrow 83.33$ km/s/Mpc
Density: $0.68 \rightarrow 0.66$ p/m³
Temperature: $3.83 \rightarrow 3.80$ K

ASTRONOMICAL OBJECTS
Galaxies: $2.57 \times 10^{13} \rightarrow 2.55 \times 10^{13}$
Stars: $1.91 \times 10^{21} \rightarrow 1.92 \times 10^{21}$
Black Holes: $1.54 \times 10^{19} \rightarrow 1.54 \times 10^{19}$
Planetary Systems: $2.58 \times 10^{21} \rightarrow 2.59 \times 10^{21}$
Worlds: $2.95 \times 10^{22} \rightarrow 2.97 \times 10^{22}$

WORLDS WITH LIFE
Simple Life: $5.23 \times 10^{17} - 2.29 \times 10^{19}$
Complex Life: $6.26 \times 10^{11} - 2.69 \times 10^{17}$
Intelligent Life: $1.92 \times 10^{8} - 2.08 \times 10^{15}$
Technological Life: $5.86 \times 10^{4} - 2.57 \times 10^{13}$
Interstellar Life: $2 - 2.57 \times 10^{12}$

It must have been quite a sight—even though no one was around to witness it—to behold the night sky from Earth shortly after its formation. While daytime was dominated by the light from our young Sun, hundreds of newborn, nearby stars would have glittered brightly and brilliantly each night. As a member of an open star cluster, the Sun was one of a few thousand young stars all born at approximately the same time. While the most massive ones likely sank towards the cluster's center, the Sun was driven towards the outskirts, like many of the more typical and lower-mass stars that formed contemporaneously. While the cores of star clusters such as this one can remain bound for up to about one billion years, most of the stars, particularly on the cluster edges, are ejected much earlier. Our Sun—and the surrounding Solar System—after about 200 million years as a member of the star cluster it was born into, is finally sent out on its own: in orbit around the galactic center.

The Sun was hardly unique at the time of its birth, as practically all of the stars that formed from the same cloud of gas possessed similar ratios of heavy elements, similar compositions for forming planets but with wide-ranging masses and brightnesses. The more massive stars are brighter and bluer in color, with dozens of stars shining more brilliantly than even Sirius will shine once it's born billions of years after this epoch. About half of these stellar siblings will wind up as members of multi-star systems, while the other half remain singlet stars like our own Sun. The most massive ones are already dying in supernova explosions and by blowing off their outer layers in planetary nebulae, further contributing to the dissociation of the star cluster.

Other stars can be seen trailing away from the cluster: some ahead of the Sun, some in the same direction, but with many appearing to fly off at random as well. As stars leave the cluster, it becomes less massive, and with less gravity to hold it together, low-mass and intermediate-mass stars are cast out more frequently. Earth's night sky evolves tremendously during this epoch, as the original star cluster that birthed it becomes less luminous, less blue, contains fewer stars, and begins to recede from view.

Although many of the stars birthed in the same stellar nursery as our Sun remain visible and begin in predictable orbits around the Milky Way, it isn't long before the other stars within the galaxy make the situation much messier. With about every million years that goes by, a passing star comes within less than 1 light-year of the Sun (and each star ejected from the Sun's original star cluster), altering the star's path slightly but significantly. As time marches on, stars move in and out of the spiral arms in our galaxy, with the Sun completing about 40% of a revolution around the Milky Way every 100 million years.

Over the span of a mere 100 million years, it becomes impossible to trace back the trajectory of these stars and determine which ones originated from the same cluster as our own Sun. The best clues we have as to which ones are truly our long-lost siblings—like age and abundance of heavy elements—are insufficient to tell us which stars were born along with us versus which ones are just look-alike impostors. Without a clear marker to identify our stellar siblings, we can only have faith that they're out there wandering through the galaxy, just like us.

The combination of violent collisions, external heating from the Sun, and its relatively small gravitational pull cause Mercury to lose not only its atmosphere but its crust and most of its mantle as well. Initially, Mercury may have been up to four times as massive as it is today, where 85% of the planet's interior, by radius, is composed of a metal-rich core. This giant impact created its Caloris Basin: a 1,500-kilometer-wide impact crater.

IMAGE BY MARK A. GARLICK

THE DIMINUTION OF MERCURY

—— *extreme conditions wear down Mercury* ——

UNIVERSE
Diameter: $2.65 \times 10^{13} \rightarrow 2.67 \times 10^{13}$ ly
  Observable: $6.61 \times 10^{10} \rightarrow 6.67 \times 10^{10}$ ly
Expansion: $83.33 \rightarrow 82.78$ km/s/Mpc
Density: $0.66 \rightarrow 0.65$ p/m³
Temperature: $3.80 \rightarrow 3.77$ K

ASTRONOMICAL OBJECTS
Galaxies: $2.55 \times 10^{13} \rightarrow 2.54 \times 10^{13}$
Stars: $1.92 \times 10^{21} \rightarrow 1.93 \times 10^{21}$
Black Holes: $1.54 \times 10^{19} \rightarrow 1.55 \times 10^{19}$
Planetary Systems: $2.59 \times 10^{21} \rightarrow 2.61 \times 10^{21}$
Worlds: $2.97 \times 10^{22} \rightarrow 2.99 \times 10^{22}$

WORLDS WITH LIFE
Simple Life: $5.17 \times 10^{17} - 2.27 \times 10^{19}$
Complex Life: $6.02 \times 10^{11} - 2.75 \times 10^{17}$
Intelligent Life: $1.85 \times 10^{8} - 2.13 \times 10^{15}$
Technological Life: $5.67 \times 10^{4} - 2.64 \times 10^{13}$
Interstellar Life: $2 - 2.68 \times 10^{12}$

The Solar System ages, and as it does, gradual changes begin to occur. The last bits of protoplanetesimal dust are blown away, as the Sun's radiation evaporates away any remaining pristine matter. The giant planets, and particularly Jupiter (with the asteroid belt) and Saturn and Neptune (with the Kuiper belt) become primarily responsible for sending large numbers of rocky and icy bodies into the inner Solar System, bombarding the rocky planets and moons, delivering water, organics, and other volatiles to their surfaces. And the inner planets, as they cool and experience these impacts, begin to develop complex surface chemistry at the boundary of where their crusts meet their atmospheres.

While Venus, Earth, and Mars all begin to accumulate water, including thick enough atmospheres to enable the possibility of that water existing in the liquid phase, Mercury experiences a vastly different fate. The other three rocky planets, being farther from the Sun, now have solid surfaces that exhibit plate tectonics: where the crust floats atop the mantle, sliding across it and colliding, giving rise to quakes, volcanoes, subduction, and more. But Mercury, with its extreme temperatures, begins to undergo a unique event within our Solar System: the evaporation of its mantle. The combined energy of impacts, along with these high temperatures and the intense solar radiation, boils away these outer materials found on our innermost planet, rendering it barren.

At the time of its formation, Mercury was more massive—up to four times as massive—as it will eventually become billions of years from now. Similar to Earth and the other planets, it had a solid inner core; a liquid outer core; a solid, thick mantle; and a thin outer crust. As it gets repeatedly struck by comets and asteroids, the crustal and mantle material gets kicked up, heated by the Sun, and, at these short distances, energetically kicked with enough force to cause them to escape from Mercury's gravity. As the bombardment continues unabated, more and more of the crust and mantle are stripped off, eventually leaving Mercury with only a thin mantle comprising the outer 400 kilometers of the world: a thin shell 15% of the total planet's diameter.

The greatest impact in its history occurs during this time, creating the mighty Caloris Basin on Mercury: an astounding 1,550 kilometers wide. Unlike Venus, Earth, and Mars, Mercury's topmost layer becomes a single, solid plate rather than a series of plates that slide over the lower layers. As a result, as Mercury cools off and contracts, enormous cracks, scarps, and ridges form on its surface: thousands of kilometers long and with steep drops that lead to abrupt changes in elevation of more than a kilometer on the surface.

The asymmetries in the shape of the planet—caused by differing mantle thicknesses and the impacts that created its largest basins—allow Mercury to slide into a resonance, where it rotates on its axis at only a slightly different rate from its revolution around the Sun. Mercury makes three complete rotations for every two full revolutions around the Sun, giving Mercury the longest-enduring "day" of any planet in the Solar System: 176 modern Earth-days pass between two successive sunrises. Its enormous core for its size means that Mercury is composed of denser elements than any planet in our Solar System; it's only the fact that Earth gravitationally compresses due to our greater mass that relegates Mercury to second place in overall density for planets in our Solar System. Without a mantle or atmosphere, however, Mercury becomes a lifeless world of extremes: with modern temperatures ranging from ~430°C (800°F) down to -200°C (-330°F).

Shortly after its formation, Mars, too, experiences a giant impact, creating not only the two moons that remain today— Phobos and Deimos—but also a larger, third, inner moon that will eventually break apart and fall back onto Mars. Here, that event coincides with Mars's watery past, as its great, still-forming volcanoes dominate the continental landscape already.

IMAGE BY MARK A. GARLICK

BORN OF VIOLENCE

an asteroid impact creates Martian moons

UNIVERSE
Diameter: $2.67 \times 10^{13} \rightarrow 2.69 \times 10^{13}$ ly
  Observable: $6.67 \times 10^{10} \rightarrow 6.73 \times 10^{10}$ ly
Expansion: $82.78 \rightarrow 82.16$ km/s/Mpc
Density: $0.65 \rightarrow 0.63$ p/m³
Temperature: $3.77 \rightarrow 3.74$ K

ASTRONOMICAL OBJECTS
Galaxies: $2.54 \times 10^{13} \rightarrow 2.52 \times 10^{13}$
Stars: $1.93 \times 10^{21} \rightarrow 1.94 \times 10^{21}$
Black Holes: $1.55 \times 10^{19} \rightarrow 1.56 \times 10^{19}$
Planetary Systems: $2.61 \times 10^{21} \rightarrow 2.62 \times 10^{21}$
Worlds: $2.99 \times 10^{22} \rightarrow 3.00 \times 10^{22}$

WORLDS WITH LIFE
Simple Life: $5.10 \times 10^{17} - 2.24 \times 10^{19}$
Complex Life: $5.77 \times 10^{11} - 2.81 \times 10^{17}$
Intelligent Life: $1.77 \times 10^8 - 2.19 \times 10^{15}$
Technological Life: $5.46 \times 10^4 - 2.73 \times 10^{13}$
Interstellar Life: $2 - 2.8 \times 10^{12}$

Our Solar System—like practically every similar system where planets, moons, and other objects all revolve around a single, central star—has always been a violent place. But in the early days, in the first few hundred million years after its creation, the number of massive, energetic collisions experienced by each world was even greater than it will be billions of years into the future. As the giant planets gravitationally tug on asteroids, Kuiper belt objects, and other massive bodies throughout the Solar System, they perturb their orbits, often either ejecting them or sending them hurtling towards the inner planets. In one seemingly random event about 300 million years after the birth of the Solar System, a large, energetic body about 500 kilometers in diameter collided with a young Mars. The aftermath of this collision would forever change the landscape of the red planet.

Throughout the Universe, there are only three known ways that planets can gain/acquire moons. The first is the way the large moons of gas giant worlds form: through instabilities in a circumplanetary disk that surrounds a massive planet, either from that planet's formation or from debris that arises later from a destroyed satellite. The second method is to gravitationally capture an interloping body: an asteroid, comet, or something else. But the third is perhaps the most common for rock- and/or ice-rich worlds: from an energetic impact. Although the Earth-Theia collision that occurred a few hundred million years earlier was far more energetic, this early Martian impact was sufficiently large that the evidence for its occurrence will persist for billions of years.

In the aftermath of this collision, large volumes of material were kicked up into orbit around Mars, similar to the earlier synestia that formed around Earth. However, while the debris cloud around Earth coalesced into only a large, single satellite, the one around Mars likely gave rise to three moons: a massive, innermost one; a less massive one exterior to that; and then the least massive one of all even farther out. The remainder of the kicked-up debris either fell back down onto Mars or escaped the gravitational pull of this newly formed Martian system.

This configuration, however, wouldn't last all that long. With a massive moon orbiting at such a short distance from Mars, it would experience a small but significant amount of atmospheric drag. Early Mars, just like Earth and Venus, had a substantial atmosphere around it. Unlike popular pictures, where a thin atmosphere all at once gives way to the abyss of empty space, real atmospheres gradually taper off, with no hard boundary at all. Even a small amount of friction, from this large moon passing through the Martian exosphere, could lead to orbital decay after only a few million or tens of millions of years.

As it spirals in towards Mars, the innermost moon breaks apart, getting stretched out into a ring. As the ring decays and rains its debris down onto the planet, it creates an enormous basin on one side of the red planet, where Mars's northern hemisphere is approximately 5 kilometers lower in elevation than the rest of the planet. Meanwhile, the two remaining moons—co-orbiting with Mars and orbiting within 1° of Mars's rotational plane—will persist for billions of years to come: inner Phobos and outer Deimos. This one event, an impact early on in our Solar System's history, will give rise to multiple moons and a permanently scarred Mars. Although oceans and billions of years of subsequent impacts will affect the red planet, they cannot erase the evidence of this ancient collision: an example of our violent past.

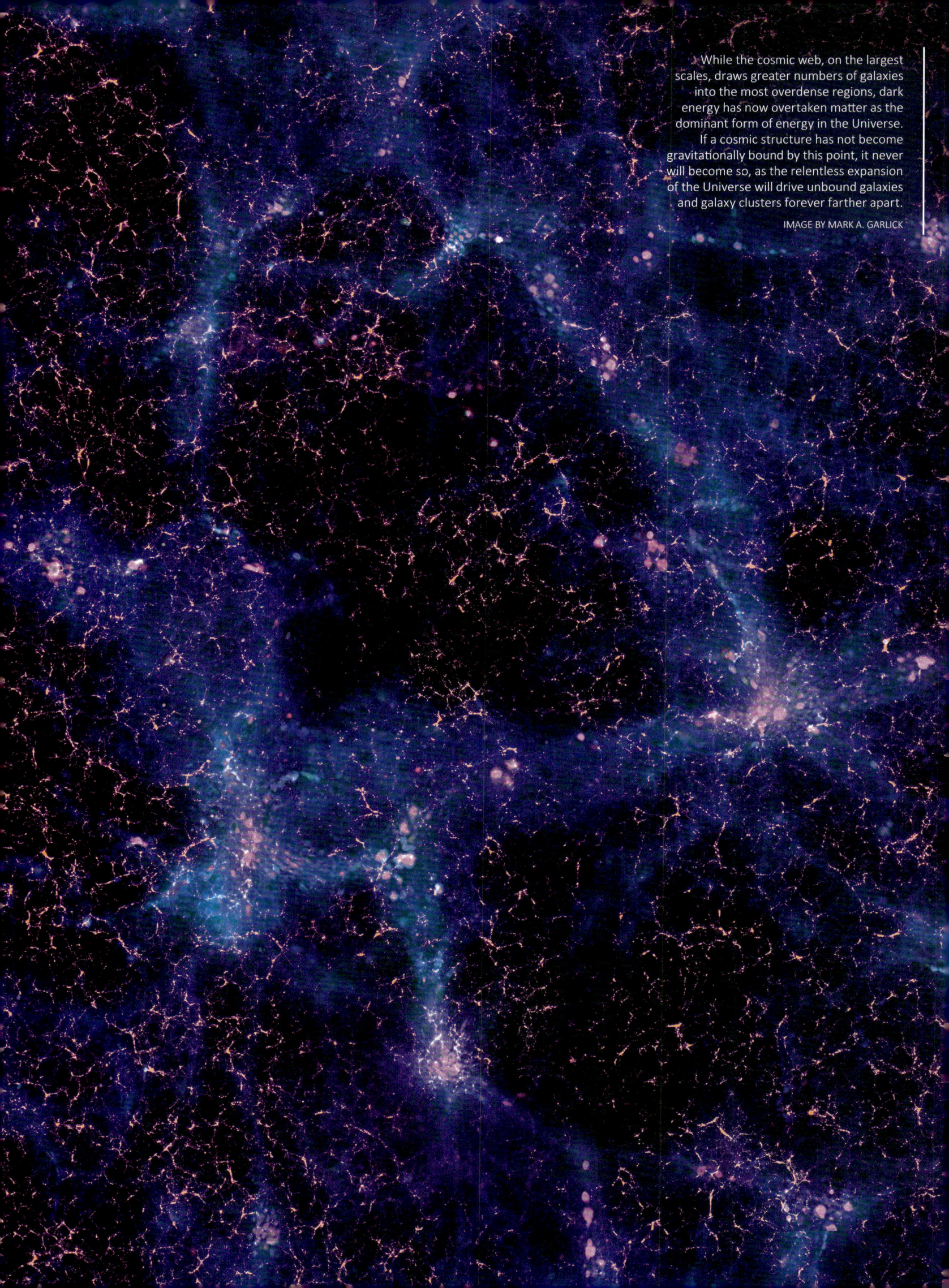

While the cosmic web, on the largest scales, draws greater numbers of galaxies into the most overdense regions, dark energy has now overtaken matter as the dominant form of energy in the Universe. If a cosmic structure has not become gravitationally bound by this point, it never will become so, as the relentless expansion of the Universe will drive unbound galaxies and galaxy clusters forever farther apart.

IMAGE BY MARK A. GARLICK

POINTS OF NO RETURN

dark energy pulls unbound structures apart

UNIVERSE
Diameter: $2.69 \times 10^{13} \rightarrow 2.71 \times 10^{13}$ ly
  Observable: $6.73 \times 10^{10} \rightarrow 6.78 \times 10^{10}$ ly
Expansion: $82.16 \rightarrow 81.63$ km/s/Mpc
Density: $0.63 \rightarrow 0.62$ p/m³
Temperature: $3.74 \rightarrow 3.71$ K

ASTRONOMICAL OBJECTS
Galaxies: $2.52 \times 10^{13} \rightarrow 2.51 \times 10^{13}$
Stars: $1.94 \times 10^{21} \rightarrow 1.95 \times 10^{21}$
Black Holes: $1.56 \times 10^{19} \rightarrow 1.57 \times 10^{19}$
Planetary Systems: $2.62 \times 10^{21} \rightarrow 2.63 \times 10^{21}$
Worlds: $3.00 \times 10^{22} \rightarrow 3.02 \times 10^{22}$

WORLDS WITH LIFE
Simple Life: $5.03 \times 10^{17} - 2.22 \times 10^{19}$
Complex Life: $5.55 \times 10^{11} - 2.86 \times 10^{17}$
Intelligent Life: $1.71 \times 10^{8} - 2.24 \times 10^{15}$
Technological Life: $5.27 \times 10^{4} - 2.80 \times 10^{13}$
Interstellar Life: $2 - 2.92 \times 10^{12}$

he hot Big Bang, looking back from this moment, is now nearly 10 billion years in the rearview mirror. During the earliest stages, only matter, both normal and dark, plus radiation, in the form of photons and neutrinos, had any appreciable consequences for the Universe's evolution. The hot Big Bang began with an initially rapid expansion, an expansion that was almost perfectly counterbalanced by the gravitational influence of matter and radiation. As the Universe aged and continued to expand, the density of the matter and radiation within it plummeted, causing the expansion rate to slow. But once it drops below a certain threshold, a new form of energy—never seen before— begins to influence our cosmic expansion: dark energy. After nearly 10 billion years, dark energy finally becomes the dominant form of energy in the Universe, overtaking dark matter and sealing our cosmic fate.

The struggle between expansion and gravitation goes all the way back to the dawn of the hot Big Bang. In many ways, it's the ultimate cosmic race: between the initial expansion, which drives gravitationally unbound objects apart, and gravitation, which strives to bring objects together and cause a complete cosmic recollapse. Early on, matter and radiation dominated the energy content of the Universe, counteracting the initially rapid expansion and slowing the expansion rate. However, as the Universe expands and gets larger, its volume increases, which leads to both matter and radiation—which are composed of a fixed number of particles—dropping in their energy densities. As the overall density drops below a certain value, it becomes apparent that matter and radiation aren't everything. In addition, there's also a form of energy whose density won't drop as the Universe expands but remains constant. That novel form of energy, behaving as though it were inherent to space itself, is what we know today as dark energy.

In the regions born with slightly more matter than average—the overdense regions of the Universe— gravitation drew more and more matter into them, while the underdense regions similarly gave up their matter to their denser surroundings. The Universe assembled its constituent particles into cosmic structures, first small-scale structures and then larger ones, as the gravitating matter dominated our cosmic story for a time. Gas clouds collapsed, forming clusters of stars, which merged to form galaxies. Small galaxies merged into larger ones, which collected to form galaxy groups and clusters. Eventually, even superclusters began to take shape, but right around the same time, dark energy began to rear its head. Because of dark energy's constant energy density, and the fact that it doesn't drop even as the Universe continues to expand, it keeps the expansion rate large and positive, which drives gravitationally unbound objects farther and farther apart, at faster and faster speeds, as time goes on.

Even though it's now been nearly two billion years since distant, unbound objects began speeding up in their recession from one another, the cosmic struggle continues between matter's gravitational pull and dark energy's repulsive push. Only now, at long last, does dark energy become the dominant component of energy in the Universe, passing dark matter. This seals our fate: All gravitationally unbound structures at this moment will forever remain unbound, doomed to accelerate away from all bound structures. Even superclusters—the largest apparent structures in space—are mere transients, destined to be ripped apart into individual groups and clusters of galaxies as time goes on. Here at home, the Local Group, dominated by Andromeda, the Milky Way, and over 100 smaller galaxies, is as large as it will ever be. Everything beyond it will only recede from us farther and farther, driven inexorably away by dark energy: the arbiter of our ultimate cosmic fate.

While Earth may be the prototypical "blue planet" as seen from space, it's only 0.02% water by mass. By contrast, many worlds, including Pluto, Titan, Callisto, Ganymede, and Europa, are composed of between 9% and 46% water. With no solid landmasses and surfaces covered in ocean and ices, these aqueous environments represent true "water worlds" in the Universe.

IMAGE BY MARK A. GARLICK

NO BLUE, NO GREEN

—— *water is conducive to life* ——

UNIVERSE
Diameter: $2.71 \times 10^{13} \rightarrow 2.73 \times 10^{13}$ ly
  Observable: $6.78 \times 10^{10} \rightarrow 6.84 \times 10^{10}$ ly
Expansion: $81.63 \rightarrow 81.10$ km/s/Mpc
Density: $0.62 \rightarrow 0.60$ p/m³
Temperature: $3.71 \rightarrow 3.68$ K

ASTRONOMICAL OBJECTS
Galaxies: $2.51 \times 10^{13} \rightarrow 2.50 \times 10^{13}$
Stars: $1.95 \times 10^{21} \rightarrow 1.96 \times 10^{21}$
Black Holes: $1.57 \times 10^{19} \rightarrow 1.57 \times 10^{19}$
Planetary Systems: $2.63 \times 10^{21} \rightarrow 2.65 \times 10^{21}$
Worlds: $3.02 \times 10^{22} \rightarrow 3.03 \times 10^{22}$

WORLDS WITH LIFE
Simple Life: $4.97 \times 10^{17} – 2.20 \times 10^{19}$
Complex Life: $5.33 \times 10^{11} – 2.91 \times 10^{17}$
Intelligent Life: $1.65 \times 10^{8} – 2.29 \times 10^{15}$
Technological Life: $5.09 \times 10^{4} – 2.88 \times 10^{13}$
Interstellar Life: $2 – 3 \times 10^{12}$

ust half a billion years after the birth of our Solar System, the major planets and moons within it have largely taken shape. In addition to the four gas giants, a number of planets and moons make compelling candidate locations for biological activity to begin to take place. As comets from the outer Solar System continue to bombard the rocky worlds, progressively larger amounts of water begin to accumulate on their surfaces and in their atmospheres. While Earth, Mars, and even Venus continue to be bombarded by water-rich bodies from the asteroid and Kuiper belts, adding to whatever stores of primeval water they were born with, other worlds possess far greater amounts of water. Many of the outer worlds in the Solar System, including Pluto, as well as Saturn's moon Titan and three of Jupiter's moons—Europa, Ganymede, and Callisto—possess at least a zettaliter, or 10^{21} liters, of water. (Earth, for comparison, contains 1.34 zettaliters of water.) Although liquid water may not be absolutely necessary for life, these wet worlds stand out as life-friendly environments in the young Solar System.

In the inner reaches of the Solar System, the combination of volcanic activity from inside the planets along with the Sun's external heating provides the tantalizing possibility of liquid water on the surfaces of Mars, Earth, and Venus. Along with water and other volatile ices, organic compounds—sugars, amino acids, and other carbon-rich molecules—are constantly being brought to the surface as a result of cosmic impacts. Precipitation gives rise to stores of fresh water when it collects on high-altitude landmasses, while saltwater oceans cover substantial portions of the lowland regions. Whenever an aqueous environment, organic molecules, and an energy gradient occur together, there's the opportunity for complex chemistry, metabolic activity, and molecular reproduction. When this occurs, it's possible that life will arise at some point.

On the more distant worlds, either around a gas giant or off in the far recesses of the Solar System's outskirts, a different set of possibilities arises. While those far-off worlds might not receive nearly the same amount of energy from the Sun as the inner planets do, they have two other factors working in their favor. One factor is particularly important for the water-rich worlds of Jupiter and Saturn: the effects of tidal forces from the large planet these satellites revolve around. Tidal forces arise when a massive body pulls more strongly on the "near side" of an orbiting companion than on the "far side," leading to tidal bulges, internal heating, and seismic activity. For colder, more distant but still water-rich worlds like Europa and Ganymede, this provides an additional heat source that could admit liquid water. In addition, many worlds in the outer Solar System are largely composed of water. Whereas Earth is just ~0.12% water, by volume, worlds like Pluto, Europa, Callisto, Titan, and Ganymede range from 9% to 46%. Even if it's beneath a thick layer of ice or rock, oceans of liquid water are common occurrences.

As time marches forward, the rate of impacts on these young worlds begins to dwindle, setting the stage for a relatively peaceful time in the Solar System's history. With liquid water, energy inputs from a combination of internal and external sources, and a copious helping of organic molecules, the same opportunities that have arisen all throughout the cosmos now arise within our own backyard: chances for life. From these early precursors and opportunities, stories of both biological successes and failures are about to unfold.

Several billion years ago, Pluto and Eris weren't the largest objects in the Kuiper belt. Rather, that honor went to Triton, more than 60% heavier than the next most massive Kuiper belt object. In time, Triton's repeated encounters with Neptune will lead to its capture, at the expense of ejecting nearly all objects within Neptune's primordial lunar system.

IMAGE BY JON LOMBERG

REACHING FOR THE MOON

Neptune captures Triton

UNIVERSE
Diameter: $2.73 \times 10^{13} \rightarrow 2.76 \times 10^{13}$ ly
  Observable: $6.84 \times 10^{10} \rightarrow 6.90 \times 10^{10}$ ly
Expansion: $81.10 \rightarrow 80.52$ km/s/Mpc
Density: $0.60 \rightarrow 0.59$ p/m³
Temperature: $3.68 \rightarrow 3.64$ K

ASTRONOMICAL OBJECTS
Galaxies: $2.50 \times 10^{13} \rightarrow 2.49 \times 10^{13}$
Stars: $1.96 \times 10^{21} \rightarrow 1.97 \times 10^{21}$
Black Holes: $1.57 \times 10^{19} \rightarrow 1.58 \times 10^{19}$
Planetary Systems: $2.65 \times 10^{21} \rightarrow 2.66 \times 10^{21}$
Worlds: $3.03 \times 10^{22} \rightarrow 3.05 \times 10^{22}$

WORLDS WITH LIFE
Simple Life: $4.90 \times 10^{17} - 2.18 \times 10^{19}$
Complex Life: $5.10 \times 10^{11} - 2.96 \times 10^{17}$
Intelligent Life: $1.58 \times 10^{8} - 2.34 \times 10^{15}$
Technological Life: $4.89 \times 10^{4} - 2.96 \times 10^{13}$
Interstellar Life: $2 - 3.16 \times 10^{12}$

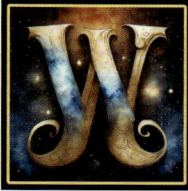

herever a star like our Sun forms, the giant planets that form around it don't just wind up with solid, metal-and-rock cores with a large gas envelope surrounding them, but with a rich system of moons as well. Just as stars are born with their own star-surrounding disks that fragment into planets, asteroids, a Kuiper belt, and more, each giant planet is born with what's known as a circumplanetary disk, which gives rise to a rich lunar system. Over time, these moons experience strong tidal forces, collisions from asteroids and comets, and gravitational interactions from one another. Although the large moons formed from the circumplanetary disks around Jupiter, Saturn, and Uranus will remain intact for many billions of years to come, Neptune isn't so lucky. From the recesses of the Kuiper belt, the largest, most massive object within it—Triton—has a close gravitational encounter with the Neptunian system, becoming captured and ejecting almost all of Neptune's primordial moons in the process.

While Jupiter's gravitational influence is most notable in perturbing the objects in the asteroid belt, frequently sending asteroids throughout both the inner and outer Solar System, Neptune is far less heavy but far closer to the Kuiper belt than any other substantial mass. Its gravitational force sends many Kuiper belt objects into resonances with it—meaning they complete either two or three orbits around the Sun for every three or four, respectively, that Neptune completes—but also influences those orbits so that they become quite eccentric: with large differences between aphelion (their farthest point from the Sun) and perihelion (their closest approach to the Sun). In many cases, such as with Pluto's orbit, the closest approach can even take these massive Kuiper belt objects interior to the orbit of Neptune.

More than three billion years ago, Pluto isn't the largest or most massive object in the Kuiper belt: That honor belongs to the icy world Triton. With a diameter of 2,700 kilometers and a mass of more than 21 quintillion tons, Triton is 14% larger and 64% more massive than Pluto. As Triton orbits the Sun, the gravitational influences of the planets pull it very close to Neptune—so close that it nearly collides with the gas giant. As Triton approaches, it begins to gravitationally interact with not only Neptune but the moons within the Neptunian system as well. As is so often the case when three or more bodies gravitationally interact, the least massive members wind up ejected, with the most massive members winding up more tightly gravitationally bound. Whatever Neptune's original lunar system looks like at this time—however many moons there are, however large and massive they are, however far away from Neptune they extend—Triton ejects almost all of them, removing all but the small, innermost moons (the largest of which is Proteus) that complete a revolution around Neptune in no more than 27 hours.

The evidence that Triton is a captured object, rather than a moon that formed around Neptune initially, will exist for the entire remainder of the Solar System. Triton orbits far out of Neptune's rotational plane and actually orbits its parent planet in the opposite direction from all other major moons. The outermost Neptunian moons all orbit in random directions and much farther out than Triton: consistent with them being icy objects that were also captured, albeit subsequently. Any observer around when the capture occurred would know when it happened and what Neptune's primordial lunar system looked like. Anyone who comes along afterward would only see the survivors.

When did life first arise on Earth? No later than this moment: 3.8 billion years before the present. These ocean-based microbes may have been extremely simple organisms, with the ability to reproduce and perhaps little else, but their remnant signatures can be found in Earth rocks even today. Once life began on Earth, it started an unbroken chain of biological activity that continues to the present day.

IMAGE BY JON LOMBERG

EARTH COMES TO LIFE

—— simple life emerges on Earth ——

UNIVERSE
Diameter: $2.76 \times 10^{13} \rightarrow 2.78 \times 10^{13}$ ly
  Observable: $6.90 \times 10^{10} \rightarrow 6.95 \times 10^{10}$ ly
Expansion: $80.52 \rightarrow 80.01$ km/s/Mpc
Density: $0.59 \rightarrow 0.57$ p/m³
Temperature: $3.64 \rightarrow 3.61$ K

ASTRONOMICAL OBJECTS
Galaxies: $2.49 \times 10^{13} \rightarrow 2.47 \times 10^{13}$
Stars: $1.97 \times 10^{21} \rightarrow 1.98 \times 10^{21}$
Black Holes: $1.58 \times 10^{19} \rightarrow 1.59 \times 10^{19}$
Planetary Systems: $2.66 \times 10^{21} \rightarrow 2.68 \times 10^{21}$
Worlds: $3.05 \times 10^{22} \rightarrow 3.07 \times 10^{22}$

WORLDS WITH LIFE
Simple Life: $4.84 \times 10^{17} - 2.15 \times 10^{19}$
Complex Life: $4.90 \times 10^{11} - 3.01 \times 10^{17}$
Intelligent Life: $1.52 \times 10^{8} - 2.39 \times 10^{15}$
Technological Life: $4.71 \times 10^{4} - 3.03 \times 10^{13}$
Interstellar Life: $2 - 3.28 \times 10^{12}$

 undreds of millions of years after its formation, early Earth is a violent place. Volcanic eruptions and earthquakes are very common, as the interior of Earth is far hotter in its early days. It rotates far more rapidly on its axis and experiences much larger tides, as the Moon is much closer. Copious major impacts continue to occur from asteroids and comets, bringing a wide variety of organic molecules to Earth's surface. And in one warm, stable pool of fresh water atop one of Earth's landmasses, a complex molecule—one capable of extracting energy by metabolizing its nutrient-rich surroundings—gains the ability to reproduce. Unlike all the similar molecules that came before it, this one isn't wiped out by changes to its environment but manages to both persist and spread to other locations. At last, some 10 billion years after the hot Big Bang, life that will sustain itself, generation after generation, has taken hold on planet Earth.

The ingredients for life's beginnings came from all across our cosmic history. Hydrogen atoms, the most numerous atoms in any living organism, were forged in the Big Bang. Heavier elements like carbon, nitrogen, oxygen, calcium, and iron were created in stars, stellar cataclysms, and from the collisions of stellar remnants. With each new generation of stars that formed from the ashes of the old, new molecules and chemical compounds were created as well. By the time our Solar System was formed, 1–2% of the elements that existed were heavier than hydrogen or helium, and the objects arising from the asteroid and Kuiper belts—the ones that continuously bombard the young Earth—contain a variety of organic molecules inside, including a wide variety of amino acids and nucleobases.

In the aqueous environments present on Earth, these amino acids run into one another over and over again, producing new molecular combinations that are far too numerous to count. These amino acid chains—peptides, to a scientist—can interact with ions in the water, where the combination of a peptide chain with an ion can produce an enzyme. The enzymes that form occasionally possess just the right configuration to interact with and extract energy from the other molecules in their vicinity: the first steps towards a metabolism. In the aqueous, polarizing environment provided by liquid water, these molecules fold in various configurations, some of which make it very easy for nucleotides—nucleobases, sugars, and phosphate groups all linked together—to align with the amino acids composing these molecules.

This leads to chains of nucleotides forming: the precursor molecules to what would evolve into RNA and DNA, which encode protein and enzyme structures for the successful, metabolically active molecules created by random chance. The early stages of life's emergence involve proteins and enzymes forming by random chance, attracting nucleotides and nucleotide chains to them, those chains peeling off and attracting amino acids to produce new copies of those proteins and enzymes, and so on. This combination—of metabolism plus reproduction in an aqueous, nutrient-rich environment—is the leading idea for how life first emerged not only on planet Earth but possibly in many other locations all across the Universe.

While many factors—changes in pH, temperature, salinity, or a much bigger catastrophe—can bring an end to this primitive, fragile, early form of life, once life makes enough copies of itself and spreads to a wide set of environments, it becomes very tenacious. Some 10 billion years after the hot Big Bang, Earth becomes a living planet: a status it has retained ever since.

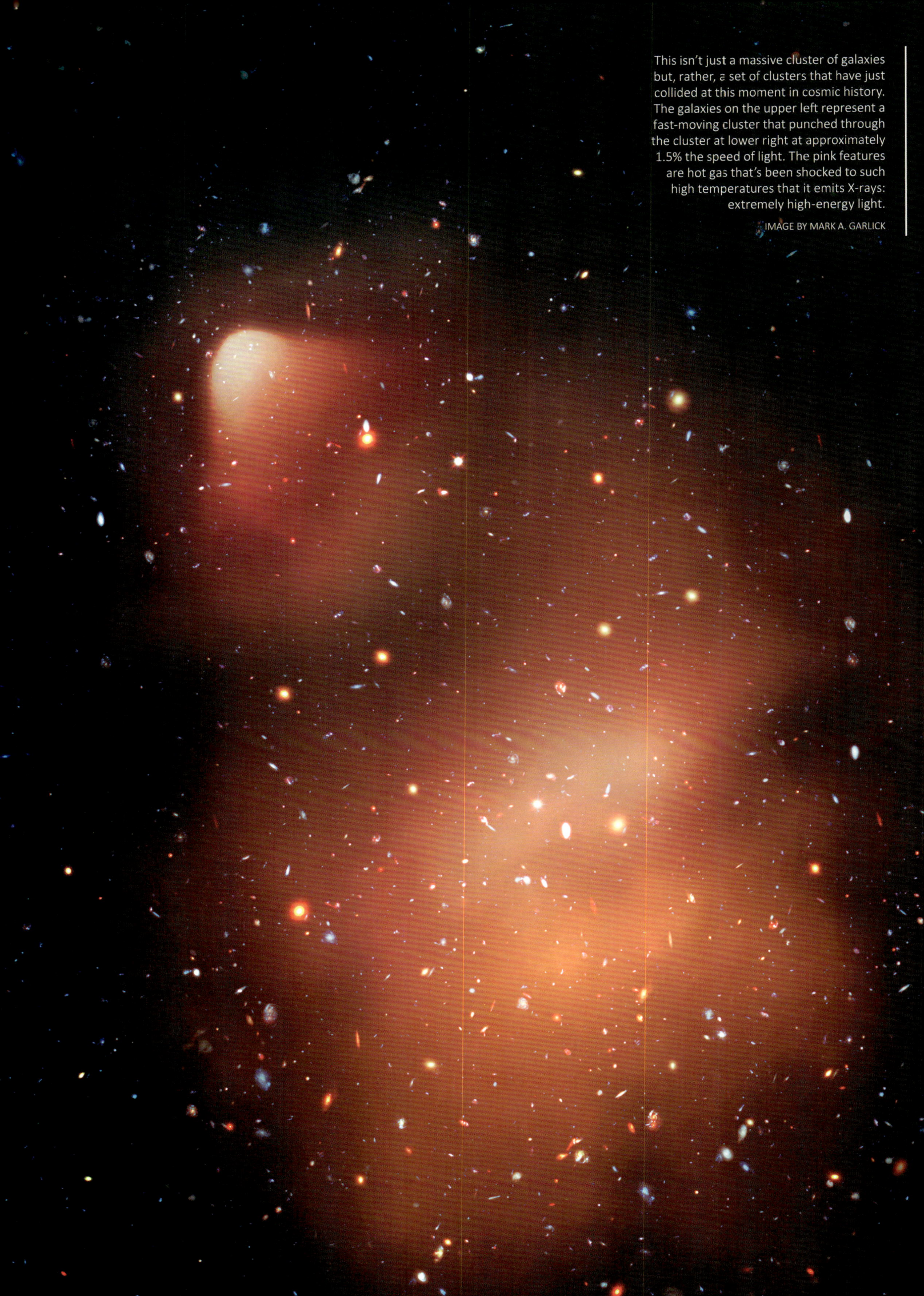

This isn't just a massive cluster of galaxies but, rather, a set of clusters that have just collided at this moment in cosmic history. The galaxies on the upper left represent a fast-moving cluster that punched through the cluster at lower right at approximately 1.5% the speed of light. The pink features are hot gas that's been shocked to such high temperatures that it emits X-rays: extremely high-energy light.

IMAGE BY MARK A. GARLICK

THE FORMS OF THINGS UNKNOWN

—— *a collision of clusters hints at dark matter* ——

UNIVERSE
Diameter: $2.78 \times 10^{13} \rightarrow 2.80 \times 10^{13}$ ly
  Observable: $6.95 \times 10^{10} \rightarrow 7.01 \times 10^{10}$ ly
Expansion: $80.01 \rightarrow 79.52$ km/s/Mpc
Density: $0.57 \rightarrow 0.56$ p/m³
Temperature: $3.61 \rightarrow 3.59$ K

ASTRONOMICAL OBJECTS
Galaxies: $2.47 \times 10^{13} \rightarrow 2.46 \times 10^{13}$
Stars: $1.98 \times 10^{21} \rightarrow 1.99 \times 10^{21}$
Black Holes: $1.59 \times 10^{19} \rightarrow 1.59 \times 10^{19}$
Planetary Systems: $2.68 \times 10^{21} \rightarrow 2.69 \times 10^{21}$
Worlds: $3.07 \times 10^{22} \rightarrow 3.08 \times 10^{22}$

WORLDS WITH LIFE
Simple Life: $4.77 \times 10^{17} - 2.13 \times 10^{19}$
Complex Life: $4.71 \times 10^{11} - 3.06 \times 10^{17}$
Intelligent Life: $1.46 \times 10^{8} - 2.44 \times 10^{15}$
Technological Life: $4.54 \times 10^{4} - 3.10 \times 10^{13}$
Interstellar Life: $2 - 3.39 \times 10^{12}$

 hile Earth's earliest life-forms begin colonizing its nutrient-rich waters, a very different event on a much larger scale is unfolding some 3.7 billion light-years away. Two massive galaxy clusters— whose combined mass is hundreds of times as great as the Milky Way—have just crashed into one another after billions of years of gravitationally drawing one another in. Colliding at speeds of 4,500 kilometers per second, about 1.5% the speed of light, these galaxy clusters smash into one another with such energy that the gas inside heats up to temperatures reaching 100 million K: far hotter than the temperatures found inside the cores of Sun-like stars.

But these clusters contain a lot more than gas; they also contain individual stars, galaxies, and copious amounts of dark matter. While the gas slows down and heats up, as the kinetic energy of the atoms and molecules inside gets converted into heat, the stars, galaxies, and dark matter mostly just pass through one another. When the light from this collision reaches Earth some 3.7 billion years later, it will provide some of the strongest, most incontrovertible evidence for dark matter's existence: the first Bullet Cluster ever discovered.

Although atoms exist all throughout the Universe, the gravitational effects on large cosmic scales—the scales of galaxies, galaxy clusters, and even the large-scale cosmic web—indicate that the amount of normal, atom-based matter alone is insufficient to explain what's observed. In principle, it could be because the laws of gravity are different than we understand them to be, but it could also be because there's a new, unseen form of matter that plays an important role on these cosmic scales. Colliding galaxy clusters are the perfect test to determine which idea is a better descriptor of the Universe.

If galaxy clusters contain only normal, atom-based matter, then there are only the stars, gas, dust, and plasmas found within each member galaxy, as well as the atomic material found in the intracluster medium: the space between the galaxies within each cluster. When the two clusters collide and the gas within them heats up, that hot gas then emits X-rays, allowing an observer to determine how much material there is, in terms of atom-based matter, in the intracluster medium. The galaxies themselves, however, behave like two handfuls of pebbles that are thrown at one another from across a yard: While two individual pebbles may randomly collide, most of the pebbles will miss one another, meaning that the galaxies will appear to pass straight through.

The overwhelming majority of the atoms present are found within the X-ray emitting gas, not the individual galaxies. But when anyone looks at this colliding set of galaxy clusters from afar, they can measure the effects of their gravity by determining how much they bend and distort the fabric of space: through the effects of gravitational lensing. If there were only atom-based matter, the strongest distortion would come from where the gas is: in the space between the two galaxy clusters. But that's not what's actually seen. Instead, the overwhelming majority of the mass is found not only with but slightly ahead of the galaxies themselves: as though there were an invisible form of mass that didn't slow down at all when the two clusters collided. That extra gravitational effect indicates that there was an extra type of matter that didn't heat up and slow down—what we'll come to know as dark matter—far in excess of any atoms. When that light arrives at Earth, billions of years in the future, it will settle the issue, proving dark matter's existence once and for all.

For a billion years or more in the early stages of the Solar System, Earth was joined by Mars as a water-rich planet containing continents, oceans, and perhaps life. It remains possible that life arose on Mars independently of life on Earth, but it's also possible that life began on one of those worlds and, through a violent impact, was transported to the other, where it then took hold. Ancient Earthlings and Martians may share a common origin.

IMAGE BY MARK A. GARLICK

STRANGERS IN A STRANGE LAND

—— impacts can spread life between planets ——

UNIVERSE
Diameter: $2.80 \times 10^{13} \rightarrow 2.83 \times 10^{13}$ ly
 Observable: $7.01 \times 10^{10} \rightarrow 7.07 \times 10^{10}$ ly
Expansion: $79.52 \rightarrow 79.03$ km/s/Mpc
Density: $0.56 \rightarrow 0.54$ p/m³
Temperature: $3.59 \rightarrow 3.56$ K

ASTRONOMICAL OBJECTS
Galaxies: $2.46 \times 10^{13} \rightarrow 2.45 \times 10^{13}$
Stars: $1.99 \times 10^{21} \rightarrow 2.00 \times 10^{21}$
Black Holes: $1.59 \times 10^{19} \rightarrow 1.60 \times 10^{19}$
Planetary Systems: $2.69 \times 10^{21} \rightarrow 2.70 \times 10^{21}$
Worlds: $3.08 \times 10^{22} \rightarrow 3.10 \times 10^{22}$

WORLDS WITH LIFE
Simple Life: $4.71 \times 10^{17} - 2.11 \times 10^{19}$
Complex Life: $4.52 \times 10^{11} - 3.11 \times 10^{17}$
Intelligent Life: $1.40 \times 10^{8} - 2.48 \times 10^{15}$
Technological Life: $4.37 \times 10^{4} - 3.17 \times 10^{13}$
Interstellar Life: $2 - 3.51 \times 10^{12}$

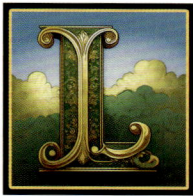

L iquid water and organic molecules, wherever they are found in combination, provide that strange, rare, and exciting opportunity for life to arise from nonlife. In many locations, something that can be described as "alive" might never come into existence; in others, the fragile living organisms that do arise quickly go extinct. However, wherever life arises, survives, thrives, and begins to diversify and occupy a variety of ecological niches, it's likely to develop a certain level of hardiness and resilience. Already on Earth, that's been the story for hundreds of millions of years, while elsewhere in the Solar System—on Venus, Mars, Europa, Titan, Ganymede, Triton, and more—it's uncertain whether life ever arose. However, one thing is definite: On all of these worlds, Earth included, significant impacts regularly occur. In many cases, material from the surface, including many of the simple forms of life present on any such world, is not only kicked up but escapes from the home world's gravity, spreading throughout the remainder of the Solar System and beyond.

Here on Earth, a small fraction of all the meteorites that have ever been found come from these other worlds, with meteorites of a Martian origin—perhaps due to a combination of its close proximity and relatively small mass—being the most common. There's no doubt that on Mars, as well as on other rock- and ice-rich worlds in the Solar System, material from Earth has traveled across the interplanetary distances to land there, delivering not just crustal rock but also any organic material present there as well. It's easy to imagine what happens in most cases: The organic material is killed before departing its home world, or burns up upon atmospheric reentry, or dies upon landing, or simply finds itself in a completely inhospitable environment that it cannot survive in.

But in this version of a planetary lottery, all it takes is one single success—one single organism—to withstand the impact that kicked it off its home world, make the interplanetary journey to a new home, and arrive in an environment where it finds conditions in which it is capable of survival. While all the water within an organism is destined to freeze solid in the vacuum of space, many such frozen organisms can then return to a life-friendly environment, thaw out, and continue their typical biological processes without suffering any ill effects. When Earth-based life comes to Mars, its biological seeds can take root in Martian waters. In this new environment, with no competitors, it's free to consume resources, reproduce and make copies of itself, and eventually give rise to a wholly new tree of life.

Even in the early stages of the Solar System, life's transfer from one inhabited world to another world that can support its life processes raises a tremendous set of possibilities. Perhaps there were, or even still are, multiple inhabited worlds seeded by the first among them to develop life. Perhaps many instances of life on multiple planets within a stellar system have a common origin and a common ancestor, even though they're now found on wholly separate worlds. Quite possibly, there was an Earthly origin for the life-forms that once inhabited Mars—and maybe even elsewhere—during the period when both worlds possessed abundant liquid water on their surfaces. And just maybe, we've gotten this part of the story completely backwards: The life that exists on Earth didn't originate here but, rather, began on Mars or a different world entirely. Once life takes hold on a planet, its seeds will inevitably spread elsewhere, throughout its home system and beyond. Wherever they land, if the conditions are favorable, there's a chance for that chain of life to continue anew, even on otherwise uninhabited worlds.

Now that Earth's moon has cooled and taken on a familiar appearance during a relatively quiet volcanic period, one might be surprised to see this view of Earth, with barely any oceans at all. Yet Earth, owing to its moving plates and active geology, frequently undergoes phases with supercontinents and superoceans. Here, the first known supercontinent on Earth, Vaalbara, covers nearly 40% of the planet's surface, creating this unfamiliar sight.

IMAGE BY MARK A. GARLICK

BEHOLD, VAALBARA

—— *the first supercontinent forms on Earth* ——

UNIVERSE
Diameter: $2.83 \times 10^{13} \rightarrow 2.85 \times 10^{13}$ ly
  Observable: $7.07 \times 10^{10} \rightarrow 7.12 \times 10^{10}$ ly
Expansion: $79.03 \rightarrow 78.54$ km/s/Mpc
Density: $0.54 \rightarrow 0.53$ p/m³
Temperature: $3.56 \rightarrow 3.53$ K

ASTRONOMICAL OBJECTS
Galaxies: $2.45 \times 10^{13} \rightarrow 2.42 \times 10^{13}$
Stars: $2.00 \times 10^{21} \rightarrow 2.00 \times 10^{21}$
Black Holes: $1.60 \times 10^{19} \rightarrow 1.61 \times 10^{19}$
Planetary Systems: $2.70 \times 10^{21} \rightarrow 2.71 \times 10^{21}$
Worlds: $3.10 \times 10^{22} \rightarrow 3.11 \times 10^{22}$

WORLDS WITH LIFE
Simple Life: $4.65 \times 10^{17} - 2.08 \times 10^{19}$
Complex Life: $4.33 \times 10^{11} - 3.15 \times 10^{17}$
Intelligent Life: $1.35 \times 10^{8} - 2.53 \times 10^{15}$
Technological Life: $4.21 \times 10^{4} - 3.24 \times 10^{13}$
Interstellar Life: $2 - 3.63 \times 10^{12}$

For as long as planet Earth has been around, its outermost solid layer—the crust—has been an evolving, changing place. From deep within Earth, heat is generated as the planet gravitationally cools and as radioactive elements within it decay. While the surface of Earth might be cool enough to be friendly to life, one only needs to go down a relatively short distance to feel the heat within our planet's interior. Descending just 1.5 kilometers into the crust, you'd find temperatures that are warmer than that at Earth's surface: 25°C (77°F). Descending to the base of the crust, where the mantle begins, temperatures reach ~1000°C (1800°F). As the lithosphere, which includes the crust and uppermost layer of the mantle, slides over the next layer down, the aesthenosphere, magma pockets form, which can erupt through the crust in oceanic ridges and continental volcanoes. The forces of these tectonic plates, along with their effects on the crust, cause continents to form, drift, separate, and collide. For the first time, a large set of continental landmasses collides and fuses, forming Earth's first supercontinent: Vaalbara.

With multiple tectonic plates rather than just one—a rarity, at least within our Solar System—Earth's surface geology never remains constant. Where plates converge, collide, and run into one another, mountains form. Just as the atmosphere "floats" atop the denser oceans and the oceans float atop the crust, Earth's crust floats atop the denser mantle, which floats atop the other, more interior layers. The crust behaves similar to how large icebergs poke out of Earth's oceans but the majority of their mass is submerged beneath the waters. At the deepest ocean depths, the crust is thinnest: less than ~5 kilometers. But underneath the highest mountains, the crust can attain thicknesses of up to ~70 kilometers, all "floating" atop the mantle.

Not all colliding plates lead to the formation of mountains, however. Sometimes, one plate slides beneath another, leading to the phenomenon of subduction: where the lower plate gets driven back into Earth's interior. While mountain formation is common over the thicker continental crust, subduction is much more common when thin, oceanic plates are involved. If a plate contains both oceanic and continental portions, when the oceanic part subducts, the thick-crusted continental component can follow, recycling them into the mantle.

However, there are also locations where plates diverge, forming and expanding the oceans while also causing the phenomenon known as continental drift. Although the motion of continents across Earth's surface is small on annual timescales—the relative distances between continents change by only 2.5 centimeters (1 inch) per year—the locations of Earth's landmasses change significantly over timescales of tens to hundreds of millions of years. Earth regularly forms new landmasses as older ones erode and occasionally subduct. Globally, Earth also goes through periods when its surface consists of many irregular, well-separated continents, as well as periods when all the landmasses collide to form a giant one: a supercontinent.

The first supercontinent ever identified on Earth—the oldest known single large landmass to form—came into existence no later than 10.3 billion years after the Big Bang: when planet Earth was merely one billion years old. The evidence from crater impacts, microfossils, and greenstone belts found across modern-day Africa, Australia, and North America all point to this ancient era as a time when a single supercontinent reigned supreme. Persisting for anywhere from 400 million to one billion years, Vaalbara's legacy will live on for as long as Earth continues to exist.

HARVESTERS OF LIGHT

—— life evolves to feed on sunlight ——

UNIVERSE
Diameter: $2.85 \times 10^{13} \rightarrow 2.87 \times 10^{13}$ ly
  Observable: $7.12 \times 10^{10} \rightarrow 7.18 \times 10^{10}$ ly
Expansion: $78.54 \rightarrow 78.07$ km/s/Mpc
Density: $0.53 \rightarrow 0.52$ p/m³
Temperature: $3.53 \rightarrow 3.50$ K

ASTRONOMICAL OBJECTS
Galaxies: $2.42 \times 10^{13} \rightarrow 2.41 \times 10^{13}$
Stars: $2.00 \times 10^{21} \rightarrow 2.01 \times 10^{21}$
Black Holes: $1.61 \times 10^{19} \rightarrow 1.61 \times 10^{19}$
Planetary Systems: $2.71 \times 10^{21} \rightarrow 2.72 \times 10^{21}$
Worlds: $3.11 \times 10^{22} \rightarrow 3.12 \times 10^{22}$

WORLDS WITH LIFE
Simple Life: $4.59 \times 10^{17} - 2.06 \times 10^{19}$
Complex Life: $4.16 \times 10^{11} - 3.19 \times 10^{17}$
Intelligent Life: $1.30 \times 10^{8} - 2.57 \times 10^{15}$
Technological Life: $4.04 \times 10^{4} - 3.31 \times 10^{13}$
Interstellar Life: $2 - 3.75 \times 10^{12}$

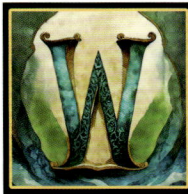

When life first took hold on Earth, several hundreds of millions of years before the current epoch, the entire planet was a nutrient-rich buffet. Rich in sugars, amino acids, and a wide variety of elements, ions, and chemical compounds, the earliest organisms had absolutely no competition for these resources. But as life survived and thrived, the various inhabitable ecological niches became occupied, and eventually ecosystems became saturated with living organisms. Suddenly, there weren't enough resources for all of Earth's living organisms to simultaneously thrive, and competition necessarily ensued. Unicellular (or single-celled) organisms that possessed membranes—conferring the ability to gather, store, and preserve resources—gained an early advantage, but nutrient resources were still finite.

However, if an organism could take advantage of a previously untapped source of energy, it could have a unique edge in survival, as it harnesses resources that no competitor can use. While unicellular bacteria were the dominant form of life at this time, they rapidly evolutionarily diverged. Archaea arose, capable of surviving in the deep ocean waters around hydrothermal vents. Plasmids, which carry genes responsible for novel abilities, arose as independent DNA molecules, unattached to the bacterial chromosome itself. And just over one billion years after the formation of Earth, the first fully photosynthetic organisms come to be. At last, life is feeding off of a virtually limitless source of energy—the Sun—powering its own life processes and generating nutrients that could serve as food for other organisms.

Photosynthesis occurs on Earth via a number of different pathways, but they all have some commonalities. Sunlight, which itself is light composed of a wide variety of different wavelengths from the ultraviolet through the visible and well into the infrared, is the driving force behind all photosynthesis. When light of a particular wavelength strikes a molecule that can absorb it, the molecule enters an excited state where one of the electrons within it gets "kicked up" to a higher energy level. That excited molecule can then have its energy extracted by the organism: either to be used directly in life processes or to create a molecule where that energy can be stored in chemical bonds until it's needed.

So long as an electron is properly excited within a molecule, there are a variety of pathways to photosynthetic success. Green and purple bacteria evolved to use hydrogen, sulfur, and a variety of acids to provide electrons for their reactions. These classes of organism use two identical proteins bound together, known as a homodimer, to absorb sunlight and initiate photosynthesis. Shortly thereafter, and perhaps completely independently, distinct organisms leverage other pathways and use two different proteins bound together—a heterodimer—with water as the electron source for photosynthesis. This latter path leads to a unique biochemical waste product that no organism had ever previously created: molecular oxygen.

At this time, the atmosphere largely consists of nitrogen, water vapor, and carbon dioxide. This epoch marks not only the first appearance of photosynthesis on Earth but also the first appearance of oxygen in Earth's atmosphere. Over the next few hundred million years, cyanobacteria—better known as blue-green algae and possibly the first organisms to utilize the molecule chlorophyll—will come into existence: leveraging the heterodimer pathway to become the only surviving class of oxygen-producing photosynthetic prokaryotes. Arriving in the aftermath of photosynthesis, these atmospheric changes will forever alter the trajectory of life on Earth.

This water-rich world from the early Solar System, surprisingly, is the planet Venus, which had a water-rich past, likely possessing oceans for hundreds of millions or even a billion years or more. However, as the oceans boil away on Venus, they increase not only the water vapor content of the atmosphere but also cloud cover, intensifying and accelerating the greenhouse effect on this doomed planet.

IMAGE BY MARK A. GARLICK

VENUSIAN GREENHOUSE

—— *the hottest planet in the Solar System* ——

UNIVERSE
Diameter: $2.87 \times 10^{13} \rightarrow 2.89 \times 10^{13}$ ly
  Observable: $7.18 \times 10^{10} \rightarrow 7.24 \times 10^{10}$ ly
Expansion: $78.07 \rightarrow 77.61$ km/s/Mpc
Density: $0.52 \rightarrow 0.51$ p/m³
Temperature: $3.50 \rightarrow 3.47$ K

ASTRONOMICAL OBJECTS
Galaxies: $2.41 \times 10^{13} \rightarrow 2.40 \times 10^{13}$
Stars: $2.01 \times 10^{21} \rightarrow 2.02 \times 10^{21}$
Black Holes: $1.61 \times 10^{19} \rightarrow 1.62 \times 10^{19}$
Planetary Systems: $2.72 \times 10^{21} \rightarrow 2.73 \times 10^{21}$
Worlds: $3.12 \times 10^{22} \rightarrow 3.13 \times 10^{22}$

WORLDS WITH LIFE
Simple Life: $4.53 \times 10^{17} - 2.04 \times 10^{19}$
Complex Life: $3.97 \times 10^{11} - 3.24 \times 10^{17}$
Intelligent Life: $1.25 \times 10^8 - 2.62 \times 10^{15}$
Technological Life: $3.89 \times 10^4 - 3.38 \times 10^{13}$
Interstellar Life: $2 - 3.86 \times 10^{12}$

In the early stages of the Solar System, three of our planets—Venus, Earth, and Mars—may all have possessed similar conditions. With thin atmospheres, chemically enriched surfaces, and copious amounts of water, all of them may have offered the potential for life. Despite its much closer proximity to the Sun than Earth and Mars, Venus might still have had liquid water oceans, as all Sun-like stars are less luminous in their early days than they are later on, once a substantial amount of their core's hydrogen fuses into helium. However, as the parent star brightens and becomes more energetic, planets and worlds that were once habitable can cross a critical threshold where water vapor in the atmosphere can no longer "rain down" to create liquid water on the surface. As the surface water evaporates and boils, converting a onetime liquid into the gaseous phase, any biological activity taking place within them grinds to a halt. Right around this moment in cosmic history, Venus's oceans are entering the gaseous phase for the final time.

In the case of Venus, it isn't the Sun that's entirely to blame; it's the combination of the Sun's increasing heat and also how Venus's atmosphere evolves. As volcanic activity occurs on Venus, large amounts of compounds such as water, carbon dioxide, and sulfur dioxide are added to the atmosphere, where the latter two are particularly heavy gases. While the Sun delivers energy to Venus mostly in the visible part of the light spectrum, Venus re-radiates that energy primarily in the infrared portion of the spectrum. Water can combine with sulfur dioxide and carbon dioxide to make sulfuric and carbonic acids: compounds that can condense and form heat-trapping clouds just like water vapor. Meanwhile, carbon dioxide and water

are effective at absorbing and re-radiating infrared light, while simultaneously being transparent to visible light. These heat-trapping effects combine to work as an ever-thickening blanket: heating the surface of Venus to much greater temperatures than it could achieve in the absence of cloud cover or these greenhouse gases.

It could have taken over a billion years from the formation of Venus to when its oceans began to boil away, but that process only makes the situation worse. As the oceans begin to disappear, the water vapor content in Venus's atmosphere rises, causing what's traditionally known as "the greenhouse effect" to play an even more severe role. As a result, Venus's temperatures rise, increasing the rate at which the oceans disappear and leading to an even greater concentration of these heat-trapping gases and clouds, creating a runaway greenhouse effect. In short order, the oceans fully disappear, leaving the high-altitude clouds as the possible last refuge for life on what quickly becomes the Solar System's hottest world: hotter than even airless Mercury, which receives nearly four times as much radiation from the Sun.

Over time, the Sun will ionize most of the water and strip away the hydrogen and oxygen from Venus, leaving only a barren hellscape behind. One of the last remaining clues to Venus's water-rich past will come from looking at the ratio of deuterium (or heavy hydrogen, with a proton and a neutron in its nucleus) to regular hydrogen (with one proton in its nucleus). With double the mass of its more common isotope, deuterium is more difficult to strip away from Venus's atmosphere than hydrogen, so seeing an enhancement in this ratio could point to a water-rich past. Billions of years later, this ratio will be ~100 times greater than it is on Earth, allowing future inhabitants to reconstruct Venus's ancient history.

Mars, with oceans, continents, a thick atmosphere, and possibly life, looks like a thriving planet in this scene; however, these conditions are fleeting. With its small size and low mass, Mars's outer core has ceased to convect, bringing an end to its protective magnetic dynamo. Without it, solar wind and radiation begin stripping away the Martian atmosphere, bringing a gradual end to this life-friendly stage.

IMAGE BY MARK A. GARLICK

THE CLOVEN SHIELD OF MARS

—— *Mars loses its magnetic field* ——

UNIVERSE
Diameter: $2.89 \times 10^{13} \rightarrow 2.92 \times 10^{13}$ ly
  Observable: $7.24 \times 10^{10} \rightarrow 7.30 \times 10^{10}$ ly
Expansion: $77.61 \rightarrow 77.15$ km/s/Mpc
Density: $0.51 \rightarrow 0.50$ p/m³
Temperature: $3.47 \rightarrow 3.44$ K

ASTRONOMICAL OBJECTS
Galaxies: $2.40 \times 10^{13} \rightarrow 2.39 \times 10^{13}$
Stars: $2.02 \times 10^{21} \rightarrow 2.03 \times 10^{21}$
Black Holes: $1.62 \times 10^{19} \rightarrow 1.62 \times 10^{19}$
Planetary Systems: $2.73 \times 10^{21} \rightarrow 2.74 \times 10^{21}$
Worlds: $3.13 \times 10^{22} \rightarrow 3.15 \times 10^{22}$

WORLDS WITH LIFE
Simple Life: $4.46 \times 10^{17} - 2.01 \times 10^{19}$
Complex Life: $3.82 \times 10^{11} - 3.28 \times 10^{17}$
Intelligent Life: $1.20 \times 10^{8} - 2.66 \times 10^{15}$
Technological Life: $3.74 \times 10^{4} - 3.45 \times 10^{13}$
Interstellar Life: $1 - 3.98 \times 10^{12}$

owever rocky planets form, the end result is always a differentiated sphere: where a core of denser elements is surrounded by a less dense mantle and an even less dense crust, with oceans and atmosphere buoyantly floating atop that. In the center, a solid core composed largely of heavy elements like iron, nickel, cobalt, and even heavier atoms finds itself surrounded by a liquid metallic outer core, which sloshes around with an ever-changing molecular structure. This flowing, metallic layer can generate planet-wide magnetic fields through a mechanism called a dynamo: the cause of magnetic fields on many worlds, including planet Earth. For the first part of Mars's history, it had a very similar dynamo to Earth and a similarly protective magnetic field. But now, about 1.3 billion years after the formation of Mars, its magnetic dynamo goes silent. This event, although it was inevitable for Earth's celestial neighbor, signals the beginning of the end for the habitability of Mars.

On the surface of planets, certain minerals—like lodestone—are found in a naturally magnetized state. These minerals, rich in magnetizable elements like iron or cobalt, were once hot, allowing their interior components to align with the ambient magnetic field of their environment. They then cooled, "freezing in" that magnetization permanently. This creates ferromagnets, or permanent magnets, that will retain their magnetization even after the underlying field changes or goes away.

On Earth, we can determine what the direction and strength of our planet's magnetic field was in the past by uncovering magnetized rocks of various ages and reconstructing our geomagnetic history. Throughout the history of Earth, our planet's outer core has always generated a relatively strong magnetic field, and the process is as follows. The outer core—a liquid layer between the solid inner core and solid mantle—has a large amount of iron and cobalt in it: magnetizable materials. There's also a temperature gradient throughout this liquid layer, as the interface with the inner core is thousands of degrees hotter than the interface with the mantle. As a result, convection occurs, where the hotter, less dense liquid rises, then cools, becomes denser, and falls back down. As long as there's a metallic, liquid layer convecting inside the planet, it continuously generates a planet-wide magnetic field.

Planets also come with all sorts of rotational speeds, with Earth and Mars spinning about their axes hundreds of times per orbit, but with Venus and Mercury barely spinning at all. Since the dawn of the Solar System, this rapid rotation combined with a warm, liquid metallic layer deep inside Earth and Mars have kept their magnetic dynamos active, leading to a planet-wide magnetic field that serves many purposes, including shielding the atmospheres of both worlds from the ionized particles emanating from the Sun—the solar wind. Without these magnetic fields to deflect those particles away, the atmosphere would eventually be depleted.

However, unlike Earth, Mars is tiny: half of Earth's diameter but only 11% of Earth's mass. Due to this relative smallness, the interior of Mars cools down far more quickly than Earth's interior, allowing its core to stratify. Rather than remaining a mix of elements and compounds that could convect and create a magnetic dynamo, the densest materials sink: joining the solid, inner core. As the planet cools and the inner core grows, the remnant outer core ceases to convect in the same way, dooming Mars's magnetic processes. Although the already-magnetized minerals will persist throughout the remainder of Mars's lifetime, they're insufficient, without a core dynamo, to protect Mars from the solar wind. Now, the atmosphere is destined to be stripped away, and its best chances for complex, advanced life will become stripped along with it.

In the great gravitational dance of galaxies, one of the Milky Way's neighboring satellites, the Sagittarius Dwarf Elliptical Galaxy, passed through the plane of the Milky Way three billion years prior, creating a gargantuan stellar stream. Under the relentless pull of gravity, it now begins approaching the Milky Way once again, doomed to eventually be absorbed by its larger, cannibalistic neighbor.

IMAGE BY JON LOMBERG

THE DANSE MACABRE II

—— *the Milky Way merges with SagDEG* ——

UNIVERSE
Diameter: 2.92 × 10¹³ → 2.94 × 10¹³ ly
  Observable: 7.30 × 10¹⁰ → 7.35 × 10¹⁰ ly
Expansion: 77.15 → 76.71 km/s/Mpc
Density: 0.50 → 0.48 p/m³
Temperature: 3.44 → 3.42 K

ASTRONOMICAL OBJECTS
Galaxies: 2.39 × 10¹³ → 2.37 × 10¹³
Stars: 2.03 × 10²¹ → 2.03 × 10²¹
Black Holes: 1.62 × 10¹⁹ → 1.63 × 10¹⁹
Planetary Systems: 2.74 × 10²¹ → 2.75 × 10²¹
Worlds: 3.15 × 10²² → 3.16 × 10²²

WORLDS WITH LIFE
Simple Life: 4.40 × 10¹⁷ – 1.99 × 10¹⁹
Complex Life: 3.67 × 10¹¹ – 3.32 × 10¹⁷
Intelligent Life: 1.15 × 10⁸ – 2.70 × 10¹⁵
Technological Life: 3.59 × 10⁴ – 3.51 × 10¹³
Interstellar Life: 1 – 4.10 × 10¹²

Every large galaxy that exists in the late-time Universe, many billions of years after the hot Big Bang took place, is the end result of all the gravitational growth that occurred over its history. A combination of steady accretion from the intergalactic medium and mergers with smaller galaxies gives rise to the modern, large galaxies visible at present. However, mergers aren't an all-at-once event but can take several billion years to complete, often participating in a complex and drawn-out gravitational and star-forming dance in the process. Just under three billion years prior to this moment, a small galaxy known as the Sagittarius Dwarf Elliptical Galaxy crashed through the plane of the Milky Way, triggering a burst of star formation. But instead of being swallowed by the Milky Way, the dwarf galaxy left a stream of gas in its wake, stretched out by the tidal and frictional forces arising from the matter within the Milky Way. At this moment, that galaxy's remaining core is hundreds of thousands of light-years away from the Milky Way's center: as far from our galaxy's core as it will ever get. Despite its great distance, the gravity of the Milky Way is too significant, and it now begins falling back towards our own galaxy, destined to be cannibalized like so many galaxies before it.

Ever since dark energy came to dominate the expansion of the Universe, it primarily ensured one thing: The galaxies, galactic groups, and galaxy clusters that were already gravitationally bound will remain so, but the ones that are not yet bound together will never become so. The largest gravitationally bound structure that the Milky Way is a part of is the Local Group, extending for some four to five million light-years in diameter. All the galaxies within it are bound together and will never escape. If one were to wait long enough, every galaxy within it will eventually merge together into one solitary remnant galaxy.

However, each of the more than 100 galaxies within the Local Group has its own properties: its own stellar population, its own set of gas and dust, its own halo of dark matter, etc. When galaxies enter into the vicinity of one another and their gravitational interactions become substantial, they not only attract one another; they experience a variety of internal interactions as a result. Gases within the galaxies experience tidal forces, creating overdensities that trigger the formation of new stars. If the small galaxy passes very close to the center of the larger one, tidal forces combined with gas interactions between the galaxies can strip material out of the small galaxy, creating a stream of matter that will form new stars, creating a stellar trail. Yet the core of the small galaxy can survive not only the initial collision but many subsequent ones; small galaxies can pass through large ones several times before they're gobbled up entirely.

For approximately three billion years prior to this epoch, the core of the Sagittarius Dwarf Elliptical Galaxy has been speeding away from the Milky Way after its first such close encounter, with a large stream of stars trailing behind it. However, gravity is a relentless force, and the smaller galaxy is unable to escape from the Milky Way's pull. After a journey of several hundred thousand light-years, the dwarf elliptical galaxy now begins to hurtle back towards the Milky Way, where it will pass through the galactic plane again and again. Its remnants will persist for billions of years, but eventually, its fate will be like all the rest: to be devoured by the Milky Way, along with its stars, gas, globular clusters, and more.

The lunar maria, or dark "seas" visible on the near side of the Moon, formed from ancient lava flows that filled in the lowest-elevation regions on Earth's nearest neighbor. Although the Moon has been volcanically inactive for more than a billion years, periodic intervals where lava flows resurface the lunar lowlands, like a cosmic Zamboni, are an integral part of its geological history.

IMAGE BY MARK A. GARLICK

SEAS OF TRANQUILITY

—— lunar maria are ancient lava beds ——

UNIVERSE
Diameter: $2.94 \times 10^{13} \rightarrow 2.97 \times 10^{13}$ ly
  Observable: $7.35 \times 10^{10} \rightarrow 7.41 \times 10^{10}$ ly
Expansion: $76.71 \rightarrow 76.30$ km/s/Mpc
Density: $0.48 \rightarrow 0.47$ p/m³
Temperature: $3.42 \rightarrow 3.33$ K

ASTRONOMICAL OBJECTS
Galaxies: $2.37 \times 10^{13} \rightarrow 2.36 \times 10^{13}$
Stars: $2.03 \times 10^{21} \rightarrow 2.04 \times 10^{21}$
Black Holes: $1.63 \times 10^{19} \rightarrow 1.64 \times 10^{19}$
Planetary Systems: $2.75 \times 10^{21} \rightarrow 2.76 \times 10^{21}$
Worlds: $3.16 \times 10^{22} \rightarrow 3.16 \times 10^{22}$

WORLDS WITH LIFE
Simple Life: $4.34 \times 10^{17} - 1.97 \times 10^{19}$
Complex Life: $3.51 \times 10^{11} - 3.35 \times 10^{17}$
Intelligent Life: $1.10 \times 10^{8} - 2.74 \times 10^{15}$
Technological Life: $3.45 \times 10^{4} - 3.58 \times 10^{13}$
Interstellar Life: $1 - 4.22 \times 10^{12}$

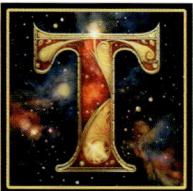

The small, airless worlds throughout the cosmos might seem like the most hopeless places for life and organic processes. But if an advanced enough civilization ever encountered one, they'd find resources and features that are exceedingly rare in more biologically friendly habitats. The internal heat of these worlds, particularly if they're in orbit around a more massive neighboring world that can create significant tides, is substantial, with liquid layers of rock inside many of them. Whenever impacts from asteroids and comets occur, they leave permanent marks on those worlds, scarring the surface and creating new craters atop any old ones that were left previously. But when an internally hot world experiences an energetic enough impact, that lava can reach the surface, often emerging in the lowest-elevation areas. These airless worlds, owing to these lava flows, can experience substantial "resurfacing" events, where molten rock from the world's interior flows into the planetary lowlands, erasing whatever geological history may have been written.

On a planet like Mercury, with a very thin mantle layer, the outer layers cool relatively quickly, causing the world to contract. As soon as the crust fully sealed off any pathways for that internal magma to make it to the surface, not only resurfacing events but any type of volcanic activity ceased on that world. Already, by this point in time, Mercury hasn't had an active volcano in some 500 million years, and the impact craters that exist all across its surface have layers of newer craters found atop older ones.

On the other extreme is a world like Io, the innermost large moon of Jupiter. As the fourth-largest moon in the Solar System—behind only Ganymede, Titan, and Callisto but ahead of Earth's moon—Io generates a tremendous amount of internal heat due to the tidal forces on it from Jupiter. While tides might more familiarly lead to small planetary quakes and oceanic bulges, more severe tidal forces can cause giant fractures within a world, creating pathways for magma to flow up and flood the surface with lava. For all of history up until now and far into the future, craters will survive no more than a few thousand years on Io, as lava flows across the surface will erase any features that would endure on most other worlds. But on Io, constant resurfacing events are all but inevitable.

The Moon of Earth, however, has been rich in volcanic activity ever since its formation. With both highlands and lowlands all across it, as well as a near-side/far-side dichotomy in terms of crustal thickness and mineral compositions, eruptions had long been more common on the lower-elevation near side. After cooling off for some 1.5 billion years, however, the final major lava flows are appearing on the Moon—this time, on the near side alone. The lowest-elevation regions are almost exclusively found on the near side, and a large impact anywhere on the world would have been capable of triggering volcanic activity, flooding the lowest-elevation areas of the Moon with lava.

During the current 100-million-year interval, the Moon experiences the formation of the last of its lunar maria: the "dark regions" that arose due to these lava flows and other, related volcanic activity that filled in and resurfaced any craters found in these low-lying regions. While short periods of minor volcanic activity would sporadically occur on the Moon over the next two billion or so years, the final formation of the lunar maria—rich in basalt and other volcanic rocks and less heavily cratered than all the highland regions of the Moon—occurs some 10.7–10.8 billion years after the Big Bang.

Just a few hundred million years after its core ccoled and its magnetic dynamo ceased, the atmosphere of Mars has now become thin and tenuous enough that the last vestiges of liquid water either freeze or evaporate from its surface. W thout a sufficient atmospheric pressure, water cannot achieve a liquid phase, and as this final pool of water evapora es from the Martian lowlands, the chain of life on Mars finally breaks.

IMAGE BY MARK A. GARLICK

REQUIEM FOR MARTIANS

—— life ends on Mars ——

UNIVERSE
Diameter: $2.97 \times 10^{13} \rightarrow 2.99 \times 10^{13}$ ly
  Observable: $7.41 \times 10^{10} \rightarrow 7.47 \times 10^{10}$ ly
Expansion: $76.30 \rightarrow 75.89$ km/s/Mpc
Density: $0.47 \rightarrow 0.46$ p/m³
Temperature: $3.39 \rightarrow 3.37$ K

ASTRONOMICAL OBJECTS
Galaxies: $2.36 \times 10^{13} \rightarrow 2.35 \times 10^{13}$
Stars: $2.04 \times 10^{21} \rightarrow 2.05 \times 10^{21}$
Black Holes: $1.64 \times 10^{19} \rightarrow 1.64 \times 10^{19}$
Planetary Systems: $2.76 \times 10^{21} \rightarrow 2.77 \times 10^{21}$
Worlds: $3.16 \times 10^{22} \rightarrow 3.18 \times 10^{22}$

WORLDS WITH LIFE
Simple Life: $4.29 \times 10^{17} - 1.95 \times 10^{19}$
Complex Life: $3.37 \times 10^{11} - 3.39 \times 10^{17}$
Intelligent Life: $1.06 \times 10^{8} - 2.77 \times 10^{15}$
Technological Life: $3.32 \times 10^{4} - 3.63 \times 10^{13}$
Interstellar Life: $1 - 4.32 \times 10^{12}$

Since losing its magnetic field, any life processes on Mars were proceeding on borrowed time. Over the past few hundred million years, a combination of ultraviolet radiation from the Sun as well as solar wind particles have been bombarding Mars, and without a magnetic field to protect it, the molecules making up the Martian atmosphere have slowly but steadily been stripped away. Made up largely of carbon dioxide and water—which can dissociate into carbon, oxygen, and hydrogen atoms—Mars's original atmosphere has thinned substantially since the loss of its core dynamo and its planet-wide magnetic field. At last, more than 1.5 billion years into the history of Mars, it experiences its last wet day: the final moment at which liquid water can stably exist on the Martian surface. Once just a few more atoms are stripped away from the atmosphere, the pressure will no longer be sufficient for water to exist in the liquid phase. From anywhere on the Martian surface, even at the lowest point in elevation, water can only take on the solid or gaseous phases, dependent solely on temperature.

Throughout the history of Mars, life has always faced an uphill battle. On a smaller planet than Earth located farther from the Sun, a substantial, thick atmosphere was required to hold in enough heat to make liquid water possible. Mars was lucky enough to possess a sufficiently thick early atmosphere for liquid water to have been copious on that world; if one were to spread out all of Mars's water evenly over a sphere, it would have made a layer between 130 and 500 meters thick—enough to create a world full of oceans and continents. Although life never achieved the level of success on Mars that it did on Earth, many features found on Mars suggest that life could have arisen and thrived in the Martian waters. But as the atmosphere slowly but steadily gets stripped away, the end of biological activity cannot be avoided.

Two factors play a major role in stripping the atmosphere away from Mars: planetary weather and space weather. Windstorms are known to frequently occur on Mars, and when they do, they can carry gases up to extremely high altitudes. This far up, it's relatively easy to escape from Mars's gravitational pull: Both ultraviolet radiation and solar wind particles can cause oxygen, carbon, and hydrogen atoms to escape. Particles are stripped away from the Martian atmosphere during planet-wide windstorms at 10 times the rates seen during non-windy periods, identifying a major culprit in the story of how Mars lost its atmosphere. But the other culprit is the Sun, which generates solar flares, coronal mass ejections, and other forms of space weather rich in fast-moving, ionized particles. When these space weather events strike Mars, they can increase the rate at which particles are lost to space by more than a factor of 20. Putting these factors together, a single day where a solar flare hits Mars during a windstorm can remove as much atmosphere as a typical year on a quieter Mars.

While the carbon dioxide largely escapes to space, only some of Mars's water does the same. A large amount of Mars's water is locked up into the solid phase: in polar ice caps and underground ice stores. Many minerals both on and beneath the Martian surface—where most of Mars's original water that isn't lost to space winds up—are likely well hydrated. As the last liquid water evaporates away, biological activity grinds to a halt. While some extreme organisms may adapt to the briny, underground environments where liquid water can occasionally still arise, life on Mars most likely died out due to thirst. If a thicker atmosphere and a protective magnetic field were ever reintroduced onto Mars, even billions of years down the road, perhaps life could someday thrive there once again.

What types of methods might a sufficiently advanced alien species use to harness energy? Perhaps building a swarm of structures around an active black hole is one option, as the accelerated particles and high-energy radiation both carry tremendous amounts of energy that could provide a source of power far greater than any mere star. Choosing a supermassive black hole would be the most ambitious option of all.

IMAGE BY MARK A. GARLICK

DYSON STRUCTURES

—— *advanced life may build megastructures* ——

UNIVERSE
Diameter: $2.99 \times 10^{13} \rightarrow 3.01 \times 10^{13}$ ly
  Observable: $7.47 \times 10^{10} \rightarrow 7.53 \times 10^{10}$ ly
Expansion: $75.89 \rightarrow 75.46$ km/s/Mpc
Density: $0.46 \rightarrow 0.45$ p/m³
Temperature: $3.37 \rightarrow 3.34$ K

ASTRONOMICAL OBJECTS
Galaxies: $2.35 \times 10^{13} \rightarrow 2.34 \times 10^{13}$
Stars: $2.05 \times 10^{21} \rightarrow 2.06 \times 10^{21}$
Black Holes: $1.64 \times 10^{19} \rightarrow 1.65 \times 10^{19}$
Planetary Systems: $2.77 \times 10^{21} \rightarrow 2.78 \times 10^{21}$
Worlds: $3.18 \times 10^{22} \rightarrow 3.20 \times 10^{22}$

WORLDS WITH LIFE
Simple Life: $4.23 \times 10^{17} - 1.92 \times 10^{19}$
Complex Life: $3.24 \times 10^{11} - 3.42 \times 10^{17}$
Intelligent Life: $1.02 \times 10^{8} - 2.81 \times 10^{15}$
Technological Life: $3.19 \times 10^{4} - 3.70 \times 10^{13}$
Interstellar Life: $1 - 4.43 \times 10^{12}$

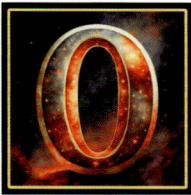

On every inhabited world, whenever one class of organism comes to exist in great enough numbers for long enough periods of time, it has the potential to initiate global changes. Single-celled organisms consume gases while producing others, changing the contents of a planet's atmosphere over sufficiently long timescales. Waste products produced from biological activity can alter the pH of even the deepest oceans, while simply overconsuming a necessary resource can cause an entire ecosystem to collapse. But once intelligent, technologically advanced life arises, even more substantial worldwide changes can ensue. For many of the worlds inhabited by intelligent life, these global transformations have already occurred.

The earliest transformations that ensue are likely small-scale and exploratory. Civilizations that launch fleets of satellites into orbit can gather energy from space, heat or cool their planet with arrays of reflective panels, and begin mining other worlds in their stellar system for the elements necessary to power their technologies on increasingly larger scales. Over time, satellite fleets can be grown into swarms, where those civilizations can harvest a substantial fraction of their parent star's energy output and put it towards good uses.

When enough materials have been excavated, collected, and refined, planetary-scale permanent structures can be constructed. Space elevators can be built along a rotating world's equator, bringing large amounts of mass into space and back down onto the world's surface with ease. Those elevators can serve as supports for a stable, connected structure: a planet-encircling ring that can be used as both a habitat and also as a hub for space-based infrastructure. Planets that are inhabited by "ring builders" in this fashion may seek one another out; perhaps a search for ringed exoplanets is a standard part of the search for extraterrestrial intelligence among advanced civilizations.

Once a civilization gains the ability to truly undertake extremely long-term tasks, they can begin to transform not only their home world, but their entire star system. Asteroids within an asteroid belt can have low-power thrusters attached to them, causing them to migrate and merge together into a single mass. That action would protect all the other worlds within the system from any potentially hazardous asteroid strikes. Similarly, an array of well-equipped spacecraft sufficiently far from the system's parent star—perhaps at the location of the outermost major planet—can serve as a defense array, protecting the inner system from comets and even interstellar strikes.

But perhaps the grandest superstructure one could envision is also one of science fiction's wildest ideas: a Dyson sphere. By building a solid structure that surrounds not only a star but the entire planetary system that orbits it, a full 100% of the star's energy can be harnessed. Perhaps lab-grown structures like carbon nanotubes or graphene-like materials could actually withstand the substantial forces that would act on such a configuration, preventing it from collapsing or otherwise breaking. And perhaps these objects are already out there, just waiting for a civilization with the right infrared capabilities to come along and find them. After all, every advanced, intelligent species ever to arise in this Universe must have wondered, "Are we alone?" right up until the moment they discovered that there were others. Just possibly, it's the very act of imagining what an advanced, successful civilization would accomplish that's the key to detecting and revealing its presence.

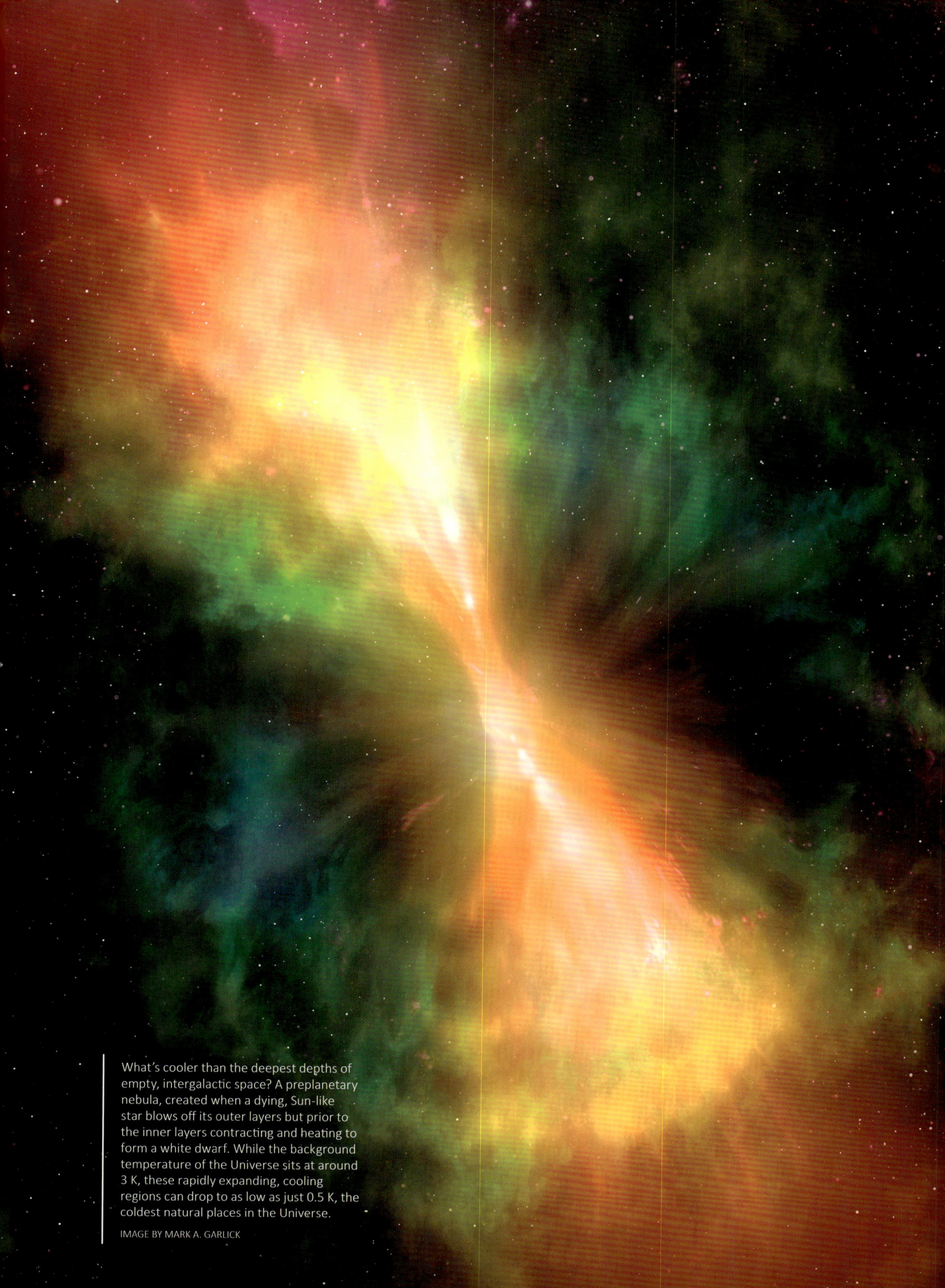

What's cooler than the deepest depths of empty, intergalactic space? A preplanetary nebula, created when a dying, Sun-like star blows off its outer layers but prior to the inner layers contracting and heating to form a white dwarf. While the background temperature of the Universe sits at around 3 K, these rapidly expanding, cooling regions can drop to as low as just 0.5 K, the coldest natural places in the Universe.

IMAGE BY MARK A. GARLICK

UNFATHOMABLE COLDNESS

—— *the coldest place in the Universe* ——

UNIVERSE
Diameter: $3.01 \times 10^{13} \rightarrow 3.03 \times 10^{13}$ ly
 Observable: $7.53 \times 10^{10} \rightarrow 7.59 \times 10^{10}$ ly
Expansion: $75.46 \rightarrow 75.05$ km/s/Mpc
Density: $0.45 \rightarrow 0.44$ p/m³
Temperature: $3.34 \rightarrow 3.31$ K

ASTRONOMICAL OBJECTS
Galaxies: $2.34 \times 10^{13} \rightarrow 2.32 \times 10^{13}$
Stars: $2.06 \times 10^{21} \rightarrow 2.06 \times 10^{21}$
Black Holes: $1.65 \times 10^{19} \rightarrow 1.65 \times 10^{19}$
Planetary Systems: $2.78 \times 10^{21} \rightarrow 2.79 \times 10^{21}$
Worlds: $3.20 \times 10^{22} \rightarrow 3.20 \times 10^{22}$

WORLDS WITH LIFE
Simple Life: $4.17 \times 10^{17} - 1.90 \times 10^{19}$
Complex Life: $3.10 \times 10^{11} - 3.46 \times 10^{17}$
Intelligent Life: $9.74 \times 10^{7} - 2.85 \times 10^{15}$
Technological Life: $3.06 \times 10^{4} - 3.76 \times 10^{13}$
Interstellar Life: $1 - 4.55 \times 10^{12}$

Left over from the very early stages of the hot Big Bang, a bath of relic radiation fills the Universe: the photons now making up the cosmic microwave background. Over the past ~11 billion years, as the Universe has expanded, the wavelength of those photons has been stretched to progressively longer distances. These photons possess a temperature corresponding to the equilibrium temperature that any object within that region would achieve after being heated by its surroundings. At the present epoch, that temperature is a mere 3.3 K—just 3.3 degrees above absolute zero. Far away from any other sources of radiation, in the deepest depths of intergalactic space, only this background of cosmic radiation matters. And yet, there are places in the Universe that are even colder, with a rare class of objects known as preplanetary nebulae achieving temperatures of ~1 K or less.

Within a typical galaxy, there are many sources of energy, including (most prominently) stars and stellar remnants such as white dwarfs, neutron stars, and black holes. These cosmic engines emit radiation and particles, substantially heating up whatever matter enters their vicinity. Within any stellar system, planets, moons, asteroids, and other objects are heated to temperatures ranging from tens of degrees at the outskirts all the way up to hundreds or even thousands of degrees close to the star. Even at distances several light-years away from the nearest star, the atoms and ions in interstellar space are heated to temperatures of 10–30 K, significantly hotter than the background radiation from the Big Bang.

However, the Universe has a way of rapidly and significantly changing the temperature of any system of particles, such as a gas, by either rapid contraction or expansion. The "rapid" component is essential, because if expansion or contraction happens slowly, heat can enter and/or leave the system. When gas clouds contract, they heat up, where systems made of hotter particles are more difficult to compress than systems made of colder ones. In order for stars to form, a gas cloud needs to be relatively cold at the start and must cool efficiently as it contracts; otherwise the gas will heat up and re-expand instead.

The reverse is also true: When a system of hot particles, like a gas or plasma, rapidly expands, it cools down. This is why, when you exhale with your mouth wide open, the exhaled air remains at your internal body temperature, but when you make only a small opening with your mouth, the air rapidly expands and cools off. When stars evolve from the standard, hydrogen-burning phase of their lives into red giants, the outer layers of the star expand and drop to a lower temperature, cooling by many thousands of degrees in the process. Because red giants continue to fuse lighter elements into heavier ones in their core, they continuously inject extra heat into the system, preventing those particles from cooling off too severely.

When red giant stars begin to run out of fuel in their cores, however—especially within the lower-mass giants that are similar in mass to our Sun—the radiation within their cores drops, and they continue to evolve. The giant star's core begins to gravitationally contract, beginning its evolution into a white dwarf and leading to a series of pulsations. These pulses expel the less dense, most tenuously held outer layers, which then rapidly expand. As a result, thick shells of gas are created around these stars, which cool quickly. This gas creates a structure known as a preplanetary nebula, a phase that lasts only a few thousand years but contains the coldest places in the known Universe. The gas achieves internal temperatures as low as ~0.5 K, making these preplanetary nebulae colder than even the emptiest depths of intergalactic space.

Although Earth is known as the blue planet owing to the colors of its oceans and skies, the evolution and rise to prominence of cyanobacteria, the blue-green algae that for billions of years were Earth's dominant oxygen producers, begins the greening of planet Earth.

IMAGE BY MARK A. GARLICK

RISE OF THE CYANOBACTERIA

—— cyanobacteria evolve on Earth ——

UNIVERSE
Diameter: $3.03 \times 10^{13} \rightarrow 3.06 \times 10^{13}$ ly
 Observable: $7.59 \times 10^{10} \rightarrow 7.65 \times 10^{10}$ ly
Expansion: $75.05 \rightarrow 74.64$ km/s/Mpc
Density: $0.44 \rightarrow 0.43$ p/m³
Temperature: $3.31 \rightarrow 3.29$ K

ASTRONOMICAL OBJECTS
Galaxies: $2.32 \times 10^{13} \rightarrow 2.31 \times 10^{13}$
Stars: $2.06 \times 10^{21} \rightarrow 2.07 \times 10^{21}$
Black Holes: $1.65 \times 10^{19} \rightarrow 1.66 \times 10^{19}$
Planetary Systems: $2.79 \times 10^{21} \rightarrow 2.80 \times 10^{21}$
Worlds: $3.20 \times 10^{22} \rightarrow 3.21 \times 10^{22}$

WORLDS WITH LIFE
Simple Life: $4.11 \times 10^{17} - 1.87 \times 10^{19}$
Complex Life: $2.95 \times 10^{11} - 3.49 \times 10^{17}$
Intelligent Life: $9.33 \times 10^{7} - 2.88 \times 10^{15}$
Technological Life: $2.93 \times 10^{4} - 3.81 \times 10^{13}$
Interstellar Life: $1 - 4.66 \times 10^{12}$

On Earth, photosynthesis has already been taking place for some 750 million years, largely among bacteria living on the surfaces of water and around deep-sea hydrothermal vents. A large variety of photosynthetic pigments—complex molecules that reflect most wavelengths of light but absorb light of a specific frequency, causing the pigment molecule to change shape—have now arisen, including carotenoids (which absorb violet, blue, and green light), phycobilins (which absorb red, orange, yellow, and green light), and the famously green-hued chlorophyll (which absorbs blue/violet and red/orange light). Although six unique types of chlorophyll molecules will eventually come to exist on Earth, three of them are already found within one class of organism alive at this time: cyanobacteria, or blue-green algae.

Having arisen in the fresh waters found atop continental landmasses, cyanobacteria easily rise high into the atmosphere from any spray-producing water source, such as a waterfall. While the first cyanobacteria to arise only had a simple cell membrane, the passage of time allowed them to develop a thick, gelatinous structural layer outside of that membrane: a cell wall. Once the cell wall develops, a kicked-up cyanobacterium precipitates down onto the salty ocean surface, where it survives, thrives, and reproduces. In short order, these photosynthetic organisms, some of which are as small as 500 nanometers across (a single wavelength of cyan-colored light), grow to almost unfathomable numbers. Once cyanobacteria begin thriving on the surfaces of Earth's oceans, they quickly saturate that environment, as no other photosynthetic competitors presently occupy that niche. Approximately 100,000 cyanobacteria can be found in every milliliter of surface water, implying a total cyanobacteria population of several octillion (10^{27}) over the entirety of Earth.

Some of these cyanobacteria are compact, ellipsoidal-shaped organisms, capable of thriving in isolation or as part of a large colony, forming a "mat" on the water's surface. Others are elongated, filamentary specimens: the precursors of the first multicellular organisms that will later appear on Earth. Although cyanobacteria can easily serve as food for other organisms that coexist along with them (and are often preyed upon by a special class of virus: the bacteriophage), they have no competitors for the ecological niche they occupy. This lack of competitors, combined with the tremendous biological diversity exhibited by cyanobacteria, ensure their ecological stability, even over long periods of time. With a practically unlimited source of water in Earth's oceans, a practically unlimited source of energy in sunlight, and a rich source of both nitrogen and carbon dioxide from the oceanic surface interfacing with Earth's atmosphere, these cyanobacteria can easily obtain almost everything they need for their life processes simply by floating atop the waters of Earth's oceans.

However, as the only known photosynthetic prokaryotic organism, cyanobacteria have now come to populate Earth in great abundance. As they undergo their biological processes, using chlorophyll to convert sunlight and carbon dioxide into energy-storing carbohydrates, they begin to abundantly produce a waste product that's only ever existed in trace amounts on Earth prior to now: molecular oxygen (O_2). Although Earth's atmosphere is largely composed of nitrogen and carbon dioxide, with small amounts of water vapor, methane, and argon, this marks the first widespread appearance of oxygen. With each 100 million years that passes, the combined effect of all the cyanobacteria on Earth increases the oxygen content of our atmosphere by somewhere between 0.5% and 1%. At long last, a distant intelligent observer, seeing these atmospheric changes occur, would identify planet Earth as a living world.

The largest volcanic structure in the known Solar System is Olympus Mons, found on the Tharsis region of Mars. Only from space can its scale be fully appreciated, as from base to summit, it gains over 21 kilometers of elevation but spans roughly 600 kilometers from end to end. From the base of the mountain, Mars's curvature is so great that the summit cannot be seen.

IMAGE BY MARK A. GARLICK

BEHOLD, OLYMPUS MONS

the largest volcano in the Solar System

UNIVERSE
Diameter: $3.06 \times 10^{13} \rightarrow 3.08 \times 10^{13}$ ly
  Observable: $7.65 \times 10^{10} \rightarrow 7.70 \times 10^{10}$ ly
Expansion: $74.64 \rightarrow 74.29$ km/s/Mpc
Density: $0.43 \rightarrow 0.42$ p/m³
Temperature: $3.29 \rightarrow 3.26$ K

ASTRONOMICAL OBJECTS
Galaxies: $2.31 \times 10^{13} \rightarrow 2.30 \times 10^{13}$
Stars: $2.07 \times 10^{21} \rightarrow 2.08 \times 10^{21}$
Black Holes: $1.66 \times 10^{19} \rightarrow 1.66 \times 10^{19}$
Planetary Systems: $2.80 \times 10^{21} \rightarrow 2.81 \times 10^{21}$
Worlds: $3.21 \times 10^{22} \rightarrow 3.23 \times 10^{22}$

WORLDS WITH LIFE
Simple Life: $4.06 \times 10^{17} - 1.85 \times 10^{19}$
Complex Life: $2.86 \times 10^{11} - 3.51 \times 10^{17}$
Intelligent Life: $8.98 \times 10^{7} - 2.91 \times 10^{15}$
Technological Life: $2.82 \times 10^{4} - 3.86 \times 10^{13}$
Interstellar Life: $1 - 4.76 \times 10^{12}$

Volcanic activity plays an important role in shaping the surfaces of Earth and the Moon, yet neither one of those familiar worlds is home to the largest planetary volcano in the Solar System. That honor instead goes to Olympus Mons, with a height of 21.9 kilometers as measured from base to peak. This is about 250% as high as the highest mountains on Earth and is rivaled only by mountains formed from crater impacts on far less massive, lower-density worlds. Although Olympus Mons began forming several hundred million years ago and will continue to exhibit periodic lava flows for another ~2.5 billion years, this epoch marks the moment when it becomes the largest planetary volcano in the Solar System.

Much like Earth, Mars has a large planetary mantle underneath the surface crust, and the farther down one ventures into the mantle, the hotter the temperature gets. Relative to Earth, however, Mars has a significantly smaller size, which means it loses its internal heat more quickly than Earth. This process has already caused Mars to lose its core dynamo, its planet-protecting magnetic field, most of its atmosphere, and its capacity to have liquid water on its surface. However, the remaining internal heat continues to radiate away, and volcanism—where magma rises from the mantle up through weak points in the crust and erupts onto the surface—is one of the primary mechanisms by which planets release their internal heat.

The type of material that emerges from volcanic eruptions is largely dependent on the mass of the world on which they occur. On Earth, magmas typically rise, cool, and mix with rock and other batches of magma. As a result, the lava that erupts as a component of volcanic activity is enriched with silica, volatile compounds, and other light elements. On the other hand, lower-mass worlds like the Moon have very primitive, unmixed magmas, making basaltic lava flows rich in magnesium and iron. Because Mars is more massive than the Moon but less massive than Earth, its magmas are somewhere in between the two: partially mixed and enriched but still rich in magnesium and iron.

Although Mars experienced most of its volcanic activity early in its history, forming a wide variety of very tall mountains, Olympus Mons is the last of its great volcanoes to come into existence and is made of thousands of basaltic lava flows that have all combined together. Beginning its formation several hundred million years before the current epoch, Olympus Mons has grown in height ever since but now starts reaching a new point in its evolution. Just as is the case on Earth, the Martian crust floats atop the Martian mantle. As Olympus Mons grows in height, this region of the crust begins pressing down on the mantle with greater force. Unlike on Earth, there are no active plate tectonics on Mars, so the internal "hot spot" beneath the mountain remains in the same location, causing the mountain to simply grow and grow over time. Forces between the mantle and crust, as well as internal forces within the crust itself, cause the volcanic mountain to spread out instead, leading to a very wide mountain with a deep depression surrounding the mountain's base. Despite its volcanic origin, it's a gently sloping mountain, with an average grade of only 5%.

With a much lower surface gravity than Earth's, Mars's Olympus Mons rises to much greater heights than any mountain on Earth can achieve. The atmospheric pressure atop this enormous mountain is only 12% as great as it is at the mountain's base. When Mars becomes enveloped in its periodic planet-wide dust storms, the summit of Olympus Mons remains one of the only surface features that can still be viewed, unobscured, from space.

Is this an artifact from an alien civilization, or a device used to "seed" uninhabited worlds with primitive life-forms? If technologica ly advanced civilizations have a penchant for colonization, life throughout a galaxy, or even beyond it, may yet have a common origin that was deliberately put in place by an intelligent species.

IMAGE BY JON LOMBERG

THERE IS A WORLD ELSEWHERE

—— life may seek to colonize the Universe ——

UNIVERSE
Diameter: $3.08 \times 10^{13} \rightarrow 3.10 \times 10^{13}$ ly
  Observable: $7.70 \times 10^{10} \rightarrow 7.76 \times 10^{10}$ ly
Expansion: $74.29 \rightarrow 73.89$ km/s/Mpc
Density: $0.42 \rightarrow 0.41$ p/m³
Temperature: $3.26 \rightarrow 3.24$ K

ASTRONOMICAL OBJECTS
Galaxies: $2.30 \times 10^{13} \rightarrow 2.28 \times 10^{13}$
Stars: $2.08 \times 10^{21} \rightarrow 2.08 \times 10^{21}$
Black Holes: $1.66 \times 10^{19} \rightarrow 1.67 \times 10^{19}$
Planetary Systems: $2.81 \times 10^{21} \rightarrow 2.82 \times 10^{21}$
Worlds: $3.23 \times 10^{22} \rightarrow 3.24 \times 10^{22}$

WORLDS WITH LIFE
Simple Life: $4.00 \times 10^{17} - 1.83 \times 10^{19}$
Complex Life: $2.72 \times 10^{11} - 3.54 \times 10^{17}$
Intelligent Life: $8.60 \times 10^{7} - 2.95 \times 10^{15}$
Technological Life: $2.70 \times 10^{4} - 3.92 \times 10^{13}$
Interstellar Life: $1 - 4.87 \times 10^{12}$

In all the Universe, no matter how many planets there are where life arises, survives, and thrives over long periods of time, there are bound to be a far greater number of worlds where something goes awry and the great cosmic chance for intelligent beings dies out. Many worlds are born with all the necessary ingredients for life: a rich supply of organic molecules, copious amounts of flowing liquid on the surface, multiple sources of energy, etc. Perhaps life even got its start and survived for a time on many of these worlds, but eventually, it all died away. Some worlds eventually became too hot, boiling their oceans away. Perhaps others got too cold, froze over, and lost their atmospheres to space. These now-lifeless worlds litter the cosmos, with trillions upon trillions of them roaming each one of the Milky Way–sized galaxies out there.

Somewhere, among the sextillions of stars within the observable Universe, one intelligent civilization now arises that takes a great interest in these worlds, studying their varied properties and histories, and gains an understanding of the physics that governs them. They're then classified into a variety of categories based on their temperatures, compositions, atmospheres, plus other astronomical and geological properties. For one such class of world, this civilization figures out something remarkable: a pathway to transform such a world, uninhabited at the moment, step-by-step into one that is not only teeming with life but that will eventually be colonizable by members of that civilization itself.

Perhaps it's a frozen world, not so different from a late-time version of Mars in our own Solar System. Instead of an internal dynamo to create a magnetic field, a robotic mission could install a large power array that creates an enormous, planet-wide circuit, capable of creating a protective magnetic field. A device that digs down into the planet's mantle could trigger the release of volcanic gases, which would work to replenish the atmosphere, eventually enabling warmer temperatures and surface liquids once again. Subsequently, an interplanetary mission with the right "seed organisms" to take hold on the surface can begin to create a biosphere. With a slow buildup of missions over time and a mix of technological and biological milestones to achieve, this civilization—one step at a time—transforms nonliving worlds into future sites of habitability for themselves and their descendants.

It's likely that such a civilization will begin with worlds within their own stellar system and will then move on to nearby stars, but once they have a repeatable formula, they will begin to expand into a truly galaxy-wide network. Unlike the first *Encyclopaedia Galactica* that arose in the Universe, which was purely a comprehensive source of information about the worlds and star systems within a galaxy, this rapidly growing colony of living, inhabited worlds serves to become the Universe's first galactic scale civilization. In a span of only a few tens of millions of years, thousands of previously uninhabited worlds are now brimming with intelligent life.

But no matter how successful this civilization becomes, even if they manage to expand to a few other nearby galaxies, any ambitions they had of colonizing the entire Universe are quickly squashed by an inevitable presence: dark energy. Most of the galaxies visible to this civilization already appear to be receding away at speeds greater than the speed of light, rendering them unreachable and rendering two-way communication impossible. Although the great colonizers might swiftly become the most "successful" civilization in all the cosmos, the Milky Way—and even the prospect of eventual contact with humanity—will forever be beyond their grasp.

Over hundreds of millions of years, the metabolic processes of photosynthetic organisms like cyanobacteria have been converting sunlight and carbon dioxide into energy, producing oxygen as a waste product. The large, gradual increase in atmospheric oxygen finally has a planet-wide chilling effect, leading to the Great Oxygenation Event and the onset of "Snowball Earth" conditions.

IMAGE BY JON LOMBERG

THE GREAT OXYGENATION EVENT

—— life transforms the atmosphere of Earth ——

UNIVERSE
Diameter: $3.10 \times 10^{13} \rightarrow 3.13 \times 10^{13}$ ly
  Observable: $7.76 \times 10^{10} \rightarrow 7.82 \times 10^{10}$ ly
Expansion: $73.89 \rightarrow 73.55$ km/s/Mpc
Density: $0.41 \rightarrow 0.40$ p/m^3
Temperature: $3.24 \rightarrow 3.22$ K

ASTRONOMICAL OBJECTS
Galaxies: $2.28 \times 10^{13} \rightarrow 2.27 \times 10^{13}$
Stars: $2.08 \times 10^{21} \rightarrow 2.09 \times 10^{21}$
Black Holes: $1.67 \times 10^{19} \rightarrow 1.67 \times 10^{19}$
Planetary Systems: $2.82 \times 10^{21} \rightarrow 2.83 \times 10^{21}$
Worlds: $3.24 \times 10^{22} \rightarrow 3.25 \times 10^{22}$

WORLDS WITH LIFE
Simple Life: $3.95 \times 10^{17} - 1.81 \times 10^{19}$
Complex Life: $2.63 \times 10^{11} - 3.57 \times 10^{17}$
Intelligent Life: $8.27 \times 10^7 - 2.97 \times 10^{15}$
Technological Life: $2.60 \times 10^4 - 3.98 \times 10^{13}$
Interstellar Life: $1 - 4.97 \times 10^{12}$

For several hundred million years, one of the dominant forms of life on Earth has been cyanobacteria. These photosynthetic bacteria use sunlight to transform carbon dioxide and water into sugars, producing oxygen (O_2) as a waste product. With a complete life cycle lasting only a few hours and a whopping 10^{27} such organisms present on Earth at any moment, they wind up adding nearly 1 million tons of molecular oxygen into the atmosphere each year. At first, the excess oxygen gets absorbed by rocks and decaying life-forms on the planet's surface, oxidizing any surface iron and aiding the decomposition of organisms. But far more oxygen is produced, and over far too long a timescale, for those sinks to effectively absorb most of the O_2. As time progresses, oxygen begins to build up in Earth's atmosphere, and as the oxygen concentration rises and rises, it reaches around 3–4% during this epoch for the first time in our planet's history. A series of planet-wide changes then begins to ensue.

One of the first things to happen is that the remaining methane in Earth's atmosphere begins to react with the now-abundant oxygen. When methane and oxygen react, they produce carbon dioxide and either hydrogen gas or water molecules. Although both methane and carbon dioxide are greenhouse gases—serving to trap heat in Earth's atmosphere and raise the overall planetary temperature—methane is far less abundant than carbon dioxide, and hence the climate is far more sensitive to its presence or absence. As the oxygen concentration rises in Earth's atmosphere, the methane concentration plummets, and Earth becomes less efficient at trapping heat.

Owing to the fact that the Sun is less luminous when it's younger—it gets about 8% brighter with each billion years

that pass—the loss of methane causes Earth's temperature to reach a tipping point: one where it cannot trap enough heat to keep its surface temperatures above the threshold at which water freezes. Simply from the life processes of a thriving single-celled organism in Earth's oceans, the passage of time catastrophically changes Earth's climate. As a result, the rise of oxygen in Earth's atmosphere, often called the Great Oxygenation Event or the Great Oxidation Event, triggers a series of planet-wide ice ages: the Huronian glaciation, which endures for several hundred million years.

The oxidation of the atmosphere—combined with the much colder temperature conditions—triggers a mass extinction among the dominant form of life on Earth at this time: the anaerobic bacteria. Oxygen not only reacts with the anaerobic bacteria in the atmosphere, on continental land, and on the planet's surface waters, but also with the organisms that reside deep beneath the lake, sea, and ocean waters. Oxygen, at this level, affects all organisms where it arises, but its effects are most strongly felt by the anaerobes. To them, oxygen is toxic, and the Great Oxygenation Event causes an estimated 80% of all species on Earth to die out during this period.

During the most severe of the ice ages of this epoch, ice covers not only the entirety of Earth's landmasses but also the oceans, even at equatorial latitudes. For the first time since its formation, our planet experiences a set of "Snowball Earth" conditions, where practically the entire surface of Earth becomes frozen. And yet, through it all, the cyanobacteria continue to thrive, while mass extinctions cause new ecological niches to open up. As is often the case, the catastrophic death of one class of organisms paves the way for new ones to rise.

We now understand that bright, prominent rings are temporary, transient phenomena in planetary history. Jupiter, shown here with its four Galilean moons, has likely seen many of its former moons destroyed over time, giving rise to spectacular but short-lived phases that included massive, brilliant rings. New satellites will soon form from this ringed debris, which will deplete in a few hundred million years at most.

IMAGE BY MARK A. GARLICK

RINGS OF DESTRUCTION

shattered moons become planetary rings

UNIVERSE
Diameter: $3.13 \times 10^{13} \rightarrow 3.15 \times 10^{13}$ ly
 Observable: $7.82 \times 10^{10} \rightarrow 7.88 \times 10^{10}$ ly
Expansion: $73.55 \rightarrow 73.17$ km/s/Mpc
Density: $0.40 \rightarrow 0.39$ p/m³
Temperature: $3.22 \rightarrow 3.19$ K

ASTRONOMICAL OBJECTS
Galaxies: $2.27 \times 10^{13} \rightarrow 2.26 \times 10^{13}$
Stars: $2.09 \times 10^{21} \rightarrow 2.10 \times 10^{21}$
Black Holes: $1.67 \times 10^{19} \rightarrow 1.68 \times 10^{19}$
Planetary Systems: $2.83 \times 10^{21} \rightarrow 2.84 \times 10^{21}$
Worlds: $3.25 \times 10^{22} \rightarrow 3.26 \times 10^{22}$

WORLDS WITH LIFE
Simple Life: $3.89 \times 10^{17} - 1.78 \times 10^{19}$
Complex Life: $2.50 \times 10^{11} - 3.60 \times 10^{17}$
Intelligent Life: $7.91 \times 10^{7} - 3.01 \times 10^{15}$
Technological Life: $2.49 \times 10^{4} - 4.02 \times 10^{13}$
Interstellar Life: $1 - 5.08 \times 10^{12}$

Throughout time, a typical glimpse of our Solar System will reveal our eight planets—four rocky worlds and four gas giants—with the larger planets possessing many moons and only a thin, tenuous system of rings. The surface features may change, as Earth is now a white, ice-covered planet and Jupiter has no "great spots" of any color at the moment, but the planets themselves all continue to remain intact. However, moon-destroying events can occur from a combination of major impact events, a gradual gravitational migration, and the tidal influences of other moons and the parent planet. The smaller and lower in mass a moon is, the easier it is to destroy, and although these destruction events are rare, several have occurred over our Solar System's history.

Whenever a significant moon is destroyed, it gives rise to a brilliant but cosmically short-lived system of rings, where several new moons and moonlets arise from within the wreckage of the ringed system. It's during this time period, after more than half a billion years of relative peace, that the then fifth-largest moon of Jupiter gets struck by an impacting body with enough energy to completely blast the world apart. What was once a single moon that orbited Jupiter at around six times the distance of the largest Galilean satellite, Callisto, has swiftly been transformed into a great ringed system of debris that now orbits Jupiter at a significant angle: inclined by about 28° to the orbit of the more prominent inner moons.

Unlike a more modest, typical ring system around a gas giant, where the rings might extend ~100,000 kilometers or so from the edge of the planet in question, these temporary Jovian rings are enormous, several million kilometers wide and orbiting 10–12 million kilometers from Jupiter itself. The matter from the formerly large moon gets drawn out into a large, thick, circumplanetary ring, which then develops instabilities and begins to form clumps within it. Just as planets form out of a protoplanetary disk around newborn stars and moons initially form out of the material in circumplanetary disks, this new system of rings will lead to the formation of a new set of moons out of the remnant material.

During this epoch in our Solar System's history, a large moon is destroyed, a giant system of circumplanetary rings appears around Jupiter, and that ring material begins, in a few places, to coalesce into major clumps of matter. These clumps of matter grow into four substantial moons—Leda, Himalia, Lysithea, and Elara—ranging in diameter from 20–150 kilometers. Himalia, the largest one by a wide margin, actually becomes Jupiter's fifth most massive moon upon re-forming. Most of the remaining material, by the time this epoch reaches its end, is very low in mass and density, and is clustered together in three rings: between Leda and Himalia, between Himalia and Lysithea, and out beyond the orbit of Elara.

Although small moonlets repeatedly form from the remnant ring material, they're easily destroyed by tidal and gravitational interactions. As a result, only a handful of small, asteroid-sized moons will eventually, in the epochs to come, survive from this ringed system, and the largest of those will be less than 1% as massive as Leda, the smallest of the four substantial moons formed from Jupiter's onetime rings. As the rings evaporate away entirely, our Solar System reverts to its most common appearance: four rocky and four giant planets, with rich systems of moons but without substantial ring systems around any of them.

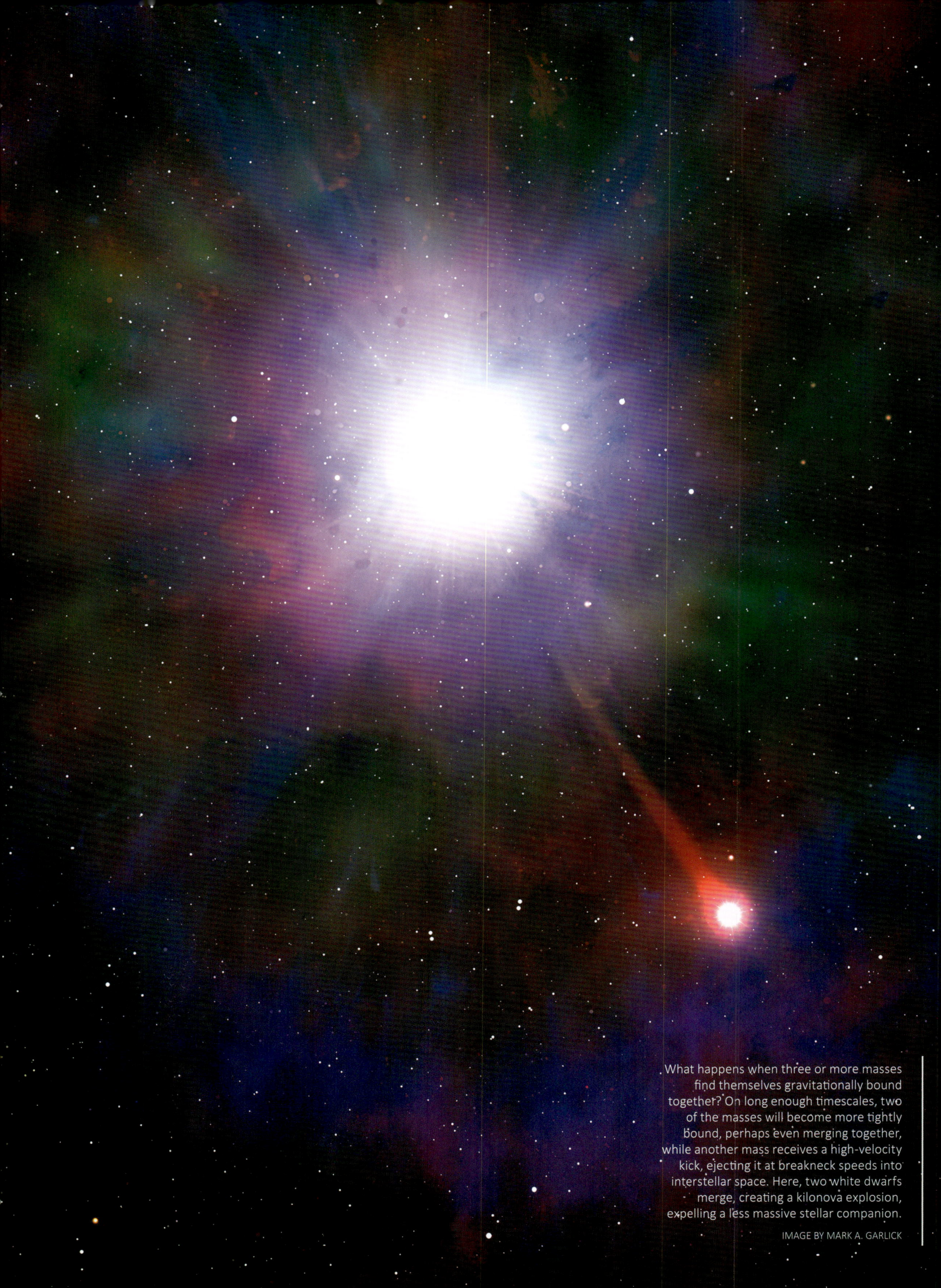

What happens when three or more masses find themselves gravitationally bound together? On long enough timescales, two of the masses will become more tightly bound, perhaps even merging together, while another mass receives a high-velocity kick, ejecting it at breakneck speeds into interstellar space. Here, two white dwarfs merge, creating a kilonova explosion, expelling a less massive stellar companion.

IMAGE BY MARK A. GARLICK

THREE-BODY PROBLEMS

three-body systems often eject a body

UNIVERSE
Diameter: $3.15 \times 10^{13} \rightarrow 3.17 \times 10^{13}$ ly
  Observable: $7.88 \times 10^{10} \rightarrow 7.94 \times 10^{10}$ ly
Expansion: $73.17 \rightarrow 72.84$ km/s/Mpc
Density: $0.39 \rightarrow 0.38$ p/m³
Temperature: $3.19 \rightarrow 3.17$ K

ASTRONOMICAL OBJECTS
Galaxies: $2.26 \times 10^{13} \rightarrow 2.24 \times 10^{13}$
Stars: $2.10 \times 10^{21} \rightarrow 2.10 \times 10^{21}$
Black Holes: $1.68 \times 10^{19} \rightarrow 1.68 \times 10^{19}$
Planetary Systems: $2.84 \times 10^{21} \rightarrow 2.84 \times 10^{21}$
Worlds: $3.26 \times 10^{22} \rightarrow 3.27 \times 10^{22}$

WORLDS WITH LIFE
Simple Life: $3.84 \times 10^{17} - 1.76 \times 10^{19}$
Complex Life: $2.42 \times 10^{11} - 3.62 \times 10^{17}$
Intelligent Life: $7.61 \times 10^{7} - 3.03 \times 10^{15}$
Technological Life: $2.40 \times 10^{4} - 4.06 \times 10^{13}$
Interstellar Life: $1 - 5.18 \times 10^{12}$

By this point in cosmic history, the effects of dark energy are not only undeniable but overwhelming. While the overdense regions notably possess stars, galaxies, and clusters of galaxies, it's the underdense regions, or cosmic voids, that are now growing rapidly. As the Universe continues to expand, the dense clumps of matter within it remain gravitationally bound, but the space between individual, bound clumps increases forevermore. And yet, as vast as the space between the galaxies and clusters of galaxies is, it isn't truly empty. In addition to the sparse gases and rarefied plasmas that persist there, there are also intergalactic "pinballs"—black holes, white dwarfs, stars, planets, and even entire stellar systems—roaming through the Universe

Inside every sufficiently massive galaxy in the Universe, the same types of objects can always be found. There are stars, born with a wide distribution of masses and evolving into black holes, neutron stars, and white dwarfs, given enough time. There are planets, including those persisting around stars, those ejected from their home system, and those that formed without a parent star at all. There are black holes, including not only the stellar mass ones that arose from massive stars that lived and died but also supermassive ones, present at the heart of almost every large galaxy. And yet, despite how massive and substantial these objects are, as well as how strong the gravitational pull of the galaxies they initially inhabit are, a subset of all of them are destined to be ejected, where they'll roam the depths of intergalactic space for all eternity thereafter.

Whenever you have a three-body interaction—that is, where three objects of comparable masses are all gravitationally bound together—what inevitably happens is that two of them wind up being much more tightly bound at the expense of the third, usually lightest, member. That third object often winds up getting ejected, sometimes with velocities so substantial that it can escape the gravitational pull of its host galaxy. Whether we're looking at planets, stars, or stellar remnants as the dancers on this cosmic stage, all are capable of being flung out of their home galaxies and into the abyss of intergalactic space.

Additionally, while most of the supernovae in the Universe come from the death throes of evolved, high-mass stars—the core-collapse supernovae that lead to neutron stars and black holes—far greater numbers of stars that are born will die much more peacefully and gradually: gently blowing off their outer layers in a planetary nebula while their centers contract to form a white dwarf. These white dwarfs, on occasion, will either accrete mass from a companion star or collide with another white dwarf, triggering an explosion that destroys the white dwarf entirely: a type Ia supernova. These supernova explosions are so energetic that they can "kick" the companion, whether a star or another white dwarf, to speeds ~1,000 kilometers per second or greater, catapulting them into intergalactic space.

Finally, whenever two similarly massed black holes inspiral and merge, whether of the stellar mass or the supermassive variety, they'll emit gravitational waves. Both black holes not only spin about their axes but also orbit one another with a particular orientation. These spin-and-orbit orientations are fairly random, but every so often, the spins of both black holes align as they merge, creating a "super-kick" situation. With gravitational waves primarily emitted in just one direction, the post-merger black hole receives a large "super-kick" in the opposite direction of several thousand kilometers per second, ejecting them into intergalactic space. These rogue objects eventually behave as intergalactic pinballs, populating the deepest depths of otherwise empty space.

For all of Earth's history until now, organisms have primarily reproduced asexually: through mitotic cell division. However, a new form of reproduction is now arising: sexual reproduction, where two parent organisms each contribute genetic material to create a new organism that possesses significant genetic variation compared to both of its parents. Complex organisms reap the largest evolutionary benefits from this advance.

IMAGE BY MARK A. GARLICK

THE ENGINE OF DIVERSIFICATION

— sex accelerates gene sharing —

UNIVERSE
Diameter: $3.17 \times 10^{13} \rightarrow 3.20 \times 10^{13}$ ly
  Observable: $7.94 \times 10^{10} \rightarrow 8.00 \times 10^{10}$ ly
Expansion: $72.84 \rightarrow 72.48$ km/s/Mpc
Density: $0.38 \rightarrow 0.38$ p/m³
Temperature: $3.17 \rightarrow 3.14$ K

ASTRONOMICAL OBJECTS
Galaxies: $2.24 \times 10^{13} \rightarrow 2.23 \times 10^{13}$
Stars: $2.10 \times 10^{21} \rightarrow 2.11 \times 10^{21}$
Black Holes: $1.68 \times 10^{19} \rightarrow 1.69 \times 10^{19}$
Planetary Systems: $2.84 \times 10^{21} \rightarrow 2.85 \times 10^{21}$
Worlds: $3.27 \times 10^{22} \rightarrow 3.29 \times 10^{22}$

WORLDS WITH LIFE
Simple Life: $3.79 \times 10^{17} - 1.74 \times 10^{19}$
Complex Life: $2.31 \times 10^{11} - 3.64 \times 10^{17}$
Intelligent Life: $7.28 \times 10^{7} - 3.06 \times 10^{15}$
Technological Life: $2.29 \times 10^{4} - 4.11 \times 10^{13}$
Interstellar Life: $1 - 5.28 \times 10^{12}$

urviving for hundreds of millions of years beneath the ices on a cold, frozen Earth, life-forms finally get another chance to thrive. As heat builds up from volcanic activity occurring beneath the ice, so do the gases that volcanoes release: water vapor, carbon dioxide, and sulfur dioxide. When cracks appear in the ice sheets and gases rise to mix with the atmosphere, they gradually create a warming effect. At last, Earth begins to thaw, becoming a blue, ocean-rich planet once again. As the ice melts over continental land, further greenhouse gases, such as methane, get released into the atmosphere, warming the planet still further. In the ecological niches that were vacated by countless species of now-extinct bacteria, new organisms that didn't exist before rise to take their place.

Now that Earth has an atmosphere that's rich in oxygen, organisms have evolved to take advantage of its presence, and aerobic (or oxygen-based) respiration has now become widespread. Multiple different pathways have already evolved by which organisms take in nutrients from their surroundings—including nutrients that are synthesized by cyanobacteria and their evolutionary descendants—and combine them with oxygen, breaking apart molecular bonds and forming new ones, all while extracting energy in the process. Compared to their anaerobic predecessors, the aerobes are far more efficient at extracting energy from the same metabolites: Up to 18 times the energy can be extracted aerobically versus anaerobically for each glucose molecule. The ability to "do more with the same ingredients" gives the aerobes such an evolutionary advantage that they begin to outcompete their anaerobic counterparts in many niches all across the globe.

Separately from the development of aerobic respiration, a new type of reproduction has arisen: sexual reproduction. Biologically, organisms that are more

complex—that perform more functions, are more highly differentiated, have greater amounts of information in their genomes, etc.—generally require more resources to survive, thrive, and reproduce than their simpler counterparts. This often means that in the time it takes for a complex organism to spawn the next generation, a simpler organism might have divided, with a chance to mutate and evolve, dozens, hundreds, or even thousands of times.

When sexual reproduction appears, it serves as a mechanism to help these more complex organisms keep up. Organisms that simply copy their genetic information—along with whatever mutations may have occurred—and pass it on to the next generation are relatively monolithic: Even thousands of generations down the line, there's very little genetic variation among the surviving organisms. That's a limitation of asexual reproduction: It may be fast and might require few resources, but when environmental conditions change, a selection pressure that's fatal to one member of the colony may lead to the demise of the entire colony.

In contrast, sexual reproduction, and explicitly the process of meiosis, allows two different members of the same species to each contribute half of their genetic information to each member of their offspring. This creates, in a single generation, a diverse set of offspring that can be vastly different not only from one another but from each of the parent organisms. The additional variations that arise in only a single generation—along with the associated resilience—lead to a rapid growth in how complex the genomes of sexually reproducing organisms can become. As is so frequently the case, a change in conditions leads to a mass extinction, and in the aftermath, a new batch of organisms, with new capabilities, rises to fill the void.

As seen from Earth, the most "radio-loud" galaxy in all of our skies is Hercules A, a galaxy located some two billion light-years away. Powered by a central, supermassive black hole, these incredibly energetic radio jets are produced at this moment in cosmic history and, after a journey across intergalactic space, will arrive just as humans begin observing the skies with radio telescopes.

IMAGE BY MARK A. GARLICK

THE LOUDEST OF ALL SIGNALS

—— Hercules A emits the loudest radio signal ——

UNIVERSE
Diameter: $3.20 \times 10^{13} \rightarrow 3.22 \times 10^{13}$ ly
Observable: $8.00 \times 10^{10} \rightarrow 8.06 \times 10^{10}$ ly
Expansion: $72.48 \rightarrow 72.15$ km/s/Mpc
Density: $0.38 \rightarrow 0.37$ p/m³
Temperature: $3.14 \rightarrow 3.12$ K

ASTRONOMICAL OBJECTS
Galaxies: $2.23 \times 10^{13} \rightarrow 2.21 \times 10^{13}$
Stars: $2.11 \times 10^{21} \rightarrow 2.11 \times 10^{21}$
Black Holes: $1.69 \times 10^{19} \rightarrow 1.69 \times 10^{19}$
Planetary Systems: $2.85 \times 10^{21} \rightarrow 2.86 \times 10^{21}$
Worlds: $3.29 \times 10^{22} \rightarrow 3.30 \times 10^{22}$

WORLDS WITH LIFE
Simple Life: $3.74 \times 10^{17} - 1.72 \times 10^{19}$
Complex Life: $2.22 \times 10^{11} - 3.66 \times 10^{17}$
Intelligent Life: $7.00 \times 10^{7} - 3.08 \times 10^{15}$
Technological Life: $2.20 \times 10^{4} - 4.16 \times 10^{13}$
Interstellar Life: $1 - 5.37 \times 10^{12}$

From all directions and locations, signals are propagating throughout the Universe at the speed of light: from the source where they were created until they arrive at their destination. These sources span the electromagnetic spectrum; although they include visible light sources, they also include phenomena that emit at both higher and lower energies. These range from tiny gamma-ray wavelengths that are shorter than the size of a proton through X-ray, ultraviolet, visible light, infrared, microwave, and radio signals, where the longest radio waves are comparable to the sizes of planets. Although the most interesting signals are artificial—broadcasts or messages created by intelligent life-forms—natural signals arising from astrophysical phenomena are far more common. It's at this moment in cosmic history that the loudest radio transmission ever recorded is created: from a galaxy known as Hercules A.

Every stellar system in the Universe has its own set of sources that emit at radio frequencies. Stars themselves typically outshine all other sources within their system at most frequencies: from X-rays down to radio waves, inclusive. Giant planets with strong magnetic fields also emit radio waves, as electrons become trapped in the space around them. Large moons around those planets affect those signals, occasionally focusing and "beaming" them in random directions. Stellar remnants, like neutron stars, black holes, and the ashes from supernova explosions, also emit them. Galaxies possess regions that contain clouds of neutral hydrogen, carbon monoxide, and even more complex molecules, and when those regions heat up due to star formation, they emit radio waves as well.

But all of these pale in comparison to the loudest radio transmissions of all within the Universe: radio galaxies and quasars. Powered by the supermassive black holes at the centers of these galaxies, large quantities of infalling material—typically arising from galactic mergers or other examples of cosmic violence—are accelerated by these central behemoths. Although a portion of that material gets devoured, growing the size of the supermassive black hole, much more of it is accreted, forming a disk that heats up, ionizes, and generates its own magnetic field. This magnetic field then accelerates the charged particles within it, and electrons in particular, which get collimated into two jets that are emitted perpendicular to the accretion disk. These accelerated jets can emit light at a wide variety of frequencies and shine more brightly, or loudly, in radio waves than anything else in the known Universe.

Some galaxies are extremely bright in this radio light, with Hercules A, a giant elliptical galaxy, emitting the loudest continuous radio signal that will be visible from Earth some 1.9 billion years from now: when human beings arise. Its impressive radio emissions extend for several million light-years, and the material that powers these emissions appears to arise as Hercules A continues to devour a smaller, but still extremely substantial, elliptical galaxy. For as long as there's material to sustain the accretion disk, there will be particles accelerated around the black hole and emitted in these energetic, collimated jets. As they collide and interact with the material in the circumgalactic medium, they create enormous features around radio galaxies: lobes of radio emissions.

The central activity of supermassive black holes can also affect the entire galaxy, in some cases, forming either active galactic nuclei or quasars. While quasars can be more powerful than radio galaxies—quasar 3C 273 is brightest in visible light, and the ultradistant quasar PSO J352 is intrinsically the loudest in the radio—no galaxy, as seen from Earth, appears louder in radio light than Hercules A.

Some stellar cataclysms, including hypernovae and gamma-ray bursts, produce powerful, highly collimated jets of radiation and energetic particles capable of sterilizing life on any planet they come in contact with for thousands upon thousands of light-years. The brightest, most energetic gamma-ray burst ever seen occurs at this moment in cosmic history: GRB 221009A, whose light is only detected in 2022.

IMAGE BY MARK A. GARLICK

THE BRIGHTEST OF ALL LIGHTS

70x brighter than prior record holders

UNIVERSE
Diameter: $3.22 \times 10^{13} \rightarrow 3.24 \times 10^{13}$ ly
  Observable: $8.06 \times 10^{10} \rightarrow 8.11 \times 10^{10}$ ly
Expansion: $72.15 \rightarrow 71.85$ km/s/Mpc
Density: $0.37 \rightarrow 0.36$ p/m³
Temperature: $3.12 \rightarrow 3.10$ K

ASTRONOMICAL OBJECTS
Galaxies: $2.21 \times 10^{13} \rightarrow 2.20 \times 10^{13}$
Stars: $2.11 \times 10^{21} \rightarrow 2.12 \times 10^{21}$
Black Holes: $1.69 \times 10^{19} \rightarrow 1.70 \times 10^{19}$
Planetary Systems: $2.86 \times 10^{21} \rightarrow 2.87 \times 10^{21}$
Worlds: $3.30 \times 10^{22} \rightarrow 3.31 \times 10^{22}$

WORLDS WITH LIFE
Simple Life: $3.69 \times 10^{17} - 1.70 \times 10^{19}$
Complex Life: $2.13 \times 10^{11} - 3.68 \times 10^{17}$
Intelligent Life: $6.72 \times 10^{7} - 3.11 \times 10^{15}$
Technological Life: $2.12 \times 10^{4} - 4.20 \times 10^{13}$
Interstellar Life: $1 - 5.46 \times 10^{12}$

While most astrophysical phenomena that can be detected occur over long—even cosmic—timescales, a number of important events are merely transient: arising all at once, peaking rapidly, then swiftly fading away. Objects that venture too close to a black hole can be ripped and shredded apart in a tidal disruption event, creating a spectacular burst of energy when it happens to a star. Supernovae are even more energetic, emitting a great blast of light across the electromagnetic spectrum whenever a massive star's core collapses or a white dwarf explodes. And neutron star mergers create kilonova events, where large quantities of concentrated nuclear material get ejected into the Universe, emitting energy from gamma rays down to radio light. But a special type of supernova, known as either a superluminous supernova or a hypernova, can outshine them all. It's right at this time that the brightest such event ever discovered, gamma-ray burst GRB 221009A, occurs in the Universe.

There are two main ways to make a gamma-ray burst in the Universe. One is through a kilonova, where two neutron stars merge, and that's responsible for approximately 30% of the gamma-ray bursts that have been observed. But the other 70% are caused by extremely energetic supernova events, hypernovae, that don't simply collapse and explode but emit the overwhelming majority of their energy in two highly collimated jets. Most of the time, these jets aren't pointed towards whatever location is observing them, and only a brighter-than-average supernova—along with an afterglow—appears. But whenever these jets happen to intersect with a star, planet, or even a distant galaxy, a special signature can be seen: a rapid and short-lasting burst of light, peaking at the highest energies of all, gamma rays. The stronger and more tightly collimated

a hypernova jet is, the more energy will be delivered to whatever objects happen to be in its path.

At the same time, however, it's important to recognize that "brightest" isn't necessarily the same thing as "most energetic" when it comes to astrophysical events. Energy can come in many different forms: particles, gravitational waves, and electromagnetic radiation (i.e., light) all included. Additionally, that energy can be emitted over a wide variety of durations, ranging from fractions of a second to hundreds of millions of years. Active galactic nuclei and quasars put out much greater amounts of energy than any supernova ever could, but they do so over extremely long timescales rather than in a short burst. Merging supermassive black holes are the most energetic events that occur in all the cosmos, excluding the hot Big Bang itself, but practically all of their energy is carried away in the form of gravitational waves; they aren't bright. "Bright" always refers to electromagnetic emission and how something appears to an observer, rather than being about any intrinsic property.

The most energetic burst to occur in the known history of the Universe was GRB 080319B, which occurred some 6.3 billion years after the Big Bang. That event injected a total of 21 quadrillion Suns' worth of energy into the Universe, powered by an extremely luminous afterglow that could be seen from billions of light-years away with the equivalent of a naked human eye. But far brighter, especially during the early, short-lived "burst" stage of a jet-driven supernova, was GRB 221009A. With an incredibly narrow, collimated jet, it outshines all prior record holders by a factor of ~70 in terms of the brightness of its burst. If a jet from this event struck a planet within its host galaxy, it would outshine that planet's sun, potentially wiping out all life on whichever hemisphere was exposed to the blast.

Unlike the smaller, more primitive prokaryotic organisms that dominated Earth for two billion years or more, eukaryotic organisms now rise to prominence. With specialized structures and differentiated organelles within it, including a cell nucleus, these eukaryotes, beginning as single-celled protists, will eventually give rise to all plants, animals, and fungi on Earth.

IMAGE BY MARK A. GARLICK

RISE OF THE EUKARYOTES

— *eukaryotes evolve on Earth* —

UNIVERSE
Diameter: $3.24 \times 10^{13} \rightarrow 3.27 \times 10^{13}$ ly
  Observable: $8.11 \times 10^{10} \rightarrow 8.18 \times 10^{10}$ ly
Expansion: $71.85 \rightarrow 71.50$ km/s/Mpc
Density: $0.36 \rightarrow 0.35$ p/m³
Temperature: $3.10 \rightarrow 3.07$ K

ASTRONOMICAL OBJECTS
Galaxies: $2.20 \times 10^{13} \rightarrow 2.19 \times 10^{13}$
Stars: $2.12 \times 10^{21} \rightarrow 2.13 \times 10^{21}$
Black Holes: $1.70 \times 10^{19} \rightarrow 1.70 \times 10^{19}$
Planetary Systems: $2.87 \times 10^{21} \rightarrow 2.88 \times 10^{21}$
Worlds: $3.31 \times 10^{22} \rightarrow 3.32 \times 10^{22}$

WORLDS WITH LIFE
Simple Life: $3.63 \times 10^{17} - 1.67 \times 10^{19}$
Complex Life: $2.04 \times 10^{11} - 3.70 \times 10^{17}$
Intelligent Life: $6.43 \times 10^{7} - 3.13 \times 10^{15}$
Technological Life: $2.03 \times 10^{4} - 4.24 \times 10^{13}$
Interstellar Life: $1 - 5.57 \times 10^{12}$

ontaining an oxygen-rich atmosphere and a thriving ecosystem filled with photosynthetic organisms, and with sexual reproduction now commonplace, life on Earth is poised to become more complex than ever. A variety of organisms have already evolved the machinery to encode specific, specialized molecules, like various types of chlorophyll, while others have evolved to successfully accomplish specific tasks, like the synthesis of the energy-storing molecule adenosine triphosphate (ATP). What now begins to occur is that larger, more complex cells begin to engulf the small, specialized prokaryotic organisms that carry out these functions, forming a symbiotic relationship. This is the first critical step that leads to the development of eukaryotic organisms: single-celled creatures that possess well-defined internal structures responsible for carrying out a variety of life processes essential to that organism's survival.

When an organism photosynthetically converts sunlight into nutrients, an organism in need of those nutrients can simply consume them, either by taking them from the organism that produced them or through predatory practices. If a photosynthetic organism produces a large store of sugars, a predator organism can acquire those sugars by consuming and metabolizing the prey organism. When an organism gains the ability to efficiently convert sugars into ATP—an energy-storing molecule that plays a vital biochemical role—a clever predator could feed sugars to that organism, wait for the organism to metabolize those sugars into ATP, and then consume that organism, acquiring the converted ATP molecules.

But another possibility for evolutionary success, other than outright predation, is to provide a hospitable environment where the organism that does whatever the host needs can thrive. This can quickly create a symbiotic relationship: The host organism benefits from the nutrients, metabolites, or life processes that the subsumed organism produces, while the hosted organism benefits from the protection and the hospitable environment offered by the host. Over extremely long timescales—often tens or even hundreds of millions of years—these organisms can coevolve into a single, larger, more complex organism (with a more complex genome) that produces its own version of these smaller, hosted organisms inside of it: organelles.

At this particular moment in Earth's evolutionary history, it's undergoing another "supercontinent" phase, where almost all of the tectonic plates that contain high-elevation landmasses have merged together. The supercontinent that emerges at this time, Columbia, is fantastically large. At its longest, it's approximately the diameter of Earth: nearly 13,000 kilometers across. With such a large supercontinent, it enables the formation of large bodies of fresh water: ponds, rivers, lakes, and even largely freshwater seas. It's in these environments that one of the most important events in evolutionary history occurs: the development of the chloroplast as a eukaryotic organelle.

Armed with the added benefit of sexual reproduction, common among (but not exclusive to) eukaryotes, organisms with increasing complex genomes are evolving. These symbiotic relationships provide a starting place for a number of important organelles to arise, which enable organisms to not only carry out a greater variety of life processes but to become more resilient to a variety of changes in their environments. Although prokaryotic life is still the overwhelming form of life on Earth at this time, the eukaryotes have definitively arrived and begin to occupy ecological niches from which no simple prokaryotic competitor can dislodge them.

Within the Milky Way, a binary star system forms. The more massive star, shown at the center, will only live for about 600 million years before dying and becoming a white dwarf. The less massive star will persist for some two billion years and will become known to humanity as Procyon, the brightest star in the constellation of Canis Minor.

WANDERERS OF THE NIGHT

the Procyon system roams the Milky Way

UNIVERSE
Diameter: $3.27 \times 10^{13} \rightarrow 3.29 \times 10^{13}$ ly
  Observable: $8.18 \times 10^{10} \rightarrow 8.23 \times 10^{10}$ ly
Expansion: $71.50 \rightarrow 71.20$ km/s/Mpc
Density: $0.35 \rightarrow 0.34$ p/m³
Temperature: $3.07 \rightarrow 3.05$ K

ASTRONOMICAL OBJECTS
Galaxies: $2.19 \times 10^{13} \rightarrow 2.18 \times 10^{13}$
Stars: $2.13 \times 10^{21} \rightarrow 2.13 \times 10^{21}$
Black Holes: $1.70 \times 10^{19} \rightarrow 1.70 \times 10^{19}$
Planetary Systems: $2.88 \times 10^{21} \rightarrow 2.88 \times 10^{21}$
Worlds: $3.32 \times 10^{22} \rightarrow 3.33 \times 10^{22}$

WORLDS WITH LIFE
Simple Life: $3.59 \times 10^{17} - 1.65 \times 10^{19}$
Complex Life: $1.96 \times 10^{11} - 3.71 \times 10^{17}$
Intelligent Life: $6.17 \times 10^{7} - 3.15 \times 10^{15}$
Technological Life: $1.95 \times 10^{4} - 4.28 \times 10^{13}$
Interstellar Life: $1 - 5.65 \times 10^{12}$

Inside every spiral galaxy in the Universe, stars are moving relative to one another, changing their real positions in three-dimensional space. On any inhabited planet within that galaxy, the creatures living there would see the stars in their night sky change over time as well. When stars are close by, they appear to change position rapidly; when they're farther away, they appear to move slowly. As they approach one another, they appear brighter; as they recede, they appear fainter. Somewhere in the Milky Way, about 18,000 light-years away from planet Earth, gas that's been funneled from the galactic halo into one of the spiral arms collapses and begins forming stars. Inside that cloud, a binary system emerges with two stars orbiting one another: a star system known as Procyon. With masses that are somewhat heavier than the Sun's mass, however, these Milky Way stars embark on a much more compressed cosmic journey of their own.

Nearby, thousands of other stars are also being born, and these newborn stars shine brilliantly, burning off the surrounding gas. With thousands of stars in such a small region of space, gravitational interactions are common. Some star systems get ejected early on, while others become more tightly bound to the cluster's center. The most massive stars in the cluster die swiftly—after only a few millions or tens of millions of years—dying in core-collapse supernovae. The combination of gravitational interactions and a variety of stellar cataclysms progressively kicks out more and more star systems, with Procyon being ejected about 30 million years after these stars first form. Now, Procyon begins roaming the Milky Way all on its own, with only its internal members to keep it company.

As it travels the Milky Way, its two main stars are separated by about the same distance that separates Saturn from the Sun. The two stars within the Procyon system, during the star-formation phase, managed to draw in more matter than our own Solar System did: a little bit over four times as much, all told. The more massive star, Procyon B, is about 2.6 times the mass of the Sun. Procyon B shines brilliantly and with a turquoise color, illuminating its surroundings 28 times as brightly as the Sun. The less massive star, known as Procyon A, is still 1.5 times the Sun's mass, shining in white light with the brightness of about four Suns. Because more massive stars burn through their fuel much more quickly, anyone identifying Procyon at the time of its birth could predict—with great accuracy—how this star system will evolve.

After about another 450 million years, Procyon B will exhaust the hydrogen in its core, evolving into a red giant and burning helium. About 50–100 million years after that, it will run out of its core helium, and as its core contracts, its outer layers will be gently blown off. Because of the presence of its binary companion—Procyon A—a large fraction of Procyon B's original mass will get expelled into the interstellar medium. As a result, the white dwarf that remains will be small and low in mass relative to the original star, weighing in at only 60% the mass of the Sun.

The loss of so much mass causes Procyon A and B to spiral outward from one another, now separated by about four times their original distance. Procyon A continues to get brighter over its lifetime and will begin to expand when its core hydrogen fuel gets depleted. This will happen right around the time it passes close to the Sun, missing a collision by only ~10 light-years. When humans eventually arise on Earth, Procyon A will appear as the eighth-brightest star in Earth's night sky—the star Alpha Canis Minoris—orbited by a faint white dwarf remnant: all that remains of Procyon B.

The most massive grouping of galaxies within 100 million light-years of Earth is the Virgo Cluster. Although the Local Group, which contains the Milky Way and Andromeda, is being gravitationally attracted by the Virgo Cluster, the effect of the expanding Universe is still greater, and the Virgo Cluster, currently receding at more than 1,000 kilometers per second, will never draw our Local Group into it.

IMAGE BY JON LOMBERG

PARTING IS SUCH SWEET SORROW

—— *the Virgo Cluster recedes from view* ——

UNIVERSE
Diameter: $3.29 \times 10^{13} \rightarrow 3.32 \times 10^{13}$ ly
  Observable: $8.23 \times 10^{10} \rightarrow 8.29 \times 10^{10}$ ly
Expansion: $71.20 \rightarrow 70.90$ km/s/Mpc
Density: $0.34 \rightarrow 0.34$ p/m³
Temperature: $3.05 \rightarrow 3.03$ K

ASTRONOMICAL OBJECTS
Galaxies: $2.18 \times 10^{13} \rightarrow 2.17 \times 10^{13}$
Stars: $2.13 \times 10^{21} \rightarrow 2.14 \times 10^{21}$
Black Holes: $1.70 \times 10^{19} \rightarrow 1.71 \times 10^{19}$
Planetary Systems: $2.88 \times 10^{21} \rightarrow 2.89 \times 10^{21}$
Worlds: $3.33 \times 10^{22} \rightarrow 3.35 \times 10^{22}$

WORLDS WITH LIFE
Simple Life: $3.54 \times 10^{17} - 1.63 \times 10^{19}$
Complex Life: $1.88 \times 10^{11} - 3.73 \times 10^{17}$
Intelligent Life: $5.93 \times 10^{7} - 3.17 \times 10^{15}$
Technological Life: $1.87 \times 10^{4} - 4.31 \times 10^{13}$
Interstellar Life: $1 - 5.74 \times 10^{12}$

 o matter where they're located, galaxies inevitably find themselves caught up in a cosmic tug-of-war. While the expansion of the Universe serves to drive everything apart, as though bound objects were raisins inside a ball of leavening dough, gravitation attempts to pull these different masses together, with more massive and closer objects exerting the greatest gravitational forces. As the Universe evolves, it clumps into galaxies, galaxy groups, galaxy clusters, and even into superclusters of galaxies. In the vicinity of the Milky Way, the largest grouping of galaxies—the Virgo Cluster—is only about 50 million light-years away. Consisting of more than 1,000 galaxies with a combined mass approaching a quadrillion (10^{15}) Suns, it's the most massive structure we'll ever encounter.

Early on, many small galaxies were drawn into the proto-Milky Way, causing it to grow into the large, massive galaxy it is at the present epoch. The nearby Andromeda galaxy grew even larger than we did, joining us to form the dominant members of our Local Group. Beyond the Local Group, an even larger, more massive collection of galaxies is found nearby: the Virgo Cluster. For the first several billion years of cosmic history, an observer would have wondered if, someday, the Local Group might be swallowed by the Virgo Cluster. By the present epoch, the answer is clear: No, no it won't. The Virgo Cluster is destined to disappear from view, as gravitational attraction simply isn't enough to overcome the dark energy–driven expansion of the Universe.

The Local Group is indeed being pulled towards the Virgo Cluster, as are all the other galaxies and galaxy groups within about 100 million light-years of its center. The Virgo Cluster is huge: Whereas our Local Group is only about three million light-years across, the Virgo Cluster spans approximately 15 million light-years in space. When someone within the Milky Way measures the relative motions of nearby objects, they find that we're "falling" towards the Virgo Cluster at a speed of about 250 kilometers per second, or faster than the Sun orbits around the center of the Milky Way. The force of gravity alone is doing this, as the Virgo Cluster's intense gravitational pull attracts any nearby mass it encounters.

The space between the galaxies of the Virgo Cluster isn't empty but, rather, is filled with an incredibly hot, low-density plasma at a temperature of ~30 million K: a signature that appears in X-ray light. As gas-rich galaxies speed through space, that plasma strips out the gas, where it collapses to form trails of stars. As many as 10% of the stars within the Virgo Cluster no longer belong to any galaxy but, rather, inhabit the intracluster space between the galaxies. But the most spectacular event occurring within the Virgo Cluster is surely the cosmic dance of giant galaxies, passing by one another and occasionally even colliding and merging at fantastic speeds of up to 1% the speed of light. Over time, some of the closest surrounding galaxies and galaxy groups even get drawn into the Virgo Cluster itself.

But the expansion of the Universe is too great for even the Virgo Cluster's gravity to draw us in. Instead, the entire cluster recedes from the Local Group at about 1,100 kilometers per second. For every 270 million years that pass, the Virgo Cluster and the Local Group separate by another million light-years. From within the Milky Way, every galaxy of the Virgo Cluster appears fainter and more distant with the passage of time. Unless an intelligent civilization arises within the Local Group that masters intergalactic travel, all its inhabitants can do is wave goodbye to the Virgo Cluster.

These small organelles, the mitochondria, are among the most important structures found inside eukaryotic cells. This cellular machinery routinely converts sugars into the molecule adenosine triphosphate (ATP), enabling cells to metabolize nutrients to generate their own sources of energy on demand and to survive even long periods of nutrient-sparse conditions.

IMAGE BY MARK A. GARLICK

POWERHOUSE OF THE CELL

—— mitochondria evolve on Earth ——

UNIVERSE
Diameter: $3.32 \times 10^{13} \rightarrow 3.34 \times 10^{13}$ ly
   Observable: $8.29 \times 10^{10} \rightarrow 8.36 \times 10^{10}$ ly
Expansion: $70.90 \rightarrow 70.57$ km/s/Mpc
Density: $0.34 \rightarrow 0.33$ p/m³
Temperature: $3.03 \rightarrow 3.00$ K

ASTRONOMICAL OBJECTS
Galaxies: $2.17 \times 10^{13} \rightarrow 2.15 \times 10^{13}$
Stars: $2.14 \times 10^{21} \rightarrow 2.14 \times 10^{21}$
Black Holes: $1.71 \times 10^{19} \rightarrow 1.71 \times 10^{19}$
Planetary Systems: $2.89 \times 10^{21} \rightarrow 2.89 \times 10^{21}$
Worlds: $3.35 \times 10^{22} \rightarrow 3.35 \times 10^{22}$

WORLDS WITH LIFE
Simple Life: $3.48 \times 10^{17} - 1.61 \times 10^{19}$
Complex Life: $1.79 \times 10^{11} - 3.74 \times 10^{17}$
Intelligent Life: $5.66 \times 10^{7} - 3.19 \times 10^{15}$
Technological Life: $1.79 \times 10^{4} - 4.35 \times 10^{13}$
Interstellar Life: $1 - 5.84 \times 10^{12}$

Both freshwater and saltwater stores on Earth, wherever liquid water persists, now teem with a tremendous diversity of living creatures. Although simple, single-celled organisms like cyanobacteria still thrive, more complex eukaryotes are thriving alongside them. This complexity includes a series of machine-like parts inside of eukaryotic cells: organelles, specialized to perform important functions. Some organelles are used to transport nutrients. Others are used to eliminate waste products. Chloroplasts, the recently evolved organelle that enables eukaryotic photosynthesis, are now found across a wide variety of species. With the right machinery in place, organisms are now producing sugars—incredible sources of stored energy—in large quantities. All that's needed to take advantage of sugars is the right cellular machinery to metabolize them. It's at this time that the eukaryotic organelle responsible for performing exactly that function evolves: the mitochondria, also known as the powerhouse of the cell.

The precursor of the mitochondria was a prokaryotic organism: one that could metabolize a simple sugar, such as glucose. Another, larger organism—likely a eukaryote, possibly even a sugar-producing eukaryote—engulfed it, but instead of digesting it, it engaged in a symbiotic relationship. The larger organism offered protection for this prokaryote and could obtain nutrients for it: sugars. The mitochondrial precursor organism transforms those nutrients into energy stored as the molecule adenosine triphosphate (ATP), which then becomes available to the larger, host organism. Both organisms survive and thrive together, creating a symbiotic relationship.

It's then that a remarkable biological process takes place: horizontal gene transfer. Although genes are most commonly transferred from parent to offspring, it's also possible for genetic material from one organism to be absorbed and incorporated into the DNA of another organism.

The mechanisms of both gene recombination and gene insertion can accomplish this, and the result is one single, complex organism that possesses all the genes of the original organism plus an incorporated set of genes from the other.

By integrating the genes from the mitochondrial precursor into its genome, the eukaryotic organism gained the ability to create a new organelle: the mitochondria. Because the eukaryote can already take care of a great number of functions that the mitochondrial precursor previously had to accomplish on its own, the mitochondria could become simpler, more efficient, and more specialized as organelles within the eukaryotes. All of a sudden, any eukaryote without these genes would find itself at a severe competitive disadvantage when compared with one that possesses mitochondria as an organelle. As a result, a eukaryotic organism needed to either acquire that genetic information—again, likely via horizontal gene transfer—or run the risk of extinction.

In short order, that's precisely what happens in this epoch. The eukaryotes that survive and thrive all possess mitochondria, and all of the mitochondria derive from a single common ancestor. With them, eukaryotes of all types gain the ability to metabolize sugars and convert them into energy. Without them, the other eukaryotes either swiftly incorporate them or else die out. Additionally, oxygen-using (i.e., aerobic) mitochondria swiftly become the most common type of mitochondria, as they liberate more energy per molecule than any other metabolic pathway. Although eukaryotes will later arise once again without mitochondria, that will be the consequence of an evolutionary loss, where certain mitochondrial pathways are replaced via horizontal gene transfer from bacteria. From this point forward, the most complex organisms on Earth can all now metabolize sugars to liberate usable energy, granting life the power to sustain itself over substantial periods via stored nutrients alone.

The most energetic phenomena in the Universe aren't stars, stellar cataclysms, or even galaxy mergers. Instead, the greatest instantaneous releases of energy come in the form of gravitational waves: generated most strongly by merging black holes. With the capability of converting enormous quantities of mass into energy via Einstein's $E=mc^2$, these events can temporarily outshine all the stars in the Universe combined.

IMAGE BY MARK A. GARLICK

TWO DRAGONS BECOME ONE

—— *black holes merge and release vast energy* ——

UNIVERSE
Diameter: $3.34 \times 10^{13} \rightarrow 3.37 \times 10^{13}$ ly
 Observable: $8.36 \times 10^{13} \rightarrow 8.42 \times 10^{10}$ ly
Expansion: $70.57 \rightarrow 70.29$ km/s/Mpc
Density: $0.33 \rightarrow 0.32$ p/m³
Temperature: $3.00 \rightarrow 2.99$ K

ASTRONOMICAL OBJECTS
Galaxies: $2.15 \times 10^{13} \rightarrow 2.14 \times 10^{13}$
Stars: $2.14 \times 10^{21} \rightarrow 2.15 \times 10^{21}$
Black Holes: $1.71 \times 10^{19} \rightarrow 1.72 \times 10^{19}$
Planetary Systems: $2.89 \times 10^{21} \rightarrow 2.90 \times 10^{21}$
Worlds: $3.35 \times 10^{22} \rightarrow 3.36 \times 10^{22}$

WORLDS WITH LIFE
Simple Life: $3.44 \times 10^{17} - 1.59 \times 10^{19}$
Complex Life: $1.72 \times 10^{11} - 3.76 \times 10^{17}$
Intelligent Life: $5.44 \times 10^{7} - 3.21 \times 10^{15}$
Technological Life: $1.72 \times 10^{4} - 4.39 \times 10^{13}$
Interstellar Life: $1 - 5.92 \times 10^{12}$

early everything that can be detected in this Universe has one thing in common: energy. Whether a star shines or brightens, the molecules within a gas cloud collide, or the mutual orbit of two masses decays, energy is released in some form: light, heat, kinetic motion, or even in the form of gravitational waves. Although this last form of energy is exceedingly difficult to detect—as even a strong gravitational wave source might only compress and expand a planet like Earth by the width of a few protons—the events that generate them can briefly release more energy than all the stars within the observable Universe combined. During this epoch in cosmic history, two black holes, each the remnant left over by a massive blue supergiant star after its death, inspiraled and merged together. During the final ~200 milliseconds or so of this merger, it outshone, in terms of energy, all the stars in the Universe put together.

Whenever stars that contain heavy elements are born, they come in a wide variety of masses, ranging from just a few percent of the Sun's mass—the minimum mass required to ignite nuclear fusion in their cores—up to several hundred times the mass of the Sun. These stars frequently arrive as members of multi-star systems: binaries, trinaries, quaternaries, and more. Only about half of all stars exist in "singlet" stars like our Sun, and many multi-star systems have widely mismatched masses, particularly at the lower-mass end. The most massive stars, on the other hand, frequently wind up bound together, with two or more stars of very high masses orbiting relatively closely to one another. Wherever large numbers of new, massive stars are born, these high-mass, multi-star systems can be found.

In one far-flung galaxy some 1.3 billion light-years from Earth, one such episode of star formation occurs. On a cosmic scale, it's rather unremarkable: birthing several thousands of newborn stars in short order, just like millions of other galaxies. The most massive newborn stars burn through their fuel and swiftly die, collapsing to form black holes after only a few million years. This creates numerous situations where two massive black holes orbit one another at relatively short distances. In these scenarios, each black hole bends the fabric of space due to its gravity, and when the other black hole moves through that bent space, it emits energy in the form of gravitational waves. Over time, these orbits decay, the rate of gravitational wave emission increases, and eventually, these pairs of black holes inspiral and merge together.

Two such black holes engage in exactly this dance: one with 36 times the mass of the Sun and one with 29 times the Sun's mass. When they finish the inspiral phase and combine together, they wind up producing a post-merger black hole with a mass of 62 solar masses. This might appear to be bizarre, since $36 + 29 = 65$, not 62. So where did the other 3 solar masses go?

They were converted into energy—energy in the form of gravitational waves—via Einstein's most famous equation: $E = mc^2$. These gravitational waves propagate outward, in all directions, at the speed of light, and alternately compress and expand any volume-occupying object they happen to pass through. After another 1.3 billion years go by, a tiny fraction of these waves will pass through planet Earth, which will be equipped, for the first time, with gravitational wave detectors capable of seeing this signal. This event, GW150914, emitted 3 solar masses worth of energy over just a few milliseconds. It was more energetic, over that time interval, than all the stars in the Universe combined.

The first complex organisms to colonize the continental landmasses of Earth are neither plant nor animal, neither of which has yet emerged at this point, but fungi. Although the first fungi arise in aquatic environments, they swiftly develop symbiotic relationships with algae, creating lichen-like structures that can exist in sufficiently wet land-based environments.

IMAGE BY MARK A. GARLICK

RISE OF THE FUNGI

fungi evolve on Earth

UNIVERSE
Diameter: $3.37 \times 10^{13} \rightarrow 3.39 \times 10^{13}$ ly
  Observable: $8.42 \times 10^{10} \rightarrow 8.48 \times 10^{10}$ ly
Expansion: $70.29 \rightarrow 70.01$ km/s/Mpc
Density: $0.32 \rightarrow 0.32$ p/m³
Temperature: $2.99 \rightarrow 2.96$ K

ASTRONOMICAL OBJECTS
Galaxies: $2.14 \times 10^{13} \rightarrow 2.13 \times 10^{13}$
Stars: $2.15 \times 10^{21} \rightarrow 2.15 \times 10^{21}$
Black Holes: $1.72 \times 10^{19} \rightarrow 1.72 \times 10^{19}$
Planetary Systems: $2.90 \times 10^{21} \rightarrow 2.91 \times 10^{21}$
Worlds: $3.36 \times 10^{22} \rightarrow 3.37 \times 10^{22}$

WORLDS WITH LIFE
Simple Life: $3.39 \times 10^{17} - 1.57 \times 10^{19}$
Complex Life: $1.65 \times 10^{11} - 3.77 \times 10^{17}$
Intelligent Life: $5.22 \times 10^{7} - 3.22 \times 10^{15}$
Technological Life: $1.65 \times 10^{4} - 4.42 \times 10^{13}$
Interstellar Life: $1 - 6.00 \times 10^{12}$

On Earth, biological organisms are getting more and more interesting as the years tick by. As the unbroken chain of life continues, the combined factors of inheritance, random mutations, and horizontal gene transfer serve to increase the total amount of genetic information found in the genomes of the most complex organisms. This results in them gaining more specialized features, and many new characteristics begin emerging. Some organisms thrive together in colonies, with identical unicellular life-forms binding to one another to ensure that the majority of them survive and thrive. Other organisms develop multicellularity: the ability for a single organism to produce multiple component parts—cells—that all remain bound together as part of the original, parent organism. And still other organisms become differentiated, where new subcomponents develop within an organism, conferring features and abilities onto it that it didn't possess before. At this moment in time, the last of these effects leads to an entirely new kingdom of life on Earth: the fungi.

Evolving well before plants or animals arise on our planet, these early fungi likely thrive in aquatic environments and possess flagella: tiny, thread-like tails that allow them to control their motion through water. The fungi are all eukaryotic organisms that reproduce through the creation of spores, which mature atop microscopic, soft-tissue structures known as fruiting bodies. In aquatic environments, flagella are required to transport spores away from the parent body, towards locations where they can gain nutrients, thrive, and survive until they reach reproductive age.

However, many species of fungus soon adapt to thrive on land as well. On land, fungi develop filament-like structures that connect member organisms with one another, where they then form a mycelium-like network. By radiating branches outward that have the potential to connect with the branches of neighboring fungi, they create a network through a process known as anastomosis, where splitting branches recombine. Due to their small sizes and soft, easily degradable bodies, fungi only rarely fossilize, leaving little trace of their presence from so long ago. While aquatic fungi are known to arise during approximately this epoch, substantial debate exists as to when terrestrial fungi first arose, with time estimates varying by more than half a billion years.

Among the oxygen producers, cyanobacteria still dominate the biosphere. While oxygen-producing algae have already come into existence during this time, they produce only a tiny fraction of Earth's oxygen and won't rise to prominence for several hundred million years. However, small populations of terrestrial algae can—in concert with fungi—begin producing structures that resemble modern-day lichen, where algae and fungi enter into a symbiotic relationship in wet environments on land. The land itself continues to shift due to Earth's tectonic activity, with the ancient supercontinent Rodinia beginning to assemble during this time.

Although the most well-known form of fungus is the mushroom, containing the familiar cap-and-stem structure, they won't arise for more than a billion years after the first fungi appear. Instead, these early fungi are mostly single-celled, aquatic-based forms of life with flagella and spores capable of traveling great distances through their watery environs. With mitochondria operating within their cells, the machinery is already in place for multicellular, sexually reproducing eukaryotes to arise. Over the next several hundred million years, these components pave the way for the first large, complex organisms. The stage on Earth is set, at long last, for the appearance of plants and animals.

The most volcanically active planet in the Solar System is likely not Earth but our sister world: Venus. Unlike Earth, Venus experiences periodic resurfacing events - where lava flows cover nearly its entire surface, including one so massive, at this moment in time, that it wipes out nearly all of Venus's geologic record prior to this period. Unlike Earth's volcanoes, Venus's are flat and wide, producing the pancake-dome structures shown here.

IMAGE BY MARK A. GARLICK

THE CHANGING FACE OF VENUS

—— *volcanism continuously resurfaces Venus* ——

UNIVERSE
Diameter: $3.39 \times 10^{13} \rightarrow 3.41 \times 10^{13}$ y
  Observable: $8.48 \times 10^{10} \rightarrow 8.54 \times 10^{10}$ ly
Expansion: $70.01 \rightarrow 69.73$ km/s/Mpc
Density: $0.32 \rightarrow 0.31$ p/m³
Temperature: $2.95 \rightarrow 2.94$ K

ASTRONOMICAL OBJECTS
Galaxies: $2.13 \times 10^{13} \rightarrow 2.12 \times 10^{13}$
Stars: $2.15 \times 10^{21} \rightarrow 2.16 \times 10^{21}$
Black Holes: $1.72 \times 10^{19} \rightarrow 1.72 \times 10^{19}$
Planetary Systems: $2.91 \times 10^{21} \rightarrow 2.92 \times 10^{21}$
Worlds: $3.37 \times 10^{22} \rightarrow 3.38 \times 10^{22}$

WORLDS WITH LIFE
Simple Life: $3.35 \times 10^{17} - 1.55 \times 10^{19}$
Complex Life: $1.58 \times 10^{11} - 3.78 \times 10^{17}$
Intelligent Life: $5.00 \times 10^{7} - 3.24 \times 10^{15}$
Technological Life: $1.58 \times 10^{4} - 4.45 \times 10^{13}$
Interstellar Life: $1 - 6.08 \times 10^{12}$

Throughout the Solar System, the solid surfaces of planets, moons, asteroids, and Kuiper belt objects all provide a record of their histories through one key property: their crater records. On an airless world like the Moon or Mercury, even small craters can easily persist for billions of years, with younger, smaller craters superimposed atop the older, larger ones. On the other extreme, worlds like Jupiter's large, inner moon Io are so volcanically active that any craters that appear on their surfaces are quickly and completely erased, as erupting lava recoats the surface with great frequency. While craters tell a vital part of the planetary stories of Mercury, Earth, and Mars, this absolutely isn't the case for Venus. Almost none of Venus's surface from this epoch or earlier persists to the present day, however. These erased features obscure Venus's early history and provide key evidence that, unlike most of the solid-surfaced worlds in our Solar System, Venus's surface has been overwritten—perhaps repeatedly—even this late in our shared cosmic history.

Venus is sometimes known as Earth's "sister" planet, due to their comparable sizes, compositions, and masses. But owing to a few small physical differences, these two planets experienced very different fates. Being closer to the Sun, Venus receives far more heat from the Sun—about 92% more—than Earth does. Once Venus's oceans boil away, it no longer has large quantities of surface water, which could hinder many activities associated with plate tectonics. That event likely precipitated the runaway greenhouse effect that took place on Venus, creating a surface that's uniformly about 480°C (900°F) beneath the pressure equivalent of 90 Earth atmospheres.

But both Venus and Earth nonetheless remain volcanically active worlds throughout their histories, with new lava flows continuously creating, growing, and resurfacing a variety of land features on both worlds. Much like Earth, hundreds of known volcanoes can be identified at most points in time across Venus's surface. Unlike Earth, however, Venus appears to experience near-global resurfacing events periodically, including a major event that wiped out almost all of Venus's geological record prior to this time. The enormous amount of carbon dioxide present in Venus's atmosphere—along with the sulfur found in its sulfuric acid clouds—may be due entirely to gases emitted as a consequence of volcanic eruptions.

Also unlike Earth, where the terrain is shaped by a wide variety of processes, the surface of Venus—although checkered with relatively recently created craters—is dominated by volcanic features. Ninety percent of Venus's surface is covered in basalt: an igneous rock formed from rapidly cooling lava. Two-thirds of the Venusian surface is coated in volcanic lava plains: the dominant terrain found on Venus. Earth and Venus both possess shield volcanoes and lava flows, but Venus has additional structures created by lava that aren't present elsewhere. It possesses pancake domes, which are extremely wide, large-diameter volcanoes that rise less than 1,000 meters in elevation, often found near highly deformed, even folded and fractured terrain. Venus also possesses volcanic features known as scalloped margin domes, which have collapsed onto themselves and experienced landslides down their sides; from above, they look like ticks, the parasitic arachnids found on Earth.

However, Earth has a type of volcano not found on Venus: stratovolcanoes. Without large quantities of surface water or subducting tectonic plates, Venus cannot create these structures. Instead, Venusian volcanoes are very wide and not particularly high, an effect possibly caused by the intense atmospheric pressure. The global resurfacing event that happens during this epoch on Venus will be followed by others, giving Venus the youngest surface of any planet within the Solar System.

While the star-formation rate continues to drop on a cosmic scale, there are isolated bursts in galaxies all across the Universe where new stars rapidly form in great numbers. During this time interval, about one billion years before the present, a burst of star formation occurred in the Milky Way's center, producing a large number of new stars and greatly growing our galaxy's central barred structure.

IMAGE BY JON LOMBERG

A SURPRISE OF STARBURSTS

——— star formation rises in the Milky Way ———

UNIVERSE
Diameter: $3.41 \times 10^{13} \rightarrow 3.44 \times 10^{13}$ ly
 Observable: $8.54 \times 10^{10} \rightarrow 8.60 \times 10^{10}$ ly
Expansion: $69.73 \rightarrow 69.46$ km/s/Mpc
Density: $0.31 \rightarrow 0.30$ p/m³
Temperature: $2.94 \rightarrow 2.92$ K

ASTRONOMICAL OBJECTS
Galaxies: $2.12 \times 10^{13} \rightarrow 2.10 \times 10^{13}$
Stars: $2.16 \times 10^{21} \rightarrow 2.16 \times 10^{21}$
Black Holes: $1.72 \times 10^{19} \rightarrow 1.73 \times 10^{19}$
Planetary Systems: $2.92 \times 10^{21} \rightarrow 2.92 \times 10^{21}$
Worlds: $3.38 \times 10^{22} \rightarrow 3.39 \times 10^{22}$

WORLDS WITH LIFE
Simple Life: $3.30 \times 10^{17} - 1.53 \times 10^{19}$
Complex Life: $1.52 \times 10^{11} - 3.79 \times 10^{17}$
Intelligent Life: $4.81 \times 10^{7} - 3.25 \times 10^{15}$
Technological Life: $1.51 \times 10^{4} - 4.47 \times 10^{13}$
Interstellar Life: $1 - 6.15 \times 10^{12}$

Falling ever since the Universe's heyday nearly 10 billion years ago, the star-formation rate across all galaxies, on average, continues to decline. Here in our own Milky Way, our home galaxy is a microcosm of that, having formed a whopping 80% of its total stars during the first five to six billion years of cosmic history. But right around now, something peculiar and unexpected begins to happen. All of a sudden, an enormous amount of gas is funneled towards the galactic center, where it accumulates in the galactic bulge. Even though the overwhelming majority of the stars in this bulge formed in the Universe's early stages, a new "burst" of star formation takes place during this interval, accounting for around 5% of the galactic bulge's total stars. All told, nearly one billion new stars form in the galaxy's center during this brief 100-million-year interval, reshaping the Milky Way in the process.

Astronomically, the formation of new stars always leaves behind one of the most clearly identifiable signatures in the Universe. First, the gas surrounding the star-forming regions gets heated and ionized, with most of that gas being hydrogen, by composition. As the (now) free electrons run into those hydrogen nuclei, they form neutral atoms once again. When the electrons cascade down the various energy levels, they release light: infrared, visible, and ultraviolet photons. One specific transition in hydrogen atoms—from the third to the second energy level—always releases light of the same precise wavelength: 656.3 nanometers. That corresponds to red light, which explains why star-forming regions always have an optically "pink" hue to them.

Typically, those star-forming regions trace out the densest regions of gas. In the Milky Way, those occur along the spiral arms and in the central region, where gas gets funneled inward, towards the galactic center. One of the more interesting ideas is that this relatively late period of intense star formation helped form and grow the "bar" at the center of the Milky Way. While a significant proportion of spiral galaxies have prominent bar-like features at the center, it was only in the 21st century that the Milky Way's bar would be confirmed and mapped out. Prior to this star-forming event, it may have only reached some 4,000–5,000 light-years away from the galactic center; during this 100-million-year interval, it more than doubles in size, reaching between 10,000 and 13,000 light-years away from the galactic center in both directions.

But perhaps the greatest change that occurs in the galaxy is due to all the new stars created in the central, bulge-like region of the Milky Way. Although the overwhelming majority of new stars that form are small, faint, and low in mass—with around 95% of them less massive than the Sun—a small fraction of them form with enough mass to end their lives in a core-collapse supernova. With this event of rapid star formation in a concentrated area of space, hundreds of thousands or perhaps even millions of new supernova events occur in the Milky Way's central region alone during this relatively brief interval. These supernovae efficiently expel the neutral matter from the galaxy's center, removing the light-blocking material and rendering it easy to view.

Unfortunately, these conditions won't last long on cosmic timescales, as the spiral arms continue to funnel new stores of gas from the galactic outskirts into the central region. By the time complex life arises on Earth, the only evidence will be a young stellar population identifiable within the Milky Way's galactic bulge.

Although single-celled life-forms still dominate the biomass of Earth, the first plants have now arisen in the waters of Earth. Resembling algae and/or moss, they colonized both the shallow seafloor and floated on the watery surface, where they feed on solar energy and breathe carbon dioxide gas, producing oxygen as a waste product. In a few hundred million years, they'll colonize Earth's continents as well.

IMAGE BY MARK A. GARLICK

RISE OF THE PLANTS

—— marine plants evolve on Earth ——

UNIVERSE
Diameter: $3.44 \times 10^{13} \rightarrow 3.46 \times 10^{13}$ ly
  Observable: $8.50 \times 10^{10} \rightarrow 8.66 \times 10^{10}$ ly
Expansion: $69.46 \rightarrow 69.20$ km/s/Mpc
Density: $0.30 \rightarrow 0.30$ p/m³
Temperature: $2.92 \rightarrow 2.90$ K

ASTRONOMICAL OBJECTS
Galaxies: $2.10 \times 10^{13} \rightarrow 2.10 \times 10^{13}$
Stars: $2.16 \times 10^{21} \rightarrow 2.17 \times 10^{21}$
Black Holes: $1.73 \times 10^{19} \rightarrow 1.73 \times 10^{19}$
Planetary Systems: $2.92 \times 10^{21} \rightarrow 2.93 \times 10^{21}$
Worlds: $3.39 \times 10^{22} \rightarrow 3.41 \times 10^{22}$

WORLDS WITH LIFE
Simple Life: $3.25 \times 10^{17} - 1.51 \times 10^{19}$
Complex Life: $1.46 \times 10^{11} - 3.79 \times 10^{17}$
Intelligent Life: $4.61 \times 10^{7} - 3.27 \times 10^{15}$
Technological Life: $1.46 \times 10^{4} - 4.50 \times 10^{13}$
Interstellar Life: $1 - 6.23 \times 10^{12}$

To a casual observer, life on Earth might not appear to be all that different at the present epoch as it was a full one billion years prior. Single-celled organisms are still the dominant forms of life on Earth, while cyanobacteria remain the primary oxygen producers. Most of the simpler organisms reproduce asexually, while the more complex ones with longer life spans are more successful when they reproduce sexually. A few life-forms still possess multiple cells acting together for the benefit of the greater organism, with several other species of unicellular organisms banding together in colonies.

But a deeper look shows a new complexity that wasn't there before. Cyanobacteria have diversified and now occupy waters with varying salinities and acidities, including nearly all saltwater and freshwater sources. The eukaryotes now all possess mitochondria, with most retaining the typically accompanying capacity for aerobic respiration. Protists, single-celled eukaryotes with a variety of internal and external structures and functions, are ubiquitous across most aqueous environments. They possess functional genomes with 10 times as much genetic information packed inside them as their predecessors from a billion years prior. Fungi have not only arisen but have begun to diversify, creating a presence on land as well as in the water. And now, for the first time, a eukaryotic, photosynthetic organism arises with not just a cell membrane to protect it but a rigid layer of sugar molecules bound together outside of it: a cell wall. The first plant life has arrived on Earth.

The first plants were likely simple collections of single-celled organisms, similar to the algal mats that cyanobacteria form on the surfaces of rocks and sediments. In fact, it's thought that plants arose due to horizontal gene transfer occurring between a flagella-containing protist and a cyanobacterium, giving rise to a new kingdom of organisms. Multicellular plants arrived relatively quickly thereafter, with red and green algae both emerging at this time. The green algae—containing both chlorophyll a and chlorophyll b molecules, absorbing blue and red light, but reflecting green light—rapidly diversify into thousands of species. After a few million years, green algae begin to occupy not only significant niches in marine environments but in freshwater communities as well.

As a consequence of horizontal gene transfer, the genomes of complex eukaryotes—plants and fungi—have expanded significantly and now include large amounts of noncoding genetic sequences (i.e., "junk" DNA) in their genetic codes. One billion years earlier, nearly all of an organism's genome was composed of coding genetic sequences: genes that actively give instructions to produce proteins necessary for the life processes of the organism. In bacteria, for instance, 90% and upwards of their DNA encodes proteins. But when mitochondria develop within eukaryotes, their acquisition also inserts what are known as "introns" into eukaryotic DNA, which help to transcribe various types of RNA. Introns enable the insertion of new sequences of nucleotides: both protein-coding genes and also nonfunctional nucleotide sequences. For some plants and fungi, the majority of their genomes become this noncoding "junk" type of DNA. Although some argue that junk DNA may serve a function—like ensuring that chromosomes bundle correctly or helping to regulate the development of an organism—most of it really is junk: simply a string of genetic code that has no impact on an organism's fitness.

Despite their aquatic origins, plants don't remain confined to the sea for long. Algae-like plants were likely the first to emerge on land, where they thrived and formed lichen-like colonies in concert with fungi. However, not long after, the descendants of green algae would join them, forming communities of complex, multicellular, photosynthetic plant life. In only a few hundred million years, they'll colonize nearly all the land on Earth.

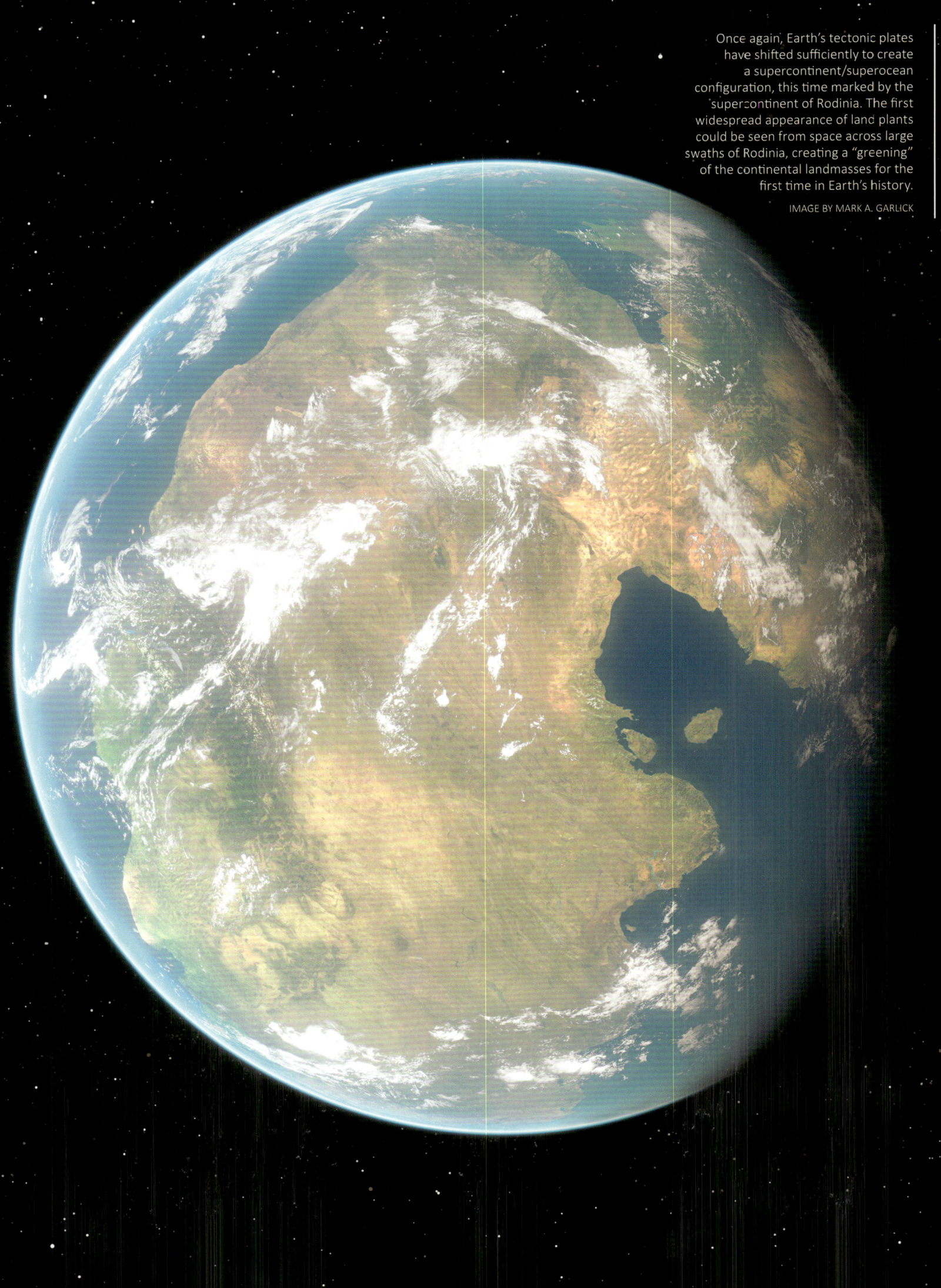

Once again, Earth's tectonic plates have shifted sufficiently to create a supercontinent/superocean configuration, this time marked by the supercontinent of Rodinia. The first widespread appearance of land plants could be seen from space across large swaths of Rodinia, creating a "greening" of the continental landmasses for the first time in Earth's history.

IMAGE BY MARK A. GARLICK

THE GREENING OF EARTH

—— *land plants evolve on Earth* ——

UNIVERSE
Diameter: $3.46 \times 10^{13} \rightarrow 3.49 \times 10^{13}$ ly
  Observable: $8.66 \times 10^{10} \rightarrow 8.72 \times 10^{10}$ ly
Expansion: $69.20 \rightarrow 68.93$ km/s/Mpc
Density: $0.30 \rightarrow 0.29$ p/m³
Temperature: $2.90 \rightarrow 2.88$ K

ASTRONOMICAL OBJECTS
Galaxies: $2.10 \times 10^{13} \rightarrow 2.08 \times 10^{13}$
Stars: $2.17 \times 10^{21} \rightarrow 2.17 \times 10^{21}$
Black Holes: $1.73 \times 10^{19} \rightarrow 1.74 \times 10^{19}$
Planetary Systems: $2.93 \times 10^{21} \rightarrow 2.94 \times 10^{21}$
Worlds: $3.41 \times 10^{22} \rightarrow 3.42 \times 10^{22}$

WORLDS WITH LIFE
Simple Life: $3.21 \times 10^{17} - 1.49 \times 10^{19}$
Complex Life: $1.40 \times 10^{11} - 3.80 \times 10^{17}$
Intelligent Life: $4.42 \times 10^{7} - 3.28 \times 10^{15}$
Technological Life: $1.40 \times 10^{4} - 4.53 \times 10^{13}$
Interstellar Life: $0 - 6.30 \times 10^{12}$

While increasingly complex and diversified forms of life are reshaping Earth's biosphere, geological processes are reshaping Earth's continents and oceans. For the previous ~360 million years before the present epoch, tectonic plates have been colliding, bringing continents together in the aftermath of the breakup of the Columbia supercontinent. But it's now some 12.9 billion years after the Big Bang and about 3.6 billion years after the creation of Earth that a new supercontinent finishes forming: Rodinia. Composed of all seven of the modern-day continents, Rodinia was surrounded by a superocean: Mirovia. With roughly 25–30% of Earth's surface covered by Rodinia, the remaining 70–75% was simply one continuous ocean.

As seen from afar, Earth's eastern and western hemispheres couldn't be more different: one completely blue and bare, the other possessing a mountain-rich supercontinent extending some 15,000 kilometers from end to end. But it's during this time period that the color of the supercontinent begins to change. Prior to now, it was only the most water-adjacent land areas—lakes, rivers, seas, and oceans—that possessed any life-forms, like fungi and plants. But about halfway through this current 100-million-year interval, land plants begin to thrive and, for the first time, begin to produce more oxygen than the cyanobacteria, building up the oxygen concentrations in Earth's atmosphere to their highest levels in terrestrial history thus far.

As plants begin to occupy these previously empty ecological niches, they begin to slowly change the color of Earth's surface: from a dirt-like series of browns and tans into a variety of shades of green. For the first time, an alien civilization viewing planet Earth from afar could see our northern and southern hemispheres change colors, as the seasonally changing temperatures would induce periods of plant growth and dormancy.

But another dramatic change begins to happen on Earth about 25 million years before this present era ends: A superplume emerges deep in Earth's interior, extending for thousands of kilometers at the core-mantle boundary. This superplume creates, over the next several million years, a series of influential changes along the supercontinent of Rodinia. As magma upswells, it affects the continental land atop it in a variety of ways. Eruptions occur, emitting both mafic (dark, rich in magnesium and iron) and felsic (light in color and typically forming quartz and feldspar) lavas, but not mixtures of the two. Crustal arching occurs, where lateral stresses combined with subterranean heating create geological structures known as anticlines. And sedimentary basins form along rifts created from the aftermath of the superplume. As the present epoch nears its end, the supercontinent of Rodinia begins to break apart.

The geological evidence of the breakup of this supercontinent can be found all across the world, including in modern-day Australia, China, India, southern Africa, and in the Middle East. As Rodinia begins to break apart, a new ocean begins to form: an ocean that will eventually grow into the Panthalassa superocean that will encircle Earth's next supercontinent, Pangaea. However, Mirovia will remain the dominant ocean on Earth for several hundred million years to come, even as Earth's tectonic plates continue to break apart and migrate across the surface. The combination of biological and geological forces are slowly reshaping planet Earth, cementing its status as a thriving, living world.

WINTER IS COMING

—— *the Sturtian glaciation begins* ——

UNIVERSE
Diameter: $3.49 \times 10^{13} \rightarrow 3.51 \times 10^{13}$ ly
 Observable: $8.72 \times 10^{10} \rightarrow 8.78 \times 10^{10}$ ly
Expansion: $68.93 \rightarrow 68.68$ km/s/Mpc
Density: $0.29 \rightarrow 0.28$ p/m³
Temperature: $2.88 \rightarrow 2.86$ K

ASTRONOMICAL OBJECTS
Galaxies: $2.08 \times 10^{13} \rightarrow 2.07 \times 10^{13}$
Stars: $2.17 \times 10^{21} \rightarrow 2.18 \times 10^{21}$
Black Holes: $1.74 \times 10^{19} \rightarrow 1.74 \times 10^{19}$
Planetary Systems: $2.94 \times 10^{21} \rightarrow 2.95 \times 10^{21}$
Worlds: $3.42 \times 10^{22} \rightarrow 3.42 \times 10^{22}$

WORLDS WITH LIFE
Simple Life: $3.17 \times 10^{17} - 1.47 \times 10^{19}$
Complex Life: $1.34 \times 10^{11} - 3.80 \times 10^{17}$
Intelligent Life: $4.24 \times 10^{7} - 3.29 \times 10^{15}$
Technological Life: $1.34 \times 10^{4} - 4.55 \times 10^{13}$
Interstellar Life: $0 - 6.37 \times 10^{12}$

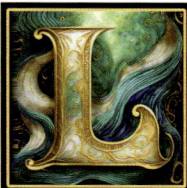

life continues to diversify in new and surprising ways, and an unprecedented combination occurs in Earth's oceans as a multicellular life-form evolves that can obtain nutrients by filtering the ocean waters—including other life-forms—through its body. This spongelike creature marks the development of the first animal on Earth: a kingdom of living organisms powered solely by the consumption of other organisms. In the protist kingdom, a type of amoeba that makes a shell evolves: the first hard-shelled organism to come into existence, populating the continental wetlands. A type of filamentous algae, dwelling in shallow water, evolves to become the most complex and differentiated form of plant life at this time, while the fungi remain mostly single-celled in Earth's waters.

The combination of cyanobacteria alongside a thriving set of plant life on Earth causes the oxygen levels to continue to rise in Earth's atmosphere, now crossing the 10% level for the first time. New rifts begin to form in the remnants of supercontinent Rodinia, while magmatism leads to relatively rapid plate motion across Earth during this time. In the span of just a few hundred million years, Earth goes from having just one supercontinent to 10 or more large, independent landmasses during this era.

As the era progresses, animals diversify to take on novel forms. Instead of simply being "filter feeders" like sponges, a new type of motile sea animal arises with an internal cavity and multiple little hairlike tendrils used for swimming. These early creatures have no bones or skeletal system and are not porous like the sponges. Instead, they travel through the waters under their own power and are the first true "predator" animals, feeding on organisms across several different kingdoms. With primitive, jellyfish-like appearances, these creatures may be extremely small—perhaps less than a millimeter in size—but can successfully feed on defenseless organisms even several times larger than themselves.

Just as all appeared to be going well for life on Earth, however, an abrupt change in Earth's climate leads to a rapid ecological change—20 million years before the close of this era, the Cryogenian era begins. Earth is plunged into a worldwide glaciation, creating either a "Snowball Earth," where all of the ocean waters are covered in a solid, frozen shell, or a "Slushball Earth," where the equatorial regions continue to flow but still remain frigid and ice-rich, while only the mid- and high latitudes are completely frozen over. Although Earth will thaw and refreeze many times over the final 20 million years of this time period, glacial deposits can be found over all of Earth's surface dating back to this time period.

As the climate on Earth changes, there's little doubt that a wide variety of extinctions occur during this time. Unfortunately, few fossils remain from this time period, as most of the sedimentary rock that forms even this late in our planet's history will metamorphose over the next ~700 million years, rendering an eventual reconstruction of the fossil record extremely difficult for future inhabitants. However, extinction of a significant fraction of species is all but assured, as so much of the ocean surface is covered in ice, causing a scarcity of resources at the base of the food chain: at the producer level. As the biomass of the creatures living in the ocean tops decreases, so must the biomass of all creatures that feed on them; we may never know just how many early plants, fungi, and animals go extinct during this time. The chain of life, however, not only remains unbroken here on Earth but has diversified to a greater degree than ever before.

Even after plants and animals have arisen, long periods of extreme heat and cold persist on the planet. Here, the Sturtian glaciation, a time period lasting 57 million years, is shown, covering most of Earth in ice and creating a "Snowball Earth" set of conditions. Once this glaciation comes to an end, algae, not cyanobacteria, emerge as the dominant oxygen producers on our planet.

IMAGE BY MARK A. GARLICK

SNOWBALL EARTH

—— *the Sturtian glaciation ends* ——

UNIVERSE
Diameter: $3.51 \times 10^{13} \rightarrow 3.54 \times 10^{13}$ ly
  Observable: $8.78 \times 10^{10} \rightarrow 8.84 \times 10^{10}$ ly
Expansion: $68.68 \rightarrow 68.43$ km/s/Mpc
Density: $0.28 \rightarrow 0.28$ p/m³
Temperature: $2.86 \rightarrow 2.84$ K

ASTRONOMICAL OBJECTS
Galaxies: $2.07 \times 10^{13} \rightarrow 2.06 \times 10^{13}$
Stars: $2.18 \times 10^{21} \rightarrow 2.18 \times 10^{21}$
Black Holes: $1.74 \times 10^{19} \rightarrow 1.74 \times 10^{19}$
Planetary Systems: $2.95 \times 10^{21} \rightarrow 2.95 \times 10^{21}$
Worlds: $3.42 \times 10^{22} \rightarrow 3.43 \times 10^{22}$

WORLDS WITH LIFE
Simple Life: $3.12 \times 10^{17} - 1.45 \times 10^{19}$
Complex Life: $1.29 \times 10^{11} - 3.81 \times 10^{17}$
Intelligent Life: $4.07 \times 10^{7} - 3.30 \times 10^{15}$
Technological Life: $1.28 \times 10^{4} - 4.57 \times 10^{13}$
Interstellar Life: $0 - 6.44 \times 10^{12}$

Earth's continents continue to drift apart, but the planet remains frozen. For an estimated 57 million years, the Sturtian glaciation continues: the longest glacial period on an Earth with plants and animals. As the glaciation continues, a battle ensues between the cyanobacteria and the various forms of algae: Which one will emerge to dominate the oceanic surface waters when the ice finally melts? About halfway through this time period, the Sturtian glaciation comes to a close, and the answer emerges: It's the algae that become the dominant oxygen producers on Earth. The algae now become a ubiquitous food source for organisms in the ocean waters, feeding the growth of zooplankton, primitive animals that simply drift with the ocean currents.

Far away from Earth, a quadruple-star system is formed, with two massive stars about 2.5 times as heavy as the Sun in a tight, binary orbit, and two red dwarf stars about half the Sun's mass bound much more loosely and orbiting significantly farther away. This star system—Capella—is located some 60,000 light-years from Earth, but approaches the Solar System as the stars of the Milky Way continue their gravitational dance. In another 600–650 million years, the two massive stars will exhaust their core hydrogen and begin expanding into giants, right around the time that they pass very close to the Sun. In time, it will become the sixth-brightest star in Earth's night sky.

But Earth, at the time Capella is being born, is experiencing some rapid, planet-altering changes. After only 15 million years in an ice-free state, a new frozen period follows: the Marinoan glaciation. Again, the entire planet becomes covered in ice, either in a Snowball or Slushball Earth state. For around 20 million years, Earth remains frozen over, with volcanic gases building up over time beneath the ice. When they finally break through, this glacial period comes to an end, as Earth becomes a blue planet once again. A new large continent—Gondwana—takes shape, representing about 50% of Earth's landmasses, while a different continent rifts apart, newly creating the Iapetus Ocean.

But on a now-ice-free Earth, the continents and oceans teem with new forms of life, marking the beginning of the Ediacaran period. In the animal kingdom, multicellular organisms are now developing distinct layers of tissues, as well as functional organs. They undergo an embryonic stage as well, developing from a single cell into an immature version of the adult organism before hatching. The early motile animals have differentiated themselves further, becoming comb jellies, which propel themselves using a series of cilia, and cnidarians, such as jellyfish, sea pens, and anemone-like creatures.

In the fungus kingdom, large colonies of lichen and mycelial structures can be found across Earth's continents as well as in the oceans and may have played a role in enhancing the oxygen content of Earth's atmosphere.

Plants remain abundant in the oceans but are still rather primitive on land, having not yet evolved a vascular system. But it's towards the very end of this era that perhaps the most important evolutionary development in history occurs: the emergence of the first wormlike animal. As the first animal with bilateral symmetry but a distinct top and bottom, it serves as the ancestor of nearly all the large animals that will come to dominate Earth's lands and oceans. Although still relatively simple, the organisms of Ediacaran Earth are developing new structures, and using them to perform new functions, at a speed never seen previously on our planet.

The diversity of animals that appear during this time period on Earth, known as the Cambrian explosion, is unprecedented in our planet's history. The first hard-shelled animals emerge, including clams, snails, and conchs. The first reef builders, the earliest arthropods, and the first evidence for eyes also appear during this time, as do the first animals with a spinal column: the first vertebrates.

IMAGE BY JON LOMBERG

BIOLOGICAL BIG BANG

— life explodes and diversifies on Earth —

UNIVERSE
Diameter: $3.54 \times 10^{13} \rightarrow 3.56 \times 10^{13}$ ly
  Observable: $8.84 \times 10^{10} \rightarrow 8.90 \times 10^{10}$ ly
Expansion: $68.43 \rightarrow 68.18$ km/s/Mpc
Density: $0.28 \rightarrow 0.27$ p/m³
Temperature: $2.84 \rightarrow 2.82$ K

ASTRONOMICAL OBJECTS
Galaxies: $2.06 \times 10^{13} \rightarrow 2.05 \times 10^{13}$
Stars: $2.18 \times 10^{21} \rightarrow 2.19 \times 10^{21}$
Black Holes: $1.74 \times 10^{19} \rightarrow 1.75 \times 10^{19}$
Planetary Systems: $2.95 \times 10^{21} \rightarrow 2.96 \times 10^{21}$
Worlds: $3.43 \times 10^{22} \rightarrow 3.44 \times 10^{22}$

WORLDS WITH LIFE
Simple Life: $3.08 \times 10^{17} - 1.43 \times 10^{19}$
Complex Life: $1.23 \times 10^{11} - 3.81 \times 10^{17}$
Intelligent Life: $3.90 \times 10^{7} - 3.31 \times 10^{15}$
Technological Life: $1.23 \times 10^{4} - 4.59 \times 10^{13}$
Interstellar Life: $0 - 6.51 \times 10^{12}$

One of the slow, gradual changes that's been happening to Earth over the past several billion years has been the slowing of our planet's rotation, accompanied by the Moon spiraling outward to greater and greater distances. Up until this moment, all solar eclipses on Earth have been either total or partial eclipses, as the Moon's disk has always appeared bigger than the Sun's in Earth's sky. However, now the situation changes in a variety of ways. The Moon's greater distance from Earth now makes annular eclipses possible: They occur when a solar eclipse overlaps with lunar apogee. Earth's day slowly lengthens, with the day now containing 22 hours. And the number of days in a year continues to decline, down to about 400 after a high of more than 1,000 several billion years earlier.

Geologically, the large continent Gondwana finishes assembling, now covering 20% of Earth's surface. Glaciations occur during this time, but they're short-lived, lasting for less than one million years apiece. With relatively stable, life-friendly conditions, organisms continue to become more complex and diversified, leading to an explosion of new life-forms.

Plants remain relatively small and dependent on the abundance of water. Fungi and algae continue to be the dominant forms of land life, while animals continue to thrive in both freshwater and saltwater environments. Sponges evolve to become larger and occupy extremely diverse locations on Earth, from the shallows of continental shelves to the deep ocean floor. Coral-like creatures emerge in the animal kingdom, joining the jellyfish, anemones, and other cnidarians. Wormlike creatures begin adapting to several different environments, and other animals with bilateral symmetry emerge, including starfish-like creatures. Animals with frond-like and disk-like shapes emerge, with their fossils preserved in sedimentary rock. Additionally, animal tracks are found dating back to this time period.

For the first 60 million years of this time period, however, it's mostly trace fossils—which preserve a record of an organism's activity but not the organism itself—that will survive. With soft bodies and no internal skeleton, most of the natural processes that occur are incapable of preserving an easy-to-reconstruct record of Precambrian life. But about 40 million years before this epoch ends, the Cambrian explosion occurs, marked by an enormous spike in novel organisms.

The start of the Cambrian is marked by the first animals with hard shells, including clam-like, snail-like, and conch-like creatures. Some of the marine animals begin building reefs, including varieties of sponge that go extinct just before this current epoch draws to a close. Arthropods first appear, including the mighty trilobite: a large, hard-shelled class of animal that survives on Earth for nearly 300 million years. The first primitive eyes develop, an invaluable tool for preying on other organisms as well as for avoiding predation. Microbes, including microbial mats and biofilms, become common on the beaches adjacent to the coast, leading to the formation of early soil ecosystems.

But the success of the animals comes with a cost: Animals require the consumption of oxygen to power their life processes, and oxygen exchange between the atmosphere and the ocean is limited in efficiency. In locations where animals are very successful, they starve the waters they inhabit of oxygen, creating an extinction event. Over the final ~15 million years of this time period, more than 25% of all species cease to exist. Still, good news comes along with the bad: The first true vertebrate, or backbone-containing animal, arises at this time. Our direct terrestrial ancestors are becoming ever more recognizable.

The greatest biodiversification event in history occurs in the time period following the Cambrian explosion, giving rise to thriving populations of filter feeders, insects, corals, and the first octopuses. New types of armored arthropods appear, including the menacing sea scorpions depicted here. The largest of them, *Jaekelopterus,* reaches up to 2.5 meters in length.

OCEANS TEEM WITH LIFE

—— marine life diversity peaks on Earth ——

UNIVERSE
Diameter: $3.56 \times 10^{13} \rightarrow 3.59 \times 10^{13}$ ly
  Observable: $8.90 \times 10^{10} \rightarrow 8.97 \times 10^{10}$ ly
Expansion: $68.18 \rightarrow 67\ 93$ km/s/Mpc
Density: $0.27 \rightarrow 0.27$ p/m³
Temperature: $2.82 \rightarrow 2\ 80$ K

ASTRONOMICAL OBJECTS
Galaxies: $2.05 \times 10^{13} \rightarrow 2.04 \times 10^{13}$
Stars: $2.19 \times 10^{21} \rightarrow 2.19 \times 10^{21}$
Black Holes: $1.75 \times 10^{19} \rightarrow 1.75 \times 10^{19}$
Planetary Systems: $2.96 \times 10^{21} \rightarrow 2.96 \times 10^{21}$
Worlds: $3.44 \times 10^{22} \rightarrow 3.45 \times 10^{22}$

WORLDS WITH LIFE
Simple Life: $3.03 \times 10^{17} - 1.41 \times 10^{19}$
Complex Life: $1.18 \times 10^{11} - 3.81 \times 10^{17}$
Intelligent Life: $3.74 \times 10^{7} - 3.32 \times 10^{15}$
Technological Life: $1.18 \times 10^{4} - 4.61 \times 10^{13}$
Interstellar Life: $0 - 6.57 \times 10^{12}$

Large, complex life-forms are arising in unprecedented numbers on Earth. At the start of this epoch, a new class of land plants begins to emerge: mosslike organisms without an internal vascular system. Arthropods without heavy, trilobite-like armor first venture onto land, as do sluglike creatures. In the waters, lampreys, hagfish, and eel-like creatures known as conodonts appear: animals with a spinal cord and, sometimes, sharp teeth. Among the jellyfish, the first complex nervous system arises. But perhaps the largest change comes within the plant and fungi kingdom, as liverworts and lichen-like symbiotes begin to colonize the lands of Earth.

As this new form of plant life colonizes the oceans and land, the carbon dioxide content of the atmosphere—which has been very high prior to this time—begins to drop. With a large increase in biomass on Earth, new ecological niches emerge, with animals and fungi rapidly diversifying to occupy them. Accompanied by still-rising oxygen levels, high temperatures, and high sea levels, life thrives along the large continental shelves, resulting in perhaps the greatest biodiversification event in Earth's history: the Great Ordovician Biodiversification Event. Large populations of filter-feeding animals thrive in the waters, as do new types of armored arthropods, including the first sea scorpion. Arthropods also give rise to the first insects. Reef-forming corals emerge, as do the first nautilus-like and octopus-like creatures.

As plants and animals diversify and increase in biomass, oxygen levels go up while carbon dioxide levels begin to plummet; and as they drop, so do the temperatures across Earth. As this key greenhouse gas gets depleted, Earth cannot hang on to as much of the Sun's heat. As a result, about halfway through this time period, the temperature falls, the ice caps grow, and the sea level retreats, reducing those life-rich continental shelves to barren, dry land. This creates the first of late Earth's "Big Five" mass extinction events, wiping out about 60% of all marine species on Earth.

While these events unfold here on Earth, elsewhere in the Milky Way, a new episode of star formation occurs in a nebula only about 2,000 light-years away. One of the newborn stars is a singlet like our Sun, possessing several planets around it but with fewer heavy elements, about 2.1 times as much mass, and with a bluer color and greater temperature. Shortly after forming, this star gets ejected from its parent star cluster, where it heads directly towards the Sun at a speed of around 14 kilometers per second. In another 450 million years, it will become one of the brightest stars in Earth's night sky. As Earth's axis precesses every 26,000 years, it will periodically appear as the brightest, most notable "pole" star that our planet's inhabitants will ever know: Vega. This fast-spinning star bulges at its equator, creating a temperature gradient of nearly 2,000 K from the hotter poles to the cooler equator.

Back on Earth, the planet begins to recover from this horrific glaciation. The first embryophytes, or vascular land plants, come into existence. Possessing a root system and internal transport structures to carry water, nutrients, and waste throughout the body of the plants, they begin to colonize the continents, coming to dominate the biomass of Earth. In the oceans, jawless fish become widespread and numerous, living alongside the sea arthropods. The first jawed, bony fish emerge, as do the first truly complex land animals: arachnids and centipedes. Zygote fungi evolve, where they thrive in soils and serve as the dominant decomposers on land. As this period draws to a close, only four major continents exist: Laurentia, Baltica, and Siberia, alongside the humongous Gondwana. Life on Earth is thriving once again.

Towering over the landscape and creating not only their own ecosystems but also their own weather, the earliest trees emerge during the Devonian period of Earth. Known as Wattieza, they possess fronds similar to modern ferns, rather than the leaves we normally associate with present-day trees. The highest oxygen levels ever seen on Earth arise now, coinciding with the presence of giant, flying insects.

IMAGE BY MARK A. GARLICK

RISE OF THE TREES

— trees evolve on Earth —

UNIVERSE
Diameter: $3.59 \times 10^{13} \to 3.61 \times 10^{13}$ ly
 Observable: $8.97 \times 10^{10} \to 9.03 \times 10^{10}$ ly
Expansion: $67.93 \to 67.69$ km/s/Mpc
Density: $0.27 \to 0.26$ p/m³
Temperature: $2.80 \to 2.78$ K

ASTRONOMICAL OBJECTS
Galaxies: $2.04 \times 10^{13} \to 2.03 \times 10^{13}$
Stars: $2.19 \times 10^{21} \to 2.20 \times 10^{21}$
Black Holes: $1.75 \times 10^{19} \to 1.75 \times 10^{19}$
Planetary Systems: $2.96 \times 10^{21} \to 2.97 \times 10^{21}$
Worlds: $3.45 \times 10^{22} \to 3.46 \times 10^{22}$

WORLDS WITH LIFE
Simple Life: $2.99 \times 10^{17} - 1.39 \times 10^{19}$
Complex Life: $1.13 \times 10^{11} - 3.81 \times 10^{17}$
Intelligent Life: $3.58 \times 10^{7} - 3.32 \times 10^{15}$
Technological Life: $1.13 \times 10^{4} - 4.63 \times 10^{13}$
Interstellar Life: $0 - 6.64 \times 10^{12}$

Planet Earth, particularly across its continents, now sees new forms of life arising, thriving, and populating every ecological niche. Plants undergo an explosion across the landmasses on Earth as they develop multiple different lignins: a key class of complex, organic molecules that form strong structural and support tissues in plants. The vascular plants take advantage of this and grow to unprecedented heights, leading to the first trees and plant seeds. As they take hold across the continents, the first rainforests form and mountains become littered with trees, providing new habitats for creatures. An insect species takes to the air: the first animals capable of flight. In the oceans, fish overtake the arthropods as the dominant form of life, including the lungfish, the coelacanth, and armored fishes like *Dunkleosteus,* growing up to 10 meters in length and weighing up to 4 tons. Some of the bony-finned fish, such as *Tiktaalik,* undergo changes to their front and rear fins, enabling them to move across land. The first tetrapods, or four-legged creatures, evolve from them, giving rise to the first amphibians. The continents Baltica and Laurentia merge into one: Euramerica.

About 30 million years into this epoch, Earth's second great mass extinction event occurs, followed by a lesser extinction event 10 million years later. Between the two, about 70% of all species die out, mostly in Earth's waters. The armored fishes, having been so prominent in the seas for about 30 million years, all go extinct. Trilobites, which had recovered over the past ~140 million years, suffer a great loss of diversity, as do the clam-like organisms and nearly all the reef builders. Jawless fish, in addition, are almost wiped out completely. Jawed fish, for unknown reasons, seem to survive both extinctions with few losses.

About 41 million years into this epoch, Earth enters the Carboniferous period, a high-temperature period in which plant life dominates the terrestrial landscape like never before. Tropical swamps proliferate across the land, with lignin-stiffened trees rising to heights in excess of 100 meters. The oxygen in Earth's atmosphere spikes, reaching its highest level ever: up to a maximum of 35% of our atmospheric gases. Insects, thriving under these oxygen-rich conditions, evolve to include giant predatory griffinflies as large as hawks and millipedes the size of human beings. They rapidly overrun the tropical rainforests of Earth's landmasses. Amphibians, too, adapt well to these continental wetlands, with a fraction of them evolving into amniotes, the first creatures to lay eggs with a protective membrane known as an amnion. Creatures hatched from an amniotic egg develop thicker skin rich in a structural protein known as keratin and a lung-based respiratory system dependent on the expansion and contraction of the rib cage. These creatures, the sauropsids and synapsids, include the ancestors of all modern reptiles and birds and mammals.

About 65 million years into this epoch, Euramerica and Gondwana collide, forming the overwhelming majority of an emerging new supercontinent, Pangaea. Except for Siberia and a few smaller microcontinents, Pangaea already consists of nearly 90% of Earth's landmasses. All across this landscape, enormous numbers of tall, carbon-rich trees are forming, growing, and dying, accumulating on the forest floors. Without bacteria or fungi capable of breaking down the hard lignin molecules within them, they simply get buried, becoming embedded beneath layers of Earth, where they'll eventually become the majority of the world's coal and oil deposits. The buildup of these carbon-rich graveyards gradually depletes the atmosphere's carbon dioxide, ending the Carboniferous with a glacial period and a minor extinction event as the world's rainforests collapse.

The Great Dying, also known as the Permian extinction, is the greatest mass extinction event in Earth's history. Occurring 252 million years before the present day, more than 95% of all life and some 83% of all genera go extinct all at once. And yet, the surviving life-forms reproduce and diversify, leading to large reptiles, the first dinosaurs, and even the first mammals emerging in the aftermath of this catastrophe.

IMAGE BY MARK A. GARLICK

THE GREAT DYING

—— *the greatest mass extinction on Earth* ——

UNIVERSE
Diameter: $3.61 \times 10^{13} \rightarrow 3.64 \times 10^{13}$ ly
  Observable: $9.03 \times 10^{10} \rightarrow 9.09 \times 10^{10}$ ly
Expansion: $67.69 \rightarrow 67.46$ km/s/Mpc
Density: $0.26 \rightarrow 0.26$ p/m³
Temperature: $2.78 \rightarrow 2.76$ K

ASTRONOMICAL OBJECTS
Galaxies: $2.03 \times 10^{13} \rightarrow 2.01 \times 10^{13}$
Stars: $2.20 \times 10^{21} \rightarrow 2.20 \times 10^{21}$
Black Holes: $1.75 \times 10^{19} \rightarrow 1.76 \times 10^{19}$
Planetary Systems: $2.97 \times 10^{21} \rightarrow 2.98 \times 10^{21}$
Worlds: $3.46 \times 10^{22} \rightarrow 3.47 \times 10^{22}$

WORLDS WITH LIFE
Simple Life: $2.95 \times 10^{17} - 1.37 \times 10^{19}$
Complex Life: $1.09 \times 10^{11} - 3.81 \times 10^{17}$
Intelligent Life: $3.43 \times 10^{7} - 3.33 \times 10^{15}$
Technological Life: $1.09 \times 10^{4} - 4.65 \times 10^{13}$
Interstellar Life: $0 - 6.70 \times 10^{12}$

The collapse of the earlier Carboniferous rainforests created a great expanse of continental desert: one that was relatively cool compared to the prior 100 million years. Creatures closely related to modern-day snails, octopuses, clams, and starfish thrive in the marine waters, alongside the ammonites, bony fish, and arthropods. But the trees that thrived previously begin to lose their habitat and are largely replaced by seed ferns and the first conifers: evergreen trees that bear seeds in pods and cones. In the more inland areas, with scarce supplies of water, plants lose their foothold, and the interior rocks of Pangaea become stained with iron oxides due to the Sun's intense, sustained heating of the now-barren terrain.

The land animals, however, begin to grow and diversify as never before. Reptiles and other sauropsids remain small, feeding mostly on insects. The synapsids come to dominate the land, including large herbivores like diadectids and edaphosaurs. The first megafaunal predator evolves in the form of the incredible *Dimetrodon,* a nonmammalian synapsid up to 4.6 meters long, up to 250 kilograms in mass, and with a large spine sail emerging from its back. For some 25 million years, *Dimetrodon* is the largest predatory land animal of all. Other synapsids later evolve to possess a novel set of properties: larger cavities in their skulls; more complex, powerful jaws; more vertical legs; and a more upright posture. This latter type of synapsid gives rise to the cynodonts, the dog-toothed ancestor of all modern mammals. Within only a few million years of their emergence, the earlier synapsids, including *Dimetrodon,* are all driven to extinction.

Siberia and the remaining microcontinents merge with Pangaea, forming a true, single supercontinent surrounded by a single superocean: Panthalassa. The crashing together of tectonic plates leads to a large set of volcanic eruptions where Siberia and Pangaea merge: the Siberian Traps, one of the largest volcanic features ever imprinted on Earth. The accompanying emission of carbon dioxide raises Earth's temperatures and acidifies Earth's oceans. About 48 million years into this epoch, this event kicks off the most severe extinction event in Earth's history: the Great Dying. More than 95% of Earth's life goes extinct, including all of the sea scorpions and trilobites, nearly all of the ammonites, snails, mollusks, anemones, clams, and corals, as well as many other marine animals. On land, most insect species go extinct, as do more than two-thirds of all amphibians and reptiles. All of the non-cynodont synapsids die out, as do all of the large herbivores.

But life survives and, over time, recovers. One million years after the Great Dying, the first complex marine ecosystem returns. Five million years later, the woody trees have fully recovered, with the crustaceans recovering soon after. Another three million years sees the return of corals and, soon after, calcified sponges. New aquatic reptiles emerge, including the ichthyosaurs and nothosaurs, displacing the large amphibians. On land, crocodilians and dinosaurs emerge, displacing the cynodonts, which survive as small, shrewlike scavengers: the first mammals. In the air, the pterosaurs—flying reptiles—arise.

Meanwhile, 58 million years into this time period, a binary star system forms about 460 light-years from Earth. The two stars that are formed weigh in with five and two times the Sun's mass, respectively. After only 14 million years, the brighter, more massive star evolves into a red giant and dies, running out of its core fuel and expelling its outer layers, losing 4 solar masses of material in the process. Its remnant core contracts to become a hot but faint white dwarf, while its less-massive companion, Sirius, will soon become Earth's brightest nighttime star.

In the tropical regions of Earth, the first flowering plants arise, in the same environment that the largest pterosaurs, or flying reptiles, dominate the skies. Flowering plants are also known as angiosperms and represent the "leafy" plants that dominate the continental flora of modern Earth, but back when they first arose, they were a novelty that were vastly outnumbered by the older seed-bearing plants, such as conifers.

IMAGE BY MARK A. GARLICK

RISE OF THE FLOWERS

flowering plants evolve on Earth

UNIVERSE
Diameter: $3.64 \times 10^{13} \rightarrow 3.66 \times 10^{13}$ ly
  Observable: $9.09 \times 10^{10} \rightarrow 9.16 \times 10^{10}$ ly
Expansion: $67.46 \rightarrow 67.23$ km/s/Mpc
Density: $0.26 \rightarrow 0.25$ p/m³
Temperature: $2.76 \rightarrow 2.75$ K

ASTRONOMICAL OBJECTS
Galaxies: $2.01 \times 10^{13} \rightarrow 2.00 \times 10^{13}$
Stars: $2.20 \times 10^{21} \rightarrow 2.20 \times 10^{21}$
Black Holes: $1.76 \times 10^{19} \rightarrow 1.76 \times 10^{19}$
Planetary Systems: $2.98 \times 10^{21} \rightarrow 2.98 \times 10^{21}$
Worlds: $3.47 \times 10^{22} \rightarrow 3.48 \times 10^{22}$

WORLDS WITH LIFE
Simple Life: $2.91 \times 10^{17} - 1.35 \times 10^{19}$
Complex Life: $1.04 \times 10^{11} - 3.81 \times 10^{17}$
Intelligent Life: $3.29 \times 10^{7} - 3.33 \times 10^{15}$
Technological Life: $1.04 \times 10^{4} - 4.66 \times 10^{13}$
Interstellar Life: $0 - 6.76 \times 10^{12}$

olcanic activity creates another mass extinction event on Earth—another of the "Big Five" extinction events within the most recent half billion years on our planet. The last of the jawless eel-like sea creatures, the conodonts, die out completely. Most of the armored and semiaquatic reptiles and remaining giant amphibians also go extinct, as do the aquatic placodonts and giant ichthyosaurs. Survivors include the lizard-like lepidosaurs, the crocodilians, the flying reptiles (pterosaurs), along with dinosaurs—including the hollow-boned theropods—which become the dominant land creatures. The last of the nonmammalian synapsids die off as well, while rodent-like scavengers and insect-eaters survive: the ancestors of all modern mammals. The egg-laying mammals split off from the rest, providing an ancestral line for modern animals like the duck-billed platypus and the echidna.

As large seas begin to open up within the supercontinent Pangaea towards the middle of this epoch, fernlike prairies became ubiquitous, with enormous herds of sauropod herbivores like *Diplodocus* and *Brachiosaurus* feeding on them. Gigantic predators like *Allosaurus* also thrive on land, while large plesiosaurs and a recovered population of ichthyosaurs thrive in the oceans. Giant salamanders arise in aquatic environments, with this family leading a resurgence of giant amphibians. Large forests of conifers cover the mountainous and temperate regions of the continents, and the first feathered, flight-capable birds arise, evolving from small, hollow-boned theropods. The armored Stegosaurus also thrives during this time period, as the great Atlantic seaway begins to open up, separating the Americas from the rest of Pangaea.

About 60 million years into this epoch, Earth enters the Cretaceous period, the longest and final period of the Mesozoic. It brings with it the largest land animals ever to inhabit planet Earth: the titanosaurs, like *Patagotitan* and *Argentinosaurus*. Pterosaurs grow larger as well, with many possessing wingspans ranging from 6 to 8 meters and some with fearsome, razor-sharp teeth. The flowering plants—angiosperms—arise in the tropics, swiftly covering Earth's landmasses in herbs, shrubs, and leafy trees, where they outcompete the conifers across warmer ecosystems. Insects play a major role in pollinating these plants, while mammals diversify into placental and marsupial groups, including wolverine-like predators. Fungi become the dominant decomposers across and within Earth's surface, but the familiar mushroom has yet to arise.

These highly complex and differentiated plants and animals have seen their genomes grow to enormous sizes, with several billion base pairs within their DNA. However, most of this DNA is both redundant and nonfunctional; less than 10% of the genome of the most complex plants and animals actually code for proteins within the organisms themselves. This 100-million-year epoch is part of the longest period in recent Earth history without a major mass extinction event, which will extend into the next epoch for another ~35 million years.

Elsewhere in the Milky Way, some familiar stellar faces make waves during the close of this epoch. Arcturus, now under 2,000 light-years away, begins to evolve into a red giant, expanding in surface area and brightening by a factor of more than 100. Altair, one of the bright stars anchoring the asterism known as the Summer Triangle, is born more than 10,000 light-years away, but its swift motion will bring it near Earth 100 million years from now. Aldebaran swells and begins fusing helium in its core, becoming some 400 times as luminous as the Sun; although it appears unremarkable from 20,000-plus light-years away, it approaches Earth at more than 50 kilometers per second. Along with Sirius, Vega, Capella, and Procyon, these familiar stars all tell their own unique chapter in the cosmic story.

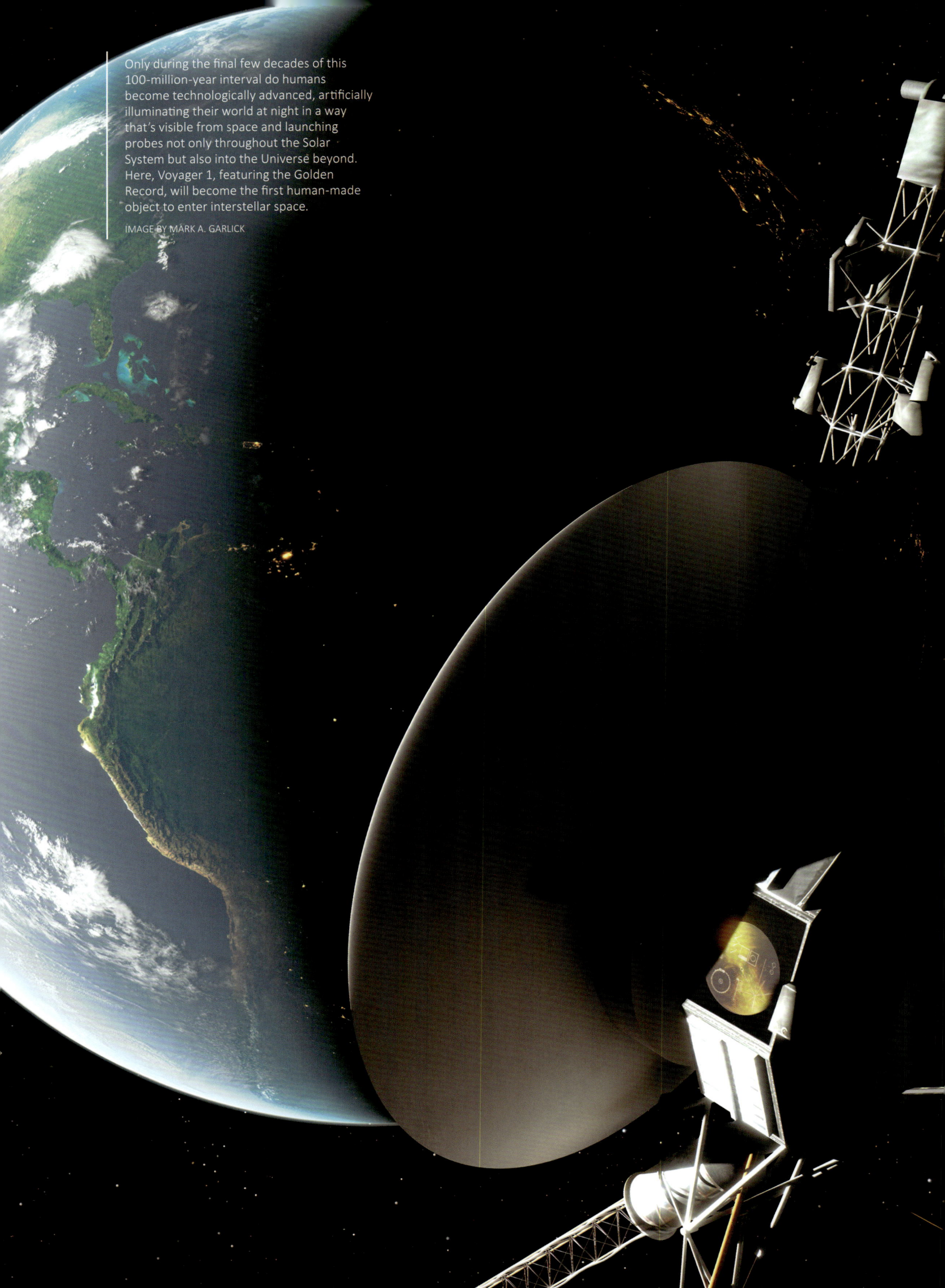

Only during the final few decades of this 100-million-year interval do humans become technologically advanced, artificially illuminating their world at night in a way that's visible from space and launching probes not only throughout the Solar System but also into the Universe beyond. Here, Voyager 1, featuring the Golden Record, will become the first human-made object to enter interstellar space.

IMAGE BY MARK A. GARLICK

THE LONG NOW

—— *spacefaring life evolves on Earth* ——

UNIVERSE
Diameter: $3.66 \times 10^{13} \rightarrow 3.69 \times 10^{13}$ ly
 Observable: $9.16 \times 10^{10} \rightarrow 9.22 \times 10^{10}$ ly
Expansion: $67.23 \rightarrow 67.00$ km/s/Mpc
Density: $0.25 \rightarrow 0.24$ p/m³
Temperature: $2.75 \rightarrow 2.73$ K

ASTRONOMICAL OBJECTS
Galaxies: $2.00 \times 10^{13} \rightarrow 1.99 \times 10^{13}$
Stars: $2.20 \times 10^{21} \rightarrow 2.21 \times 10^{21}$
Black Holes: $1.76 \times 10^{19} \rightarrow 1.76 \times 10^{19}$
Planetary Systems: $2.98 \times 10^{21} \rightarrow 2.99 \times 10^{21}$
Worlds: $3.48 \times 10^{22} \rightarrow 3.49 \times 10^{22}$

WORLDS WITH LIFE
Simple Life: $2.87 \times 10^{17} - 1.33 \times 10^{19}$
Complex Life: $9.98 \times 10^{10} - 3.81 \times 10^{17}$
Intelligent Life: $3.15 \times 10^7 - 3.34 \times 10^{15}$
Technological Life: $9.96 \times 10^4 - 4.67 \times 10^{13}$
Interstellar Life: $0 - 6.81 \times 10^{12}$

angaea continues its breakup, with the Atlantic Ocean continuing to expand, the Indian Ocean opening up to separate Asia from Africa, and a large southern continent breaking apart into Antarctica and Australia. Earth begins to cool as dinosaurs continue to dominate the landscape, with tyrannosaurs, ankylosaurs, triceratops, and duck-billed hadrosaurs sitting atop the food web. The great ichthyosaurs go extinct in the ocean, replaced by a group of large marine reptiles: the mosasaurs. Flowering plants become, for the first time, the dominant form of flora on planet Earth, and mushrooms finally appear among the fungi. Birds diversify into two main groups: a dominant group with teeth and claw-bearing wings and a subdominant group lacking both features. Mammals remain largely rodent-like, while increased volcanic activity begins to add poisonous gases to the atmosphere, driving many plants and animals into decline.

And suddenly, about 34 million years into this epoch, everything changes all at once. A large asteroid, about 5 kilometers in diameter, slams into a continental shelf in the southern Gulf of Mexico—the largest impact event in Earth's geologic history. A large cloud of ash and dust is kicked up into the atmosphere, blocking out the Sun for several weeks, depositing a layer of iridium-rich ash all across the globe, and leading to the most recent of Earth's "Big Five" mass extinctions. All of the non-avian dinosaurs go extinct, as do the flying reptiles, all toothed birds, all large marine reptiles, and all ammonites. Many species of sharks, mammals, ray-finned fishes, mollusks, insects, and lizards also go extinct; an estimated 75% of all species vanish in the aftermath of this event. Fungi, however, remain largely unaffected.

But life survives and quickly repopulates the Earth. Dense forests cover the lands, and the surviving birds and mammals thrive in them, while enormous sharks rule the oceans. Initially, mammals remain small, with early lemur-like primates and the first horse, the eohippus, being no more than 0.3 meter tall, while large, flightless birds like the *Paracrax* stand some 2.2 meters high. Midway through this epoch, a disruption in ocean currents causes a global cooling trend, enabling the growth of mammals to gargantuan proportions. Some mammals reenter the waters, becoming first semiaquatic and then fully aquatic, leading to the emergence of whales. On land, grasses emerge, quickly seeding their way across most of the continental areas. As grasslands and savannas emerge, the first elephants, cats, dogs, and modern marsupials thrive within them. Plants develop the C_4 photosynthesis pathway, granting them an evolutionary advantage in high-temperature, low-rainfall climates.

In the heavens, the bright stars Canopus, Rigel, Achernar, Betelgeuse, and Spica all form during this epoch, as the constellations begin to take their modern shapes. Sirius becomes Earth's brightest nighttime star; Vega, the fifth brightest, becomes located at Earth's north celestial pole every 26,000 years. The leftover glow from the Big Bang has cooled to just 2.725 K, and dark energy has rendered 94% of the galaxies visible in the Universe unreachable to any space traveler, even at light speed.

The first apes emerge 75 million years into this epoch, defined by the lack of a tail, splitting off from the rest of the monkeys. And 86 million years into this era, the great apes emerge. The largest ape of all time, *Gigantopithecus,* arises 91 million years into this epoch. In the final fraction of a million years of this era, a technologically advanced civilization of great apes emerges: humanity. Their activity initiates the start of a sixth great mass extinction, while they simultaneously launch probes into the Universe, seeking to answer perhaps a universal question among intelligent species, "Are we alone?"

Although human civilization will no doubt continue to advance in a great variety of technological ways, that technology brings with it a gruesome series of existential threats, from climate change to ecological collapse to food/water shortages to warfare. There are enough nuclear devices to destroy all humans on Earth already, and all it takes is one poor decision to bring about the demise of our species.

IMAGE BY MARK A. GARLICK

THE SHORT TOMORROW

— *humanity falters* —

UNIVERSE
Diameter: 3.69×10^{13} ly
  Observable: 9.22×10^{10} ly
Expansion: 67.00 km/s/Mpc
Density: 0.24 p/m³
Temperature: 2.73 K

ASTRONOMICAL OBJECTS
Galaxies: 1.99×10^{13}
Stars: 2.21×10^{21}
Black Holes: 9.28×10^{17}
Planetary Systems: 2.99×10^{21}
Worlds: 3.49×10^{22}

WORLDS WITH LIFE
Simple Life: $2.87 \times 10^{17} - 1.33 \times 10^{19}$
Complex Life: $9.98 \times 10^{10} - 3.81 \times 10^{17}$
Intelligent Life: $3.15 \times 10^{7} - 3.34 \times 10^{15}$
Technological Life: $9.96 \times 10^{4} - 4.67 \times 10^{13}$
Interstellar Life: $0 - 6.81 \times 10^{12}$

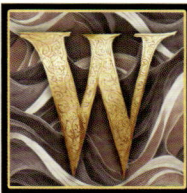

Will humanity survive our technological infancy, or will the unprecedented challenges facing us today culminate in the fall of our species before we ever realize our full potential? If we fail to both recognize and address the existential threats facing human civilization—both internal and external—we will likely become just another tombstone in the great cosmic graveyard, with no human having ever left our own Solar System. It's simultaneously fearsome and facile to envision how our entire species could collapse in only a few millennia if we turn a blind eye to the menacing obstacles troubling planet Earth today. The following is a scenario in which our species dies out only a few thousand years in the future.

In order to meet our ever-growing energy and resource needs, our species continues to explore Earth to remove resources—oil, coal, and gas reserves beneath the ground—while intensifying mining efforts and removing more of our planet's wild, natural habitats. The atmospheric carbon dioxide concentrations continue to rise, reaching levels unseen for millions of years, while the percentage of wilderness habitat shrinks to below 25%. As global temperatures rise, the oceans continue to acidify. Water and weather patterns destabilize further, forcing habitat loss for both land and marine species. Ever greater numbers of species experience complete extinction on Earth.

Increased contact between humans and wild animals, combined with a more interconnected world, leads to the emergence of novel pandemics, both lethal and highly infectious. Resource scarcity and rising inequality lead to civil unrest and war as humans scramble to gather remaining resources for their own tribes and factions. Rising sea levels lead to the abandonment of many coastal cities worldwide, displacing nearly a billion humans. Hundreds of thousands of new satellites occupy low-Earth orbit, as global surveillance and space-based weapons become the "new normal" on Earth. Eventually, war and panic lead to the launch of a single nuclear weapon, with hundreds of others soon following. Global catastrophe results in the death of billions of humans and the extinction of many other species. Earth's inhabitants are forced to reckon with a sixth great mass extinction.

When the next great solar storm strikes Earth, the first major collision between satellites in low-Earth orbit occurs, triggering a chain reaction of collisions among them. With too great a density of satellites up there, Earth becomes surrounded by a cloud of deadly debris—succumbing to Kessler syndrome—making new launches a near impossibility. As human civilization struggles to recover from disease, nuclear war, resource scarcity, and a collapsing set of ecosystems, we find ourselves completely unprepared for a possible ensuing catastrophe, when a close gravitational encounter with Jupiter sends comet Swift-Tuttle, the parent body of the Perseid meteor shower, on a collision course with Earth.

A resource-depleted, fractious human civilization can make only futile attempts to deflect this massive object—hurtling towards us with more than 25 times the kinetic energy of the vaunted asteroid that caused Earth's prior mass extinction—and even our best efforts fail. In the year 4,479, comet Swift-Tuttle strikes planet Earth, wiping out the last vestiges of humanity on Earth and bringing an end to the first intelligent, technologically advanced civilization in our planet's history. Any off-world colonies, such as on the Moon or Mars, are cut off from one another and die out without a consistent supply of terrestrial resources. While our species manages to launch a few uncrewed probes out of the Solar System and detects potential signs of life on worlds beyond our own, we never make contact with another intelligent species and never set foot on a world orbiting a star other than our own, joining so many predecessor civilizations in our isolated cosmic grave.

The lush, green, thriving soil down below isn't part of planet Earth but, rather, is part of an enormous, rotating, self-sustaining ecosystem in space that creates its own artificial gravity: an O'Neill cylinder. Humanity, if we can get our act together, has the opportunity to spread the seeds of Earth life not only to other planets around other stars but all throughout the space between stars as well.

IMAGE BY MARK A. GARLICK

THE LONG TOMORROW

— *humanity flourishes* —

UNIVERSE
Diameter: 3.69×10^{13} ly
  Observable: 9.22×10^{10} ly
Expansion: 67.00 km/s/Mpc
Density: 0.24 p/m³
Temperature: 2.73 K

ASTRONOMICAL OBJECTS
Galaxies: 1.99×10^{13}
Stars: 2.21×10^{21}
Black Holes: 9.28×10^{17}
Planetary Systems: 2.99×10^{21}
Worlds: 3.49×10^{22}

WORLDS WITH LIFE
Simple Life: $2.87 \times 10^{17} - 1.33 \times 10^{19}$
Complex Life: $9.98 \times 10^{10} - 3.81 \times 10^{17}$
Intelligent Life: $3.15 \times 10^7 - 3.34 \times 10^{15}$
Technological Life: $9.96 \times 10^4 - 4.67 \times 10^{13}$
Interstellar Life: $0 - 6.81 \times 10^{12}$

lthough the possibility of extinction awaits our civilization down many possible avenues—from ecosystem collapse to a catastrophic pandemic to nuclear war to a massive global impact event—the possibility of recognizing and rising to these challenges remains within our grasp. Perhaps the next few centuries and millennia won't be the end of our species but, rather, will represent our rise to become the spacefaring, technologically advanced species that has for so long been a part of our science fiction dreams. The following describes one such scenario.

As the pressures on modern-day humans continue to mount, our species comes to an important realization: The alternative to sustainable, long-term human activity on Earth is ecological collapse and human extinction. To avoid this fate, unprecedented funding is diverted towards basic scientific research, resulting in rapid new infrastructure developments, yielding planet-wide benefits. Earth swiftly transitions to a clean energy economy anchored by advancements in room-temperature superconductivity and nuclear fission and fusion technologies. We cease to emit a net positive amount of greenhouse gases, and with it, the recent trend of global warming ceases. The climate continues to change, but on timescales that are hundreds of times slower than what Earth currently experiences, with the Antarctic and Greenland ice sheets persisting for thousands of generations.

Future development of infrastructure and procurement of resources—here on Earth, in low-Earth orbit, and all throughout the Solar System—are all done in a fashion that considers long-term sustainability. Efforts are made to economize how humans use land and ocean resources, rewilding more than 50% of Earth's surface area. Global buy-in towards pandemic prevention occurs, and although outbreaks of new diseases routinely arise, they never devastate the human population. Resources are allocated in such a way that severe inequalities are eliminated, and all humans gain access to modern technologies and infrastructure on a global scale.

Humanity develops a comprehensive catalog of objects within the Milky Way, recording over 200 billion star systems within our home galaxy and characterizing the planets around them. Several million are found to be inhabited, and thousands of those are inhabited by some form of intelligent life. The first round-trip communication signals are exchanged between humanity and another technologically advanced species within the Milky Way; even though it takes thousands of years to send and receive a response, "First Contact" is officially made.

Evidence of past life is found on Mars, while a thriving, primitive oceanic ecosystem is discovered within the oceans of Jupiter's moon Europa. Several harmless supernovae are seen and measured from within our galaxy, as numerous Wolf-Rayet stars, as well as Antares and Betelgeuse, explode in cataclysms that are easily visible from Earth. Permanent, self-sustaining human colonies are set up on and around worlds throughout our Solar System, with thriving habitats on the Moon, Mars, and several moons of the outer planets, with various attempts made at interstellar colonization as well. Robotic probes are launched throughout the galaxy, exploring worlds on millennium-long missions and returning data to Earth.

Gradually, however, human beings begin to express new genetic traits that come about from the selection pressures inherent to these novel, technologically enhanced environments. After several hundred thousand years go by, the intelligent creatures that inhabit this futuristic version of the Solar System are no longer recognizably human, as they've evolved into a new set of animal species. Our direct descendants continue to inhabit these worlds for several million years, as these civilizations thrive with the offspring of humanity inhabiting them.

THE LONGEST TOMORROW

— *humanity endures* —

UNIVERSE
Diameter: 3.69 × 10^{13} ly
 Observable: 9.22 × 10^{10} ly
Expansion: 67.00 km/s/Mpc
Density: 0.24 p/m^3
Temperature: 2.73 K

ASTRONOMICAL OBJECTS
Galaxies: 1.99 × 10^{13}
Stars: 2.21 × 10^{21}
Black Holes: 9.28 × 10^{17}
Planetary Systems: 2.99 × 10^{21}
Worlds: 3.49 × 10^{22}

WORLDS WITH LIFE
Simple Life: 2.87 × 10^{17} − 1.33 × 10^{19}
Complex Life: 9.98 × 10^{10} − 3.81 × 10^{17}
Intelligent Life: 3.15 × 10^{7} − 3.34 × 10^{15}
Technological Life: 9.96 × 10^{4} − 4.67 × 10^{13}
Interstellar Life: 1 − 6.81 × 10^{12}

Human beings only first arose a few hundred thousand years ago. Almost all of the major technological and societal advances that shape recorded human history—agriculture, metal tools, sanitation, animal domestication, etc.—are only a few thousand years old, at most. Industrialization is only a few hundred years in our past, while electronics and rocketry are technologies that are less than a century old. Asking a modern-day human to imagine what the long-term future trajectory of our civilization would look like, even under the most optimistic of circumstances, might be like asking one of our long-extinct hominid ancestors, an *Australopithecus* perhaps, to imagine our present-day world: It would appear, as Arthur C. Clarke once wrote, "indistinguishable from magic."

And yet, future humans will no longer be bound by the current limitations of our brains. Instead, we'll possess the full scientific knowledge accessible to humanity millions of years from now, the calculational power of a quantum supercomputer, all while still maintaining the unbridled human capacity for ideation. New discoveries surely await us; as we supersede our modern understanding of fundamental physics, new breakthroughs will arise in the aftermath of uncovering the true nature of dark matter and dark energy. When we build a particle accelerator the size of the Solar System, it puts our highest-energy particle physics theories to their most direct tests ever. Such an endeavor has a chance of altering the quantum vacuum and creating an outward-propagating "bubble of destruction" that destroys our entire Local Group and more; only performing this experiment will yield the ultimate answer.

Although some humans still live and die as we do today, others, through technological enhancement, achieve a type of immortality, for our minds and perhaps even for our bodies. We spawn an entirely new species adapted to the zero-gravity environment of space—without skeletal systems or eyes—capable of spending an entire lifetime journeying between the stars. Planet Earth is no longer recognizable as a "blue marble" with continents and oceans, as we instead terraform it into a multitiered world, with humans not only living on the surface but underground, in the air, and even in space far above it.

We successfully traverse the vast interstellar distances, not only colonizing inhabitable worlds but also transforming uninhabitable ones into worlds suitable for human life. Some endeavors "clean out" the hazardous, debris-rich belts around inhabited systems, vacuuming up these individual chunks into single planet-like objects. We successfully characterize every form of life in our galaxy, discovering every inhabited planet and every intelligent species throughout the Milky Way. Galactic archaeology becomes a new science, as we reconstruct the past history of life and intelligence throughout our home galaxy and beyond, creating an *Encyclopaedia Galactica* that exceeds our current imaginings. We send spaceships to the outskirts of star-forming regions, becoming bold enough to seed novel, potentially habitable worlds with "starter life," preparing them for our eventual colonization. And we join together with all the other intelligent and technologically advanced species of our Local Group, culminating in some sort of galactic federation.

Although speculations about the future we might achieve can serve as tremendous motivators to work together for the betterment of the species, the planet, and our civilization, even these profoundly creative thoughts are insufficient to capture the full extent of what's truly possible for us. The only limits that are certain are set by the fundamental laws governing reality itself. Our imagination, resourcefulness, and capacity to learn and overcome—irrespective of the challenges faced—may prove to be our most lasting and important legacy within the cosmos.

While the fate of our Solar System will be for our Sun to die, blowing off its outer layers and contracting to form a white dwarf after engulfing the inner planets and melting the icy objects in the Kuiper belt and beyond, processes continue on longer timescales. Eventually, all the stars will die, all stellar remnants will fade to blackness, and even the most massive black holes will evaporate.

IMAGE BY MARK A. GARLICK

THE END

—— from order, randomness ——

UNIVERSE
Diameter: $3.69 \times 10^{13} \rightarrow 10^{1,000,000}$ ly
 Observable: $9.22 \times 10^{10} \rightarrow 10^{1,000,000}$ ly
Expansion: $67.00 \rightarrow 45.56$ km/s/Mpc
Density: $0.24 \rightarrow 0$ p/m³
Temperature: $2.73 \rightarrow 10^{-30}$ K

ASTRONOMICAL OBJECTS
Galaxies: $1.99 \times 10^{13} \rightarrow 10^{11}$
Stars: $2.21 \times 10^{21} \rightarrow 0$
Black Holes: $9.28 \times 10^{17} \rightarrow 10^{18}$
Planetary Systems: $2.99 \times 10^{21} \rightarrow 10^{21}$
Worlds: $3.49 \times 10^{22} \rightarrow 10^{22}$

WORLDS WITH LIFE
Simple Life: 0
Complex Life: 0
Intelligent Life: 0
Technological Life: 0
Interstellar Life: 0

In the End, there will be a return to conditions not seen since before the Beginning. Although there are a great many uncertainties about the short-term future of humanity, planet Earth, and a great many other cosmic variables, the long-term fate of our Universe is well understood. In fact, based on the laws of physics, it was arguably decided the very moment our Universe came into existence.

In the short term, life on Earth will continue to thrive for up to two billion years. As our Sun continues to evolve, however, it brightens and becomes more luminous, gradually increasing the temperature on all of its planets. After another two billion years or so, it becomes hot enough to boil Earth's oceans, effectively ending life as we know it on our world. The unbroken string of life that began shortly after our world's formation is finally cut, rendering Earth an extinct planet.

Another two billion years later, Andromeda begins merging with the Milky Way, triggering a massive episode of gas-driven star formation in the new, post-merger galaxy: Milkdromeda. The Sun transitions into a red giant, swallowing Mercury, Venus, and Earth. The Kuiper belt and Oort cloud objects melt in the process, and as the great galactic merger completes nearly three billion years after it began, the Sun blows off its outer layers and contracts to a white dwarf.

Dark energy continues to drive the expansion of the Universe, fragmenting our vast cosmic web into receding islands of structure. Individual galaxies find themselves bound into small and midsize groups as well as the occasional large cluster, but each individual group and cluster accelerates away from the others. After a further seven billion years pass, 99% of the galaxies visible to any observer have become unreachable. New stars now form only rarely, as most galactic groups have merged into one supergalaxy like Milkdromeda, exhausting the last major untapped reservoirs of hydrogen gas. Over the next 20 billion years, practically all of the Universe's stars more massive than red dwarfs finish their life cycles, dying in planetary nebula/white dwarf combinations. The remaining red dwarfs burn through their fuel far more slowly, persisting for up to 200 trillion years, at which point star formation transitions from a trickle to a mere series of drops. The cosmic microwave background has disappeared, shifted to wavelengths so long that they would require a planet-sized radio telescope to detect.

New stars arise only when two failed stars—brown dwarfs—merge together to cross that critical mass threshold, initiating core nuclear fusion and birthing a new red dwarf. When red dwarfs die, they emit no planetary nebula but simply contract to form a white dwarf made solely of helium. On timescales of about one quadrillion (10^{15}) years, white dwarfs cool down, becoming yellow, orange, red, and eventually invisible: black dwarfs. As burned-out stellar remnants within these isolated supergalaxies pass one another in their orbits, they gravitationally interact. A small fraction will collide together, triggering a cataclysmic explosion when they do, while most such interactions result in ejections: hurtling these black dwarfs into the intergalactic abyss.

After $\sim 10^{25}$ years pass, only isolated corpses persist. Black holes remain, slowly decaying away via Hawking radiation. The lightest black holes evaporate after about 10^{68} years, while the most massive ones take up to $\sim 10^{110}$ years before disappearing completely. The last vestiges of energy arise from gravitational waves, emitted wherever masses orbit one another, which cause any stellar or planetary systems to inspiral and eventually merge. When the last mergers and evaporations are complete, the Universe will have attained an equilibrium state from which no further energy can be extracted, marking the heat death of the Universe.

On macroscopic scales, the Universe appears to be completely empty, with dark energy having driven all remaining particles away from one another. However, random processes and quantum fluctuations still exist, getting stretched across the expanding Universe. This state, similar to what existed before the Beginning in many ways, may yet conceivably give rise to something novel, unexpected, and profound.

IMAGE BY WILLIAM LIDWELL

AFTER THE END

all is void and randomness

UNIVERSE
Diameter: $10^{1,000,000}$ ly → ∞
  Observable: $10^{1,000,000}$ ly → ∞
Expansion: 50 km/s/Mpc
Density: 0
Temperature: 10^{-30} K

ASTRONOMICAL OBJECTS
Galaxies: 10^{11} → 0
Stars: 0
Black Holes: 10^{18} → 0
Planetary Systems: 10^{21} → 0
Worlds: 10^{22} → 0

WORLDS WITH LIFE
Simple Life: 0
Complex Life: 0
Intelligent Life: 0
Technological Life: 0
Interstellar Life: 0

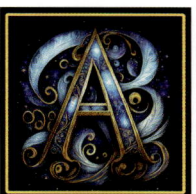

After the End, the Universe will be a vast, expanding ocean of void. The last gravitational waves will have been generated, and the last black holes will have decayed away. In isolated clumps, a few entities will persist: black dwarfs, dead neutron stars, rogue planets, as well as smaller entities, all the way down to atoms and stable, subatomic particles, like protons and electrons. Depending on the true nature of dark matter, additional quanta may also remain within this eternally expanding Universe. As dark energy has now taken over the Universe, these tiny clumps will relentlessly recede from one another, caught up in an incessantly, exponentially expanding Universe.

But this void, despite its ever-growing nature, refuses to empty out completely. Dark energy's omnipresent persistence continues to affect the quantum vacuum of empty space, even as the individual particles of matter are driven apart forevermore. A small amount of uniform radiation always arises in all directions from the presence of a cosmic horizon: a limit set by cosmic expansion and the speed of light, creating a new cosmic radiation background at the unfathomably chilling temperature of ~10^{-30} K. This radiation would require a telescope or antenna approximately 10^{28} meters long to detect (the size of the observable Universe at an age of ~13.8 billion years) but should eternally prevail. Quantum fluctuations always endure throughout space as well, appearing superimposed atop this radiation background, too. This maximal entropy state, according to our best understanding of physics, should linger for an eternity.

As the Universe's volume continues to expand, its total entropy remains constant, while its entropy density decreases to arbitrarily low values. If dark energy is truly a constant, consistent with today's best observations, the Universe's ultimate fate will be to persist in this "heat death" scenario forevermore. But if dark energy evolves, either by strengthening, decaying away, or transitioning to a lower-energy value, new possibilities arise. The Universe could rip itself apart; it could reverse its expansion and recollapse; or, perhaps enticingly, it could yet give rise to a new set of ultra-low-energy quanta all throughout space in a new creation event, just as the end of inflation in our region of the cosmos gave rise to the hot Big Bang all of those googols of years earlier.

This state of the Universe, persisting eternally into the future, bears many similarities to the initial inflationary state from which our hot Big Bang and familiar Universe arose. It's empty: devoid of matter. It contains a low-energy bath of radiation, created by the existence of quantum fields, the limiting speed of light, and the presence of a cosmological horizon. It's expanding in an exponential, relentless fashion due to a form of energy inherent to space itself: dark energy in this case, as opposed to the energy of inflation early on.

There are key differences as well: The rate of expansion and the strength of the dark energy inherent to space are far, far lower than the inflationary rate of expansion and the inflationary energy inherent to space. Whereas a transition to a lower-energy state was possible during inflation, it may be that dark energy truly represents the lowest-energy state of all: the true zero-point energy of space. If so, there will be no further transitions in the future, and no new entities will emerge; our Universe will return to a formless, empty void, with only quantum fields and their fluctuations populating it.

But even if our Universe reaches its ultimate demise in heat death, other existing universes continue to live on, and new universes will continue to emerge, live, and die throughout the cosmos, born from the original, inflationary bubble that started it all and that extends far beyond the reaches of our Universe.

ACKNOWLEDGMENTS

The authors would like to thank
the many backers of
Encyclopaedia Cosmologica,
the precursor to this edition.

Without their generous support,
this book would not have been possible.

————

Special thanks to

Jill Butler of *Stuff Creators Design*
for her creative and design support;

David Umla, Jim Gilsinan, and *Allan Hainey*
for their eagle-eyed proofing;

and

Elisa Gibson, Tyler Daswick, Michael O'Connor,
Anne LeongSon, Heather McElwain,
Susan Tyler Hitchcock, and *Lisa Thomas*
of *National Geographic Books*
for helping the Universe
share its story.

GLOSSARY

accretion disk—a disk of heated material that accumulates around a dense, collapsed object (such as a black hole) due to the interaction between the denser object and surrounding or infalling matter

active galaxy—a galaxy whose central supermassive black hole is actively feeding, producing jets, outflows, and other particle- and radiation-rich features

Andromeda—the largest and most massive galaxy within the same galaxy group as our Milky Way: the Local Group

animal—one of three kingdoms of highly complex, differentiated, and multicellular life-forms found on Earth

antimatter—the antiparticle counterpart to matter, where each species of particle (e.g., proton, electron, neutrino, quark, etc.) has a corresponding antiparticle of antimatter (e.g., antiproton, positron, antineutrino, antiquark, etc.) with the same mass but opposite electric charge

antiparticle—a particle composed of antimatter, where if it interacts with its matter counterpart, both members will annihilate into two photons; the energy of each photon (E) is given by Einstein's equation $E = mc^2$, where m is the rest mass of the particle/antiparticle that annihilates

atmosphere—the layers of gases that exist around any gravitationally-bound body that has a solid and/or liquid surface, separating the surface from the vacuum of space

Big Bang—an ambiguous term that refers to either the early, hot, dense state of our young Universe or to a theorized initial singularity that birthed space-time itself; we adopt the term "hot Big Bang" to refer solely to the first option

binary—a system where two comparable objects in size and mass (black holes, neutron stars, white dwarfs, stars, planets, etc.) orbit one another about their mutual center of gravity

biomass—the total cumulative amount of mass of all presently living organisms, combined, inhabiting a world or planet

biosphere—the thin layer of an inhabited world, including the solid and liquid surface as well as the area beneath it and the atmosphere above it, where the life processes of organisms on that world take place

black dwarf—the eventual end state of white dwarfs, attainable only after an amount of time greater than 100 trillion (10^{14}) years, after they've cooled and radiated so much of their heat away that they no longer emit any visible light at all

black hole—a region of space where so much mass is collected into such a small volume that no light can escape from within its event horizon: the location where the speed necessary for escape exceeds the speed of light

blue straggler—a relatively luminous, high-mass star found in a cluster of lower-mass stars, formed subsequently to the most recent episode of star formation by the merger of two predecessor stars

brown dwarf—a substellar object whose mass ranges between 1.3% and 7.5% the mass of the Sun, where the core temperature is high enough to initiate deuterium fusion but remains below the critical 4,000,000 K temperature needed to fuse hydrogen into helium through a chain reaction

Bullet Cluster—the first colliding cluster of galaxies ever identified to clearly show the separation between the distribution of normal matter (through the emission of X-rays) and the total amount of mass (through gravitational lensing) present within the cluster

Cambrian explosion—the event that occurred on Earth approximately 550 million years ago that led to an explosion of diversity among the complex, differentiated life-forms appearing in the fossil record

carrying capacity—the maximum population size of a particular set of organisms within an ecological niche of a specific size and with a finite amount of resources; exceeding the carrying capacity will lead to a rapid crash in the population size

cataclysm—a catastrophic event that significantly alters or even completely destroys the object experiencing it

cell—the smallest self-contained unit of life found on modern Earth, theorized to be ubiquitous on inhabited worlds all throughout the Universe

circumgalactic—referring to the environment or halo surrounding a galaxy

circumplanetary—referring to the environment or disk surrounding a planet, planetary system, or protoplanet

circumstellar—referring to the material, environment, disk, or halo surrounding a star or protostar; found most frequently around very young stars

cluster dissociation—the process by which stars bound together in large collections, such as open star clusters, are ejected over time, eventually destroying the progenitor parent collection entirely

coronal mass ejection—an event where large numbers of charged particles are rapidly expelled outward from a star's corona, typically as a result of a magnetic reconnection event occurring between the star's outer layers and the heated, rarefied material present in the outermore corona

cosmic microwave background — the leftover glow remaining from the primeval plasma phase of the early Universe, having cooled and been redshifted by the expansion of space to be at just a few degrees above absolute zero by the present time

cosmic void — a giant hole largely devoid of structures like stars and galaxies in the large-scale structure of the Universe

cosmos — all that is, all that ever was, and all there ever will be, including but by no means limited to our own Universe

cyanobacteria — a class of prokaryotic organisms that are among the earliest and most prominent photosynthesizers in Earth's history; they have persisted on our planet for at least 2.7 billion years

dark energy — a type of energy that's inherent to the fabric of space itself, which came to dominate the energy budget of the Universe only a few billion years prior to the present day

dark matter — an unknown substance whose gravitational effects can be quantified, localized, and measured, but whose nature remains unknown

differentiated — an organism that contains multiple specialized components within it, with varied structures that carry out different life processes essential to the organism's functioning

dust — small grains of neutral matter that block short-wavelength light, but allow longer-wavelength light to pass through, copiously produced by the ejecta from massive and evolved stars

dwarf galaxy — a low-mass galaxy with only thousands to millions of stars, rather than the billions (or more) of stars common to modern Milky Way–like galaxies

dwarf planet — a world that is large and massive enough to gravitationally pull itself into hydrostatic equilibrium, but that is insufficiently massive to clear its orbit at its present distance from its parent star

Dyson sphere — a theorized structure that would completely enshroud a star, capturing 100% of the star's emitted energy

E = mc² — Einstein's most famous equation, detailing the equivalence of mass (m) and energy (E) and their relation through a conversion factor (c²), enabling the creation of particle-antiparticle pairs in sufficiently energetic collisions and allowing for mass to be converted into energy through processes such as nuclear fusion

Earth — the home planet of human beings, formed roughly 4.5 billion years ago as the third planet from the Sun in our own Solar System

Earth-like — a planet comparable in size and mass to Earth, orbiting at an Earth-like distance from a star similar in mass and luminosity to the Sun

El Gordo — the largest, most massive galaxy cluster known relative to the age of the Universe at the time the cluster's now-arriving light was first created

electromagnetic force — the fundamental interaction that occurs between all electrically charged particles, giving rise to all electric and magnetic phenomena in the Universe, including light

electron — the lowest-mass particle known to possess a non-zero electric charge, and found in all atoms everywhere across the known Universe

electroweak force — a unified force that existed only briefly in the hot, dense aftermath of the Big Bang, before breaking into the electromagnetic and weak nuclear forces we experience today

elliptical galaxy — one of the most common types of galaxies in the modern Universe, typically among the largest, highest-mass galaxies of all but with only small reservoirs of gas with the potential to form new stars

Encyclopaedia Galactica — the name given to a comprehensive atlas of stars, planets, and civilizations within an entire galaxy and possibly beyond, which can only be created once one or more technologically advanced species arise within a galaxy

entropy — a fundamental measure that represents the unavailability of energy from a system to perform mechanical work or other energy-requiring tasks; a maximum entropy state equates to the lowest energy state of a physical system

epoch — a particular period of time in natural history, usually characterized by the dominance of some physical process or biological life-form

eukaryote — a complex, differentiated single-celled organism possessing specialized organelles for carrying out its essential life processes and functions; much larger and with a longer genome than their prokaryotic counterparts

evolution — the change in any system over time, including nonliving physical systems and living populations of biological systems both

expansion rate — the rate at which the distance between any two arbitrary points in space increases with time, so long as those two points are not both part of some large-scale, gravitationally bound structure

extinction event — an event that causes not only the death of an enormous number of organisms, but the outright extinction of a large percentage (up to 100%) of all species living on a particular world

filament — a cosmic structure formed by matter in an expanding universe, leading to galaxies connected by a web of invisible dark matter across distances up to two billion light-years, at present

first contact — the moment at which a technologically advanced civilization discovers the presence and existence of another intelligent civilization originating from a different star system than their own

fluctuation — an initial imperfection in an otherwise smooth, uniform structure, leading to overdense and/or underdense regions as gravitational effects accumulate over time

fungus — along with animals and plants, one of three main kingdoms of complex, differentiated, multicellular life on Earth; arguably the first one to arise

fusion — a nuclear process where the atomic nuclei of originally light elements combine to produce a heavier element, releasing energy via $E = mc^2$ in the process

galactic archaeology — the science of mapping out one's home galaxy in a variety of wavelengths of light, allowing scientists to reconstruct the detailed history of stars, gas, dust, and plasma in the galaxy

galactic cannibalism — the act of one high-mass galaxy devouring, or merging with and absorbing, a lower-mass galaxy, reducing the total number of galaxies while increasing the larger galaxy's total mass

galaxy — a large collection of normal matter and dark matter where at least some stars are present, held together through their mutual combined gravitation

galaxy cluster — a collection of anywhere between dozens and thousands of large, high-mass galaxies all gravitationally bound together, with typical masses of 100 trillion (10^{14}) solar masses and upward

gamma-ray burst — a high-energy event caused by a stellar cataclysm, such as merging neutron stars or a superluminous supernova, producing the highest-energy photons in the Universe

General Relativity — our current theory of gravity, put forth by Einstein in 1915, which relates the curvature and expansion of the Universe to the matter and energy present within it

globular cluster — a collection of anywhere from tens of thousands to several million stars, all formed in one (or very few) bursts, bound together in a sphere of just a few light-years in radius

gravitational collapse — the tendency of large collections of normal matter to shrink in volume over time, drawn gravitationally towards a central point but often resisted by factors such as radiation and the pressure exerted by atoms or other quantum particles

gravitational force — a mutually attractive force experienced by all matter- and energy-containing entities, causing those entities to accelerate towards one another

gravitational lens — a phenomenon caused when a large foreground mass intervening along our line of sight to a more distant object or sets of objects stretches, distorts, and magnifies the background light sources into arcs or rings by the gravitational curvature of space-time

gravitational wave — ripples in space-time that are caused by an inherently gravitational form of radiation generated when large masses accelerate in close proximity to one another and during cosmic inflation

Great Oxygenation Event — also called the Great Oxidation Event, this occurred more than two billion years ago in Earth's history, when photosynthetic organisms produced enough oxygen as a waste product of respiration to dramatically change Earth's climate

habitable — a planet or world with all the right conditions, including chemical and temperature conditions, to enable the emergence and persistence of life

heat death — the thermodynamic end state of a system, from which no further energy can be extracted to perform mechanical work

Herbig-Haro objects — newborn, massive stars that eject collimated jets of gas and plasma, commonly found in star-forming regions and copious producers of cosmic dust

Higgs — the symmetry, mechanism, field, and particle that are part of the Standard Model of elementary particles, conferring a nonzero rest mass to particles like quarks and the electron when that symmetry breaks

hot Big Bang — the event that marks the onset of the hot, dense, almost perfectly uniform, radiation-dominated phase that began our Universe as we know it some 13.8 billion years ago

hybrid galaxy — a galaxy possessing both a disk (like a spiral galaxy) and a halo of stars (like an elliptical galaxy), display features common to both major galaxy types

hypervelocity — a star, brown dwarf, planet, or stellar remnant moving with speeds greater than the escape velocity of the galaxy or galaxy cluster it originated from, destined to enter intergalactic space

inflation — the cosmic mechanism preceding and setting up the hot Big Bang, where space expanded quickly and relentlessly at a constant rate, driven by a large amount of energy inherent to the fabric of space itself

inhabited — a habitable world where life not only arises and persists, but where living biological organisms still reside on that world

intelligent — a species of organism that isn't just complex, multicellular, and differentiated, but that is also capable of problem solving, learning, memory, communication, and potentially consciousness

intergalactic medium — the empty space between galaxies that's largely devoid of matter, lying beyond the circumgalactic halo of matter that surrounds most individual galaxies

interstellar medium — the space between stars within a galaxy, consisting of cosmic particles, dust grains, plus large amounts of hydrogen, helium, oxygen, and carbon atoms and ions

intracluster — the space between galaxies within a galaxy cluster, often containing gas, dust, and large amounts of dark matter

ion/ionize — an atom or molecule that has a net electric charge due to the number of electrons it possesses not matching the total number of protons in its atomic nucleus/nuclei, or the process of a previously neutral atom or molecule losing one or more electrons

irregular galaxy — a galaxy that is neither spiral nor elliptical in nature, often possessing features indicating an interaction or merger with a massive companion galaxy

jet — a linear feature produced when the energetic emission of photons and/or particles travels through an extended region of space

kilonova — an energetic stellar cataclysm produced by the merger of two neutron stars, creating a gamma-ray burst, the emission of gravitational waves, and resulting in either a neutron star or a black hole as the end product

Kraken — the largest, highest-mass galaxy (relative to the Milky Way's mass) ever to have been cannibalized by the Milky Way over the entire 13.8-billion-year history of the Universe

Kuiper belt — a belt of icy and rocky bodies, largely confined to a disk, existing beyond the Solar System's last planet, with Kuiper belt–like analogs found around stars throughout the cosmos

life — defined here as any collection of molecules capable of metabolizing energy from its environment, reproducing or replicating itself, and surviving through multiple generations of reproduction

light-year — the distance covered by a massless entity, such as a photon or a gravitational wave, over the time span of one calendar year; corresponds to 9.4 trillion kilometers

Local Group — the collection of galaxies and dark matter of which the Milky Way is a member, with the Andromeda, Milky Way, and Triangulum galaxies making up the three largest members

magnetic reconnection — the phenomenon where magnetic field lines, typically within a stellar or planetary system, break and reconnect, resulting in the release of energy and the acceleration of charged particles

major merger — where two galaxies of comparable masses (generally where one is no more than three times the mass of the other) interact and merge together, often resulting in large episodes of star formation

massive overdense object — an ultramassive galaxy cluster that contains upwards of a quadrillion (10^{15}) solar masses, representing the largest, most massive gravitationally bound structures within the observable Universe

matter — used to mean normal matter, composed of protons, neutrons, and electrons, as opposed to antimatter and/or dark matter

metabolism — the ability to extract energy from nutrients, molecules, light, heat, or other resources found ubiquitously in the local environment

Milkdromeda — the name given to the galaxy that will arise from the eventual merger of the Milky Way with Andromeda, as well as, on long enough timescales, all galaxies within the Local Group

Milky Way — our home galaxy, where our Sun is located roughly 27,000 light-years from the center, consisting of roughly one trillion solar masses of matter and containing between 200 and 400 billion stars

mini-Neptune—a planet between the mass of Earth and Neptune that has a large, thick envelope of hydrogen and helium gas surrounding a solid planetary core

molecular cloud—a high-mass collection of neutral atoms, often opaque to visible light but transparent to infrared wavelengths, where gravitational collapse will someday lead to the formation of new stars and planets

neutral atom—an atomic nucleus, containing a specific number of positively charged protons, orbited by an equal number of negatively charged electrons

neutrino—the lightest massive particle in the Standard Model, coming in three flavors (electron, muon, and tau), which can be absorbed or emitted in a variety of nuclear reactions, including during the hot Big Bang

neutron star—a stellar remnant whose core is composed mainly of neutrons, and whose size is only ~10 kilometers in radius despite having upwards of a solar mass's worth of material inherent to it

Olympus Mons—the highest mountain and largest active volcano on any planet within our Solar System, found on Mars

Oort cloud—a large, diffuse cloud of mostly icy bodies found at the outskirts of our Solar System, extending from about 1,000 times the Earth-Sun distance away to more than a light-year away; Oort cloud analogs are thought to exist around most stellar systems in the Universe

opaque—a structure that light cannot freely pass through, obscuring the view of any objects behind it

open star cluster—a collection of hundreds to thousands of young stars, typically formed within the plane of a spiral/disk galaxy, and which will dissociate through mutual gravitational interactions after hundreds of millions to a few billion years

optical—the portion of the electromagnetic spectrum whose light is visible to human eyes; a synonym for visible light

organelle—a differentiated structure within a eukaryotic cell capable of performing one or more specific life processes necessary to the overall functioning of a cell, including the ability to self-destruct

organic molecule—a molecule that contains at least one carbon-hydrogen bond, including sugars, amino acids, fats, and nucleic acids

orphan planet—also known as a rogue planet, this is a planet that either formed without a parent star at all or a planet that formed around a star but was subsequently ejected into interstellar space

panspermia—the process whereby life that originated on one world spreads to another world, representing the possible origin of life on many planets, including Earth, throughout the Universe

parent star—the star that a planet orbited when that planet was first formed, as opposed to an orphan planet or a gravitationally captured planet

parsec—an astronomical unit of distance defined by how far away an object would have to be from Earth to have a parallax of one arc second when viewed six months apart; equates to about 3.26 light-years

particle—a quantum unit of matter or radiation that, if it strikes its antiparticle counterpart, will annihilate into two photons, where the energy of each photon (E) is given by Einstein's equation $E = mc^2$

Pauli exclusion principle—the principle that states that no two identical fermions, where fermions include all quarks, leptons, antiquarks, and antileptons, can occupy the same quantum state; a collection of densely packed fermions will exhibit a degeneracy pressure that resists gravitational collapse, as in a white dwarf or neutron star

periodic table—the organization of the atoms found in nature by their electron structures and, hence, chemical properties

photosynthesis—the process by which light is absorbed and converted into usable energy, either for immediate use by the organism or stored later (e.g., in the form of sugars) for use when necessary

planet—a massive body that orbits its parent star, is not a moon of another planet, and can clear its orbit of all other major masses over the lifetime of its stellar system

planetary nebula—the end state of a Sun-like star, where after evolving into a red giant, it blows off its outer layers, forming an ionized cloud of material that persists for thousands of years before fading away

plant—the largest of the three main kingdoms (along with animals and fungi) of complex, differentiated, and multicellular life, responsible for most (about 80%) of planet Earth's biomass

plasma—a hot, ionized state of matter where one or more electrons have been stripped away from the atoms in that region at extreme temperatures

Population I — stars like our Sun that formed at late times in cosmic history and are heavily enriched with heavy elements

Population II — stars that are poor in heavy elements, either forming early in cosmic history or at later times in regions that have only been lightly enriched by previous generations of stars

Population III — the first generation of stars to form after the Big Bang, pristinely made of 99.99999%-plus hydrogen and helium alone

positron — the antimatter counterpart of the electron, with an identical mass but the opposite electric charge

preplanetary nebula — an important but short-lived phase in the evolution of a giant star, having begun to blow off its outer layers but before those ejecta become ionized, which they will as the preplanetary nebula evolves to become a full-fledged planetary nebula

pristine — an environment or property that has been largely unchanged since its origin, often used to mean unaltered since the earliest stages of the hot Big Bang

prokaryote — a small, simple, undifferentiated organism that possesses no nucleus and no membrane-bound organelles, including all types of bacteria and the earliest-appearing, most primitive terrestrial life-forms

protogalaxy — a massive cloud of gas or a small collection of star clusters that will someday evolve into a full-fledged, mature galaxy, but is still in the early stages of formation

protoplanet — a large collection of mass in orbit around a star, still in the process of growing and developing into a planet

protoplanetary disk — an extended disk of material that orbits around a young star or protostar, from which planets, moons, asteroids, and other rocky and icy bodies will eventually form

protoplanetesimal — a very small clump of matter within a protoplanetary disk that may someday grow into a planet or protoplanet, but that is also likely to be destroyed by energetic collisions within that disk

protostar — a hot, dense, contracting cloud of matter that has enough mass to initiate nuclear fusion in its core, but whose conditions have not yet achieved hydrogen burning, which would signal the birth of a full-fledged star

pulsar — a rotating neutron star whose strongly magnetized environment causes the emission of regular, periodic radio pulses with every complete rotation of that stellar remnant

quantum field — an energy-containing field that cannot be removed or separated from even empty space; responsible for mediating one of the fundamental forces or interactions of nature

quark-gluon plasma — a hot, dense, energetic state of matter where stable hadrons (e.g., protons, neutrons, pions, etc.) cannot form, instead leading to an unbound, undifferentiated state of free quarks and gluons

quasar — an astronomical object, usually found at great distances, where an active supermassive black hole at a galaxy's core emits intense amounts of energetic radiation; originally an acronym for QUAsi-StellAr Radio source

radiation — any species of quanta that moves at or very close to the speed of light; applies to any massless particle or antiparticle, as well as particles/antiparticles whose kinetic energy is much greater than their rest mass energy

radio waves — the longest wavelength of electromagnetic radiation, capable of transmitting the greatest amount of information with the smallest amount of energy

red and dead — a phrase that applies to galaxies, particularly elliptical galaxies, that no longer have any neutral gas and dust within them, preventing further creation of short-lived blue stars, while long-lived red stars persist

red dwarf — the lowest-mass, lowest-temperature, and longest-lived class of star, ranging between 7.5% and 40% the mass of the Sun, in which nuclear reactions still fuse hydrogen into helium in its core, but in which no helium fusion will ever occur

red giant — a phase that all Sun-like (and more massive) stars will undergo when they run out of hydrogen fuel in their cores, swelling and becoming more luminous, while their cores contract until they reach sufficient temperatures to begin the next phase of their lives: helium fusion

replication — the capacity of a complex molecule, crystal, or life-form to make copies of or otherwise reproduce a structure identical to the original

ring galaxy — a rare type of galaxy formed by the interaction of two progenitor galaxies, resulting in a central, luminous structure surrounded by a gap, with a bright, thick ring of stars found even farther out; does not include galaxies whose light is stretched into a ring-like shape by gravitational lensing

rogue planet — *see* orphan planet

runaway process — any process where, once it begins, induces an intensification of that process without bound, until that process leads to a cataclysmic change; examples include gravitational collapse, nuclear explosions, or, in some cases, a planetary greenhouse effect

Snowball Earth — a condition where the entirety of the surface water on Earth is frozen into an icy phase, where the only liquid water on the planet is found beneath an icy crust

solar flare — an event, driven by magnetic reconnection on or near the surface of a star, where hot particles consisting of stellar plasma are accelerated and ejected away from the parent star, escaping its gravitational pull entirely

Solar System — the name given to our particular stellar system, with the Sun at the center and where Earth is the third of eight planets in orbit around it

spaghettification — the process of an object coming so close to a collapsed, dense, massive body that the tidal forces acting on the object cause it to become stretched into a long, thin, spaghetti-like strand

spallation — a process where energetic cosmic rays strike heavy atomic nuclei, splitting them apart into smaller, lower-mass nuclei; the cosmic abundances of boron, beryllium, and lithium are substantially created by this process

spectroscopy — the act of breaking light into its component wavelengths, where absorption and emission signatures stand out against the continuum background of the remaining light

spiral galaxy — a galaxy whose stars are primarily distributed in a gas- and dust-rich disk, with denser collections of stars forming a pattern of arm-like structures that spiral outward from the galactic center

star — any body with at least 7.5% of the Sun's mass that actively fuses hydrogen into helium (or heavier elements into still-heavier ones) in its interior, particularly in the core and shells surrounding the core

star formation — an active episode where clouds of molecular gas contract and fragment under the influence of gravity to create new stars and stellar systems

starburst — a phenomenon where a star-forming region is so large that it encompasses most or even all of the entire galaxy; the nearby galaxy Messier 82 (the Cigar galaxy) is a close example of a starburst galaxy

stellar corpse — the remnant core of a star that has ceased all fusion reactions in its interior; examples include white dwarfs, neutron stars, and black holes

stellar remnant — any remaining structure from an episode of stellar death or a stellar cataclysm, including not just stellar corpses but also diffuse phenomena such as supernova remnants and planetary nebulae

stellar system — any system of stars, planets, dwarf planets, moons, and more that are all gravitationally bound together; the more general term for a solar system that applies to stars other than our own

Sun — the specific star that planet Earth orbits, which formed and ignited nuclear fusion in its core roughly 4.55 billion years ago

super-Earth — a planet whose mass lies between the masses of Earth and Neptune, but that lacks the thick envelope of hydrogen and helium gas that is the defining feature of a mini-Neptune, marking a transition to a gas-rich giant planet

supercluster — a collection of clusters of galaxies that appear adjacent to one another, but that are not mutually gravitationally bound together and whose components will expand away from one another along with the expansion of the Universe

supercontinent — a large landmass where more than 50% of the entire continental mass of a planet is connected together into one unbroken structure

supergiant — the most massive and most luminous class of stars in the known Universe; typically at least 10,000 times intrinsically brighter than a Sun-like star

supermassive black hole — a black hole composed of at least one million solar masses of material, most often found at the centers of massive galaxies

supernova — a stellar cataclysm where a runaway fusion reaction occurs inside a star or stellar corpse, causing a remarkable but temporary bright outburst that gradually fades away over time; often leaves a neutron star or black hole remnant, but also some varieties leave no remnant at all

superocean — a large body of liquid water on a planetary surface that covers at least 50% of the planet's surface area; often accompanies a supercontinent phase on planets that have a mix of continents and oceans

surface mass ejection — a rare but catastrophic type of space weather where enormous sections of the stellar surface, including sections as massive as a Mars-like planet, get ejected from a very massive star

synestia—a puffy, dust-rich structure that enshrouds a world after a massive impact or collision event; often results in the production of one or more moons in orbit around the surviving planetary body

technological infancy—the first few hundred or thousand years that an intelligent civilization experiences after becoming technologically advanced; a critical time where that civilization has the power to either cause or prevent its own self-destruction

Theia—the hypothetical protoplanet that collided with Earth roughly 50 million years into our Solar System's history, leading to the eventual formation of our Moon in the collision's aftermath

Thorne-Żytkow object—a theorized structure where a giant star collides with and absorbs a neutron star; candidate Thorne-Żytkow objects have been identified but none have been confirmed to date

tidal disruption event—an event where a star or other large-mass object passes very close to a black hole, where the black hole's tidal forces tear that object apart and destroy it

tidal force—the differential force experienced by a body being gravitationally attracted by a mass, as different parts of the body feel differing strength and directions of force; tidal forces become much stronger at small separation distances from the attracting mass

tidal lock—when a massive body that orbits another mass takes equal amounts of time to complete an orbit as to rotate by a full 360°, resulting in the same "face" always pointing towards the mass it orbits

transparent—allowing light to pass through; the opposite of opaque

trinary—a stellar system containing three separate stars within it, along with any number of planets

ultraviolet—the portion of the electromagnetic spectrum that's more energetic than visible light, but is still produced in great amounts, continuously, by Sun-like and more massive stars; includes photons that are capable of ionizing a neutral hydrogen atom

Universe—our particular universe, as opposed to any of the other universes within the multiverse

violent relaxation—the process by which the lowest-mass objects gravitationally bound together in a complex structure are ejected at the expense of the remaining, heavier-mass objects becoming more tightly bound

Virgo Cluster—the closest galaxy cluster to Earth, located 55–60 million light-years away and consisting of approximately 1,000 galaxies comparable in mass to the Milky Way

visible light—*see* optical

volatile—a chemical compound that can stably exist on a world or icy body at extremely low temperatures, but that would boil or sublimate away if brought close enough to a luminous star; examples include a volatile envelope of light gases around a planet or the volatile ices found on objects originating from our Kuiper belt

water world—a world whose surface is 100% covered by liquid water, with no continents or landmasses rising above sea level

weak nuclear force—the force that governs radioactive decays and that enables particles, such as quarks and leptons, to change flavor/species from one type into another

white dwarf—the remnant of a Sun-like star that's reached the end of its life, with a core typically dominated by carbon and oxygen, weighing between 0.4 and 1.4 solar masses

world—a generic term that includes planets, moons, and other large nonstellar bodies that are typically primarily composed of ices, gases, rocks, and/or metals

X-ray—a very energetic form of electromagnetic radiation that only arises from extreme astrophysical processes, such as gas heated to millions of degrees or material accelerated by a black hole or neutron star

INDEX

G

warfare **186,** 187

see also Encyclopaedia Galactica; Intelligent life

Terra Prima 55, 87

Terrestrial planets *see* Rocky planets

Theia 197, 308

Thorne-Żytkow object 145, 163, 308

Tidal disruption events **188,** 189, 308

Tidal forces 39, 67, 189, 207, 227, 308

Tidal lock 79, 308

Titan (Saturn's moon) 109, 206, 207

Trees **280,** 281, 283

Trinary star systems 47, 145, 308

Triton (Neptune's moon) **208,** 209

Type Ia supernova **80,** 81, 113, 145, 245

U

Ultraviolet photons **22,** 23, 308

Universe

from above cloud tops **6**

after the end **296,** 297

death of 295, 297

defined 308

god's-eye view **2**

properties 9

see also specific objects, processes, and topics

V

Vaalbara (supercontinent) **216,** 217

Venus

formation 55, 197

greenhouse effect 220, 221, 265

life potential 221

rotational speed 223

swallowed by Sun 295

volcanism **264,** 265

water 201, 207, **220,** 221

Violent relaxation 43, 95, 308

Virgo Cluster **256,** 257, 308

Volcanic activity

Earth 237, 265, 271, 283, 285, 287

Earth-like planets 87

Mars **236,** 237

Mercury 227

moons 189, **226,** 227, 265

Venus **264,** 265

W

Water

Earth-like planets 87, 109

as most common liquid in Universe 109

precipitation 155

as precursor for life **58**

Water worlds

defined 308

life on **108,** 109, 156, **156,** 157, 207

Solar System **206,** 207, **220**

Weak nuclear force 13, 308

White dwarfs

collisions and mergers 27, 31, 37, 81, 171, **244,** 245

death of universe 295

defined 308

explosions **26,** 31

formation 19, 31, 81, 113, 171

low-mass X-ray binary (LMXB) system 163

mass transfer 145

tidal disruption events 189

type Ia supernova **80,** 81, 245

Worlds, defined 9, 308

X

X-ray emissions

defined 308

galaxy cluster collisions 63, 76, 77, **160,** 161, **212,** 213

low-mass X-ray binary (LMXB) system **162,** 163

Z

Zero-point energy 11

ABOUT THE AUTHORS

This book is the product of more than two years of spirited collaboration via Zoom, with author meetings every week to discuss topics, review copy and artwork, and deliberate design. This screenshot is from one of our meetings, taken on February 22, 2022.

Ethan Siegel (top left) is an astrophysicist, author, and science communicator living in Washington State. His award-winning blog, Starts With a Bang, specializes in explaining high-level concepts about the universe to general audiences. @startswithabang

Mark A. Garlick (bottom right) is an astrophysicist, scientific illustrator, and computer animator living in the United Kingdom. He is author and illustrator of several books, including *The Story of the Solar System* and *Cosmic Menagerie*. markgarlick.com

Jon Lomberg (bottom left) is a space artist living in Hawaii. He was Carl Sagan's principal artistic collaborator, chief artist for the classic television series *Cosmos,* astronomical visual consultant for the movie *Contact,* and design director for the Voyager Golden Record. jonlomberg.com

William Lidwell (top right) is a designer and researcher living in Texas. He is the author of several books, notably *Universal Principles of Design*. His lectures on design are available through The Great Courses and LinkedIn Learning. @williamlidwell

Since 1888, the National Geographic Society has funded more than 15,000 research, conservation, education, technology, and storytelling projects around the world. National Geographic Partners distributes a portion of the funds it receives from your purchase to National Geographic Society to support their mission to illuminate and protect the wonder of our world.

National Geographic Partners, LLC
1145 17th Street NW
Washington, DC 20036-4688 USA

Get closer to National Geographic Explorers and photographers, and connect with our global community. Join us today at nationalgeographic.org/joinus

For rights or permissions inquiries, please contact National Geographic Books Subsidiary Rights: bookrights@natgeo.com

Originally published in 2024 as *Encyclopaedia Cosmologica*. First National Geographic edition 2025.

Page 2 illustration of large-scale structure of the Universe based on Millennium-II Simulation (M. Boylan-Kolchin/Max Planck Institute for Astrophysics) and Millennium-XXL Simulation (R.E. Angulo/Max Planck Institute for Astrophysics).

ISBN: 978-1-4262-2443-0

Printed in China

25/RRDH/1

AN ADVENTURE INTO THE VAST OCEAN OF SPACE

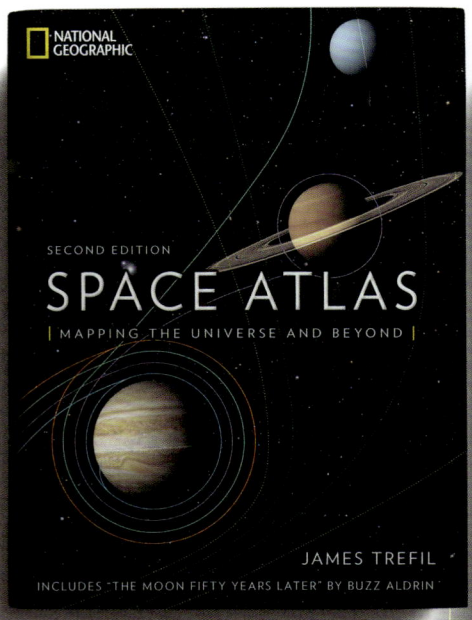

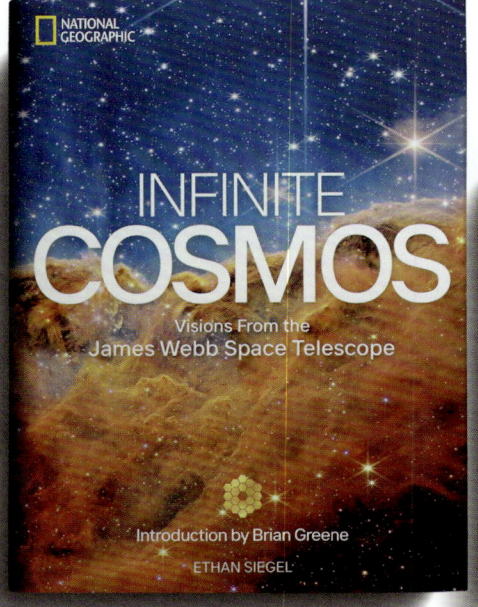

  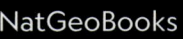